OXFORD STUDIES
IN
NUCLEAR PHYSICS

GENERAL EDITOR
P. E. HODGSON

A SHELL MODEL DESCRIPTION OF LIGHT NUCLEI

I. S. Towner

CLARENDON PRESS · OXFORD
1977

Oxford University Press, Walton Street, Oxford OX2 6DP

OXFORD LONDON GLASGOW NEW YORK
TORONTO MELBOURNE WELLINGTON CAPE TOWN
IBADAN NAIROBI DAR ES SALAAM LUSAKA ADDIS ABABA
KUALA LUMPUR SINGAPORE JAKARTA HONG KONG TOKYO
DELHI BOMBAY CALCUTTA MADRAS KARACHI

ISBN 0 19 851508 1

Printed by Thomson Litho Ltd., East Kilbride, Scotland.

PREFACE

The single most important question that has plagued nuclear physicists for the past two decades is the following: can the detailed properties of nuclei be calculated accurately using interactions deduced from nucleon-nucleon scattering data? The answer, unfortunately, is inconclusive. The optimists are divided from the pessimists over the interpretation of the word 'accurately'.

Nevertheless, the formal steps in the chain from free nucleons to finite nuclei seem to be well established. What is lacking at the present time is a demonstration from a detailed numerical calculation that would both vindicate the formal steps and enable an unqualified affirmative to be appended to the question. This is not to say that there has been any lack of effort, quite the reverse. Many detailed and sophisticated calculations have been mounted and some produce quite impressive results for the properties of finite nuclei. The difficulty is that invariably approximations are introduced somewhere along the chain, so ultimately the discussion centres on the accuracy of these approximations rather than on the question posed initially.

The central theme of this book is to trace the steps from the free nucleon-nucleon interaction to the properties of finite nuclei. The path is not straight. There are many diversions along the way, some of them extremely interesting and worthy of exploration. I have chosen to include a few such diversions, but many others have been purposely avoided. My style is to treat in as much detail as space allows the topics selected, thus making for a fairly self-contained treatise.

The natural starting point should be a discussion of the free nucleon-nucleon interaction. This in itself is a vast subject extensively covered in the literature. I have chosen not to review this field but to accept as a starting point the existence of a nucleon-nucleon interaction that fits the available two-body scattering data. I have also chosen not to discuss infinite nuclear matter, despits its historical importance in the development of this subject. Instead, the starting point will be the simplest of calculations for finite nuclei, namely the computation of the binding energy of a spherical closed-shell nucleus.

The standard technique, as borrowed from atomic physics, is a Hartree-Fock procedure, in which the gross effect of the two-body interactions is to produce an average one-body field. The resulting wavefunction is an optimum single Slater determinant of single-particle wavefunctions with no correlations between the particles. One difficulty is that the nature of the realistic nucleon-nucleon interactions, exhibiting as they do strong short-range repulsive components, leads to poor convergence, or in the extreme case of infinite hard core potentials, does not even allow the Hartree-Fock procedure to begin.

This observation pin-points the central dilemma: two-body scattering data exhibit a behaviour characteristic of strong short-range components in the nucleon-nucleon interaction, whereas nuclear structure studies indicate the effective interaction in nuclei to be weak, long-range, and attractive. This dilemma was resolved when the significant role played by the Pauli principle in smoothing out short-range fluctuations was appreciated. The stepping stone between free nucleons and finite nuclei is furnished in Brueckner theory through the construction of a reaction matrix, G. This G-matrix expresses the interaction between a pair of nucleons in a background of spectator nucleons. Hartree-Fock calculations with a G-matrix are extremely involved and time-consuming and it must have been a source of great disappointment that the results were not more encouraging. It was found that the binding energy and the root mean square charge radius of closed-shell nuclei could not be explained simultaneously without the need for yet more corrections. Nevertheless, the current state of the art is that Brueckner-Hartree-Fock calculations can be mounted for closed-shell nuclei essentially free of approximations.

Next, we turn our attention to excited states in closed-shell nuclei and consider just the one-particle one-hole states. Tamm-Dancoff and random phase approximations are discussed and although this is a departure from our main theme, considerable insight can be gained into the properties of these states by resorting to schematic rather than realistic interactions.

This concludes the discussion on closed-shell nuclei, and we turn next to the more complicated question of open-shell nuclei. Again a Hartree-Fock procedure could be invoked, but now the average one-body field will be deformed. I have chosen not to follow this avenue, not because deformations are unimportant – they are, and that may well be the source of subsequent problems – but, because Brueckner-Hartree-Fock calculations with a realistic G-matrix for open-shell nuclei are very complicated and they have not, to my knowledge, so far been attempted.

Instead we take the alternative shell model approach in which a many-body wavefunction is expanded in some chosen basis. Formally this basis spans an infinite-dimensioned Hilbert space and a characteristic shell model truncation has to be introduced, which renormalizes the operators and interactions to be used within the smaller space. These renormalizations effectively take into account the influence of neglected configurations and a perturbative formalism for their evaluation has been developed. However, the first results are hardly encouraging. It is found that the renormalizations are moderately large they depend strongly on the particular calculation performed and what is worse the perturbation series itself is likely to diverge in all cases of practical interest. The outlook appears bleak, yet to be fair not all avenues have yet been explored. The optimists remain none the less optimistic.

Underlying all this endeavour, from the very earliest shell model calculations, with phenomenological rather than realistic interactions, is the belief that the properties of light nuclei can be simply explained in terms of the mixing of a few configurations by a weak attractive two-body interaction. The problem is to determine from first principles what this interaction is, and which are the important configurations.

The last step in the chain, once an effective interaction has been constructed, is the calculation of energy levels and transition probabilities in nuclei with three, four, five, and more nucleons removed from the closed-shell situation. There is no ambiguity here. It is just a tedious exercise in Racah algebra and a continual battle in coping with the explosion in the number of active configurations. This explosion ultimately limits the practicality of any chosen calculation.

The final chapter can only be described as a whim of the author. I have for many years been fascinated by the type of nuclear structure information that can be learned from 'direct' nuclear reactions, in which one or more nucleons are transferred from a target to a residual nucleus. Naively one learns of the overlap of one nucleus with another and this is expressed in the theory by a quantity known as a spectroscopic factor. In the last chapter, shell model expressions are given for spectroscopic factors and their sum rules highlighted.

The last two chapters were the basis of a series of lectures given in Oxford in 1970. This book grew out of that experience and I am very grateful to Dr. Peter Hodgson whose gentle but firm persuasion resulted in the completion of the manuscript. I must also thank the members of the theoretical physics group at Chalk River Nuclear Laboratories for many poignant discussions, and in particular thank Dr. Malcolm Harvey whose constructive criticism resulted in several major revisions.

Malcolm Macfarlane wrote in the 1967 Varenna lecture series that the shell model, like the Roman god Janus, has two faces that look in opposite directions. One face looks towards the free-nucleon scattering data, the other towards the spectra of complex nuclei. The goal is to unite these two aspects of its personality. Nuclear physicists are still striving towards this goal.

I.S.T.

CONTENTS

1
SOME MATHEMATICAL PRELIMINARIES

1.1 OCCUPATION NUMBER REPRESENTATION

Consider the nucleus as a collection of A indistinguishable fermions describable in a non-relativistic quantum mechanical formalism and represented by a wavefunction $\Psi(\underset{\sim}{x}_1, \underset{\sim}{x}_2, \ldots, \underset{\sim}{x}_A)$. The coordinates $\underset{\sim}{x}$ specify the orbital, spin and isospin variables associated with each nucleon. Our first task is to establish an equivalence between the coordinate representation and a certain operator formalism known as 'second quantization'. This formalism will be exceedingly useful since it enables the combinatorial aspects of Fermi-Dirac statistics to be automatically taken into account.

First, we define indistinguishability by stating that particles are indistinguishable if the expectation value of any physical observable is unaltered by the exchange of particles or equivalently by a permutation of their labels. Let P_r be such a permutation operator, and let us evaluate an observable, F, in the state Ψ. Then from the statement of indistinguishability we have

$$\langle \Psi | F | \Psi \rangle \equiv \langle P_r \Psi | F | P_r \Psi \rangle$$

$$= \langle \Psi | P_r^{-1} F P_r | \Psi \rangle, \tag{1.1}$$

that is $F = P_r^{-1} F P_r$ or the commutator $[F, P_r] = 0$. This result leads to a requirement on the operator F, namely that if the expectation value of F is to be invariant under permutations of the labels of the particles, then the operator F must be a symmetric function of the particle coordinates. Most physical operators satisfy this condition. In particular, the Hamiltonian, H, is invariant under P_r, i.e. $[H, P_r] = 0$ so that the symmetry character of a state is preserved in time.

To construct an antisymmetric wavefunction from arbitrary

functions of A variables $x_1 \ldots x_A$, we introduce an antisymmetrizer

$$\mathscr{A} = \frac{1}{A!} \sum_r (-)^{p_r} P_r$$

where the summation is over all A! elements of the symmetric group of order A. The sign factor $(-)^{p_r}$ is +1 for an even permutation and -1 for an odd permutation. The antisymmetrizer has the properties: (i) it is Hermitian; (ii) it commutes with any permutation; $[\mathscr{A}, P_r] = 0$; and (iii) it is a projection operator, i.e. $\mathscr{A}^2 = \mathscr{A}$. Proofs of these statements can be found in Schweber (1962).

These properties of $\mathscr{A}$ can be used to simplify the computation of expectation values. For if $|\Phi\rangle = \mathscr{A}|\Psi\rangle$ where $|\Psi\rangle$ is a wavefunction without symmetry properties, then

$$\begin{aligned} \langle\Phi|F|\Phi\rangle &= \langle\Psi|\mathscr{A}^{\dagger} F\mathscr{A}|\Psi\rangle = \langle\Psi|F\mathscr{A}^2|\Psi\rangle \\ &= \langle\Psi|F\mathscr{A}|\Psi\rangle = \langle\Psi|F|\Phi\rangle \end{aligned} \tag{1.2}$$

since F and $\mathscr{A}$ commute. It is therefore only necessary to use one properly symmetrized function $\mathscr{A}|\Psi\rangle$ in computing the expectation value.

It is useful to characterize the A-particle system in terms of products of single-particle functions. A complete set of such single-particle functions can be obtained by solving the one-body wave equation

$$h\, \psi_\alpha(\underset{\sim}{x}) = \varepsilon_\alpha \psi_\alpha(\underset{\sim}{x}) \tag{1.3}$$

with h being a one-body Hamiltonian, and ε_α the corresponding energy eigenvalue. The subscript α labels the quantum state of the particle. The one-body Hamiltonian, h, does not have to be specified at this time, but we have in mind a Hartree-Fock Hamiltonian for which the interactions between nucleons have been incorporated in some average sense.

The next step is to construct a set of product wavefunctions in A variables of the form

$$\Psi_{\alpha\beta\ldots\nu}(\underset{\sim}{x}_1,\underset{\sim}{x}_2\ldots,\underset{\sim}{x}_A) = \psi_\alpha(\underset{\sim}{x}_1)\psi_\beta(\underset{\sim}{x}_2)\ldots\psi_\lambda(\underset{\sim}{x}_i)\ldots\psi_\nu(\underset{\sim}{x}_A).$$

These functions Ψ form a complete orthonormal set in A variables if the set ψ_λ is complete. A system of A indistinguishable fermions is then represented by a properly antisymmetrized combination of such product functions, viz. $\Phi = \mathscr{A}|\Psi\rangle$. In this wavefunction, Φ, the quantum states $\alpha,\beta\ldots\lambda\ldots$ cannot be associated with any particular particle coordinates $\underset{\sim}{x}_1,\underset{\sim}{x}_2\ldots$, instead it is sufficient to list how many quantum states λ repeat, which ones, and how often. Such specification is in terms of an infinite sequence of integers $n_\alpha, n_\beta\ldots n_\lambda\ldots$, the occupation numbers, where n_λ designates the number of times ψ_λ (irrespective of argument) occurs in any one of the product functions Ψ comprising Φ.

For fermions, the necessity of using antisymmetrized wavefunctions leads to the condition $n_\lambda \leqslant 1$, i.e. n_λ can only be equal to 0 or 1. To see this, suppose that in the product function $\Psi_{\alpha\beta\ldots}(\underset{\sim}{x}_1\ldots\underset{\sim}{x}_A)$, two of the single-particle functions, say the rth and the sth, are identical. That is,

$$\Psi_{\alpha\beta\ldots}(\underset{\sim}{x}_1,\ldots,\underset{\sim}{x}_A) = \psi_\alpha(\underset{\sim}{x}_1)\ldots\psi_\lambda(\underset{\sim}{x}_r)\ldots\psi_\lambda(\underset{\sim}{x}_s)\ldots,$$

then a simple transposition of the rth and s^{th} coordinates leaves Ψ unchanged, i.e. $P_{rs}\Psi = \Psi$. But for the correctly antisymmetrized function constructed from Ψ we have that

$$\Phi = \mathscr{A}\Psi = \mathscr{A}P_{rs}\Psi = -\mathscr{A}\Psi = 0,$$

since $\mathscr{A}P_{rs} = -\mathscr{A}$ for any simple transposition. Therefore Φ vanishes unless all the λ are different, i.e. unless $n_\lambda = 0$ or 1. This is just a statement of the Pauli Exclusion Principle - that no two fermions in a given assembly can occupy the same quantum state.

Starting, then, with a given product function Ψ, the correctly antisymmetrized function $\Phi = \mathscr{A}\Psi$ can be conveniently expressed as a Slater determinant

$$\Phi = (A!)^{-1/2} \begin{vmatrix} \psi_\alpha(\underset{\sim}{x}_1) & \psi_\alpha(\underset{\sim}{x}_2) & \cdots & \psi_\alpha(\underset{\sim}{x}_A) \\ \psi_\beta(\underset{\sim}{x}_1) & \psi_\beta(\underset{\sim}{x}_2) & & \vdots \\ & \vdots & & \vdots \\ \psi_\nu(\underset{\sim}{x}_1) & & & \psi_\nu(\underset{\sim}{x}_A) \end{vmatrix}$$

$$\equiv |n_\alpha n_\beta \ldots n_\lambda \ldots\rangle$$

$$\equiv |\alpha\, \beta \ \ldots \ \lambda \ \ldots\rangle\,. \tag{1.4}$$

We have introduced two alternative notations for the Slater Determinant. The first $|n_\alpha n_\beta \ldots n_\lambda \ldots\rangle$ is an occupation number representation and just comprises a string of zeros and ones. The second $|\alpha\beta\ldots\lambda\ldots\rangle$ is somewhat simpler and merely specifies which quantum states are occupied. In the determinantal form, it is obvious that no two ψ_λ can be the same, since if this were so, two rows would be identical and the determinant would vanish.

Let us now introduce an operator a_λ, such that

$$a_\lambda |n_\alpha n_\beta \ldots n_\lambda \ldots\rangle = (-)^{p_\lambda} n_\lambda |n_\alpha n_\beta \ldots n_\lambda - 1 \ldots\rangle \tag{1.5}$$

where $p_\lambda = \Sigma_\alpha^{\lambda-1}\, n_\alpha$ is equal to the number of occupied states up to the λth. The operator a_λ is called an annihilation operator, since it destroys a particle in the state ψ_λ. It has matrix elements

$$\langle n'_\alpha n'_\beta \ldots n'_\lambda \ldots | a_\lambda | n_\alpha n_\beta \ldots n_\lambda \ldots\rangle$$
$$= (-)^{p_\lambda} n_\lambda \delta_{n'_\alpha n_\alpha} \delta_{n'_\beta n_\beta} \ldots \delta_{n'_\lambda, n_\lambda - 1} \ldots \tag{1.6}$$

Two such operators a_λ and a_μ anticommute for

$$a_\mu a_\lambda |\ldots n_\lambda \ldots n_\mu \ldots\rangle = (-)^{p_\lambda} n_\lambda a_\mu |\ldots n_\lambda - 1 \ldots n_\mu \ldots\rangle$$
$$= (-)^{p_\lambda} (-)^{p_\mu - 1}\, n_\lambda n_\mu |\ldots n_\lambda - 1 \ldots n_\mu - 1 \ldots\rangle \tag{1.7}$$

whereas operating in the reverse order gives

$$a_\lambda a_\mu|\ldots n_\lambda\ldots n_\mu\ldots\rangle = (-)^{p_\lambda}(-)^{p_\mu} n_\lambda n_\mu|\ldots n_\lambda-1\ldots n_\mu-1\ \ldots\rangle \quad (1.8)$$

Thus $a_\lambda a_\mu = -a_\mu a_\lambda$. Note that if $\lambda=\mu$ we must have that $a_\lambda^2 = 0$, which is the operator expression of the fact that no state can have an occupation number greater than one.

Next we introduce the adjoint operator, $a_\lambda^\dagger$, defined such that

$$\langle n'_\alpha n'_\beta\ldots n'_\lambda\ldots|a_\lambda|n_\alpha n_\beta\ldots n_\lambda\ldots\rangle = \langle n_\alpha n_\beta\ldots n_\lambda\ldots|a_\lambda^\dagger|n'_\alpha n'_\beta\ldots n'_\lambda\ldots\rangle \quad (1.9)$$

then it follows by comparison with eqn (1.6) that

$$a_\lambda^\dagger|n_\alpha n_\beta\ldots n_\lambda\ldots\rangle = (-)^{p_\lambda}(1-n_\lambda)|n_\alpha n_\beta\ldots n_\lambda+1\ldots\rangle. \quad (1.10)$$

The operator $a_\lambda^\dagger$ is called a creation operator, since operating on an A-particle state it transforms it into an A+1 particle state by increasing the number of particles in the state ψ_λ by one. Notice the use of the factor $(1-n_\lambda)$ in the definition of the adjoint rather than $(n_\lambda+1)$ as suggested by the strict equality of eqn (1.6) with eqn (1.9). The two definitions are equivalent for $n_\lambda = 0$, but the former guarantees that if the state ψ_λ is already occupied, $n_\lambda=1$, a second particle cannot be put into that state. In analogy to eqns (1.7) and (1.8) it is easily shown that $a_\lambda^\dagger$ and $a_\mu^\dagger$ anticommute and that $a_\lambda^\dagger a_\lambda^\dagger = 0$.

Next consider the application of the operators $a_\lambda a_\mu^\dagger$ on the Slater determinant state:

$$a_\lambda a_\mu^\dagger|\ldots n_\lambda\ldots n_\mu\ldots\rangle = (-)^{p_\lambda+p_\mu} n_\lambda(1-n_\mu)|\ldots n_\lambda-1\ldots n_\mu+1\ldots\rangle, \quad (1.11)$$

while reversing the order of the operators leads to

$$a_\mu^\dagger a_\lambda|\ldots n_\lambda\ldots n_\mu\ldots\rangle = (-)^{p_\lambda+p_\mu-1} n_\lambda(1-n_\mu)|\ldots n_\lambda-1\ldots n_\mu+1\ldots\rangle \quad (1.12)$$

and hence an anticommutation relation

$$[a_\lambda, a_\mu^\dagger]_+ = 0. \quad (1.13)$$

In the particular case that $\lambda=\mu$, we have

$$a_\lambda a_\lambda^\dagger|\ldots n_\lambda\ldots\rangle = (1-n_\lambda)|\ldots n_\lambda\ldots\rangle, \qquad (1.14)$$

where the fact that $n_\lambda^2 = n_\lambda$, since $n_\lambda = 0$ or 1, has been used. Similarly with the operators in reverse order, we obtain

$$a_\lambda^\dagger a_\lambda|\ldots n_\lambda\ldots\rangle = n_\lambda|\ldots n_\lambda\ldots\rangle \qquad (1.15)$$

and hence the anticommutation relation

$$[a_\lambda, a_\lambda^\dagger]_+ = 1. \qquad (1.16)$$

Putting together the results in eqns (1.7), (1.8), (1.13) and (1.16), a summary set of commutation rules is now verified to be

$$[a_\lambda, a_\mu]_+ = [a_\lambda^\dagger, a_\mu^\dagger]_+ = 0$$

$$[a_\lambda, a_\mu^\dagger]_+ = \delta_{\lambda,\mu}. \qquad (1.17)$$

Note in particular that the operator, $N_\lambda = a_\lambda^\dagger a_\lambda$, has as its eigenvalue the occupation number n_λ. Therefore N_λ is known as the number operator for the λth state. It also has the additional property of being a projection operator since

$$N_\lambda{}^2 = a_\lambda^\dagger a_\lambda a_\lambda^\dagger a_\lambda = a_\lambda^\dagger(1-a_\lambda^\dagger a_\lambda)a_\lambda$$

$$= a_\lambda^\dagger a_\lambda = N_\lambda \qquad (1.18)$$

so that its eigenvalues are 0 and 1 as expected.

1.2 REPRESENTATION OF ONE-BODY OPERATORS

The next step is to obtain the representation in occupation number space of an operator, F, which does not change the number of particles and whose configuration space representation is known. Consider first the case when F is a sum of one-body operators:

$$F(\underset{\sim}{x}_1, \underset{\sim}{x}_2 \ldots, \underset{\sim}{x}_A) = \sum_{i=1}^{A} F(\underset{\sim}{x}_i). \tag{1.19}$$

By definition such an operator is capable of changing the quantum state of at most one particle in the system. Furthermore the total number of particles is conserved; thus it should be possible to write the operator in an occupation number representation as a linear combination of terms $a^\dagger_\lambda a_\mu$:

$$F = \sum_{\lambda\mu} C_{\lambda\mu} \; a^\dagger_\lambda a_\mu. \tag{1.20}$$

To find what the coefficient $C_{\lambda\mu}$ should be, we must establish the connection with the configuration space representation.

Consider the case of single-particle states, and suppose the operator F induces a transition from an initial state $\psi_\beta(\underset{\sim}{x})$ to a final state $\psi_\alpha(\underset{\sim}{x})$, then the matrix element for the transition is written

$$\int \psi^\dagger_\alpha(\underset{\sim}{x}) \; F(\underset{\sim}{x}) \; \psi_\beta(\underset{\sim}{x}) \; d\underset{\sim}{x} \equiv \langle \alpha | F | \beta \rangle. \tag{1.21}$$

The probability that the transition will occur is proportional to the square of this matrix element. This is a configuration space description.

In an occupation number representation, the single-particle states are written

$$\psi_\alpha(\underset{\sim}{x}) = a^\dagger_\alpha |0\rangle \qquad \psi_\beta(\underset{\sim}{x}) = \alpha^\dagger_\beta |0\rangle$$

where $|0\rangle$ is a vacuum state, i.e. a state containing no particles, and $a^\dagger_\alpha$ creates a particle in the quantum state α. The transition matrix element is now written

$$\begin{aligned} \langle 0 | a_\alpha \; F \; a^\dagger_\beta | 0 \rangle &= \sum_{\lambda\mu} C_{\lambda\mu} \; \langle 0 | a_\alpha a^\dagger_\lambda a_\mu a^\dagger_\beta | 0 \rangle \\ &= \sum_{\lambda\mu} C_{\lambda\mu} \; \delta_{\lambda\alpha} \; \delta_{\mu\beta} \\ &= C_{\alpha\beta}. \end{aligned} \tag{1.22}$$

The anticommutation relations, eqn (1.17), and the fact that $a_\mu|0\rangle = 0$ are used to evaluate the vacuum expectation value. Comparing eqns (1.21) and (1.22) we see the coefficient $C_{\alpha\beta}$ is just the single-particle matrix element $\langle\alpha|F|\beta\rangle$. Thus the representation of a one-body operator in an occupation number space is

$$F = \sum_{\lambda\mu} \langle\lambda|F|\mu\rangle\, a^\dagger_\lambda a_\mu . \tag{1.20}$$

As an example of how this expression is used, let us calculate the transition matrix element between the two-particle states $|\alpha\beta\rangle$ and $|\lambda\delta\rangle$ of a one-body operator. The result is

$$\begin{aligned}\langle\alpha\beta|F|\lambda\delta\rangle &= \sum_{\lambda\mu} \langle\lambda|F|\mu\rangle \langle 0|a_\alpha a_\beta a^\dagger_\lambda a_\mu a^\dagger_\delta a^\dagger_\gamma|0\rangle \\ &= \langle\alpha|F|\gamma\rangle\delta_{\beta\delta} - \langle\alpha|F|\delta\rangle\delta_{\beta\gamma} - \langle\beta|F|\gamma\rangle\delta_{\alpha\delta} + \langle\beta|F|\delta\rangle\delta_{\alpha\gamma}.\end{aligned} \tag{1.23}$$

Extending to an A-particle state, the diagonal matrix element of a one-body operator is

$$\langle\alpha\ldots\lambda\ldots\nu|F|\alpha\ldots\lambda\ldots\nu\rangle = \sum_\lambda \langle\lambda|F|\lambda\rangle, \tag{1.24}$$

and for a typical off-diagonal matrix element

$$\langle\alpha\ldots\lambda'\ldots\nu|F|\alpha\ldots\lambda\ldots\nu\rangle = \langle\lambda'|F|\lambda\rangle. \tag{1.25}$$

If the initial and final states differ by more than one quantum state then by definition the matrix element of a one-body operator vanishes between these states.

1.3 REPRESENTATION OF TWO-BODY OPERATORS

A two-body operator in a configuration space representation is of the form

$$G = \sum_{i<j} G(\underset{\sim}{x}_i, \underset{\sim}{x}_j) = \frac{1}{2} \sum_{\substack{ij \\ i\neq j}} G(\underset{\sim}{x}_i, \underset{\sim}{x}_j). \tag{1.26}$$

Such an operator is capable of changing the quantum state of at most two particles in the system. We shall only be concerned with number-conserving operators: thus, in complete analogy to the treatment of one-body operators in the last section, the representation of G in occupation number space is

$$G = \sum_{\substack{\alpha\leqslant\beta \\ \lambda\leqslant\delta}} \langle\alpha\beta|G|\gamma\delta\rangle_A \; a^\dagger_\alpha a^\dagger_\beta a_\delta a_\gamma . \tag{1.27}$$

The quantum states have been ordered in an arbitrary way so that each pair of states is counted only once in the sum. The coefficient $\langle\alpha\beta|G|\gamma\delta\rangle_A$ represents the matrix element of G between antisymmetrized states (note the presence of the subscript A), viz.

$$\langle\alpha\beta|G|\gamma\delta\rangle_A = \langle\alpha\beta|G|\gamma\delta\rangle - \langle\alpha\beta|G|\delta\gamma\rangle. \tag{1.28}$$

As discussed in connection with eqn (1.2) only the bra or the ket wavefunction needs to be explicitly antisymmetrized. Of the two terms in eqn (1.28), the first is called the direct and the second the exchange matrix element. The configuration space representation of the matrix element is

$$\langle\alpha\beta|V|\gamma\delta\rangle = \int \psi^\dagger_\alpha(\underset{\sim}{x}_1)\psi^\dagger_\beta(\underset{\sim}{x}_2)G(\underset{\sim}{x}_1,\underset{\sim}{x}_2)\psi_\gamma(\underset{\sim}{x}_1)\psi_\delta(\underset{\sim}{x}_2)d\underset{\sim}{x}_1 d\underset{\sim}{x}_2 . \tag{1.29}$$

The expression (1.27) can be written in two alternative forms with unrestricted sums over the quantum states $\alpha\beta\gamma\delta$. The first uses antisymmetric two-body matrix elements:

$$G = \frac{1}{4} \sum_{\alpha\beta\gamma\delta} \langle\alpha\beta|G|\gamma\delta\rangle_A \; a^\dagger_\alpha a^\dagger_\beta a_\delta a_\gamma \tag{1.30}$$

and the second uses unsymmetrized matrix elements

$$G = \frac{1}{2} \sum_{\alpha\beta\gamma\delta} \langle\alpha\beta|G|\gamma\delta\rangle \; a^\dagger_\alpha a^\dagger_\beta a_\gamma a_\delta \tag{1.31}$$

Eqn (1.30) is the form most frequently used.

For an A-particle state, the diagonal matrix element of a two-body operator is

$$\langle \alpha\ldots\lambda\ldots\nu|G|\alpha\ldots\lambda\ldots\nu\rangle = \frac{1}{2}\sum_{\lambda\mu}\langle\lambda\mu|G|\lambda\mu\rangle_A$$

For matrix elements non-diagonal in the quantum state of either one or two particles, we have

$$\langle \alpha\ldots\lambda'\ldots\nu|G|\alpha\ldots\lambda\ldots\nu\rangle = \sum_{\substack{\mu \\ \mu\neq\lambda,\lambda'}}\langle\lambda'\mu|G|\lambda\mu\rangle_A \tag{1.33}$$

$$\langle\ldots\lambda'\ldots\mu'\ldots|G|\ldots\lambda\ldots\mu\ldots\rangle = \langle\lambda'\mu'|G|\lambda\mu\rangle_A. \tag{1.34}$$

1.4 ANGULAR MOMENTUM

We have already mentioned that the Hamiltonian commutes with the permutation operators P_r. The Hamiltonian also commutes with the angular momentum operator $\underset{\sim}{J}$, i.e. $[H,\underset{\sim}{J}^2] = 0$ and the total angular momentum of the nucleus is preserved in time. The one-body functions $\psi_\alpha(\underset{\sim}{x})$, solutions to the Schrödinger equation, eqn (1.3), are consequently eigenfunctions of the angular momentum operators:

$$\underset{\sim}{J}^2\psi_\alpha(\underset{\sim}{x}) = j_\alpha(j_\alpha+1)\psi_\alpha(\underset{\sim}{x}) \qquad J_z\psi_\alpha(\underset{\sim}{x}) = m_\alpha\psi_\alpha(\underset{\sim}{x}).$$

Our notation for these one-body functions will be

$$\psi_\alpha(\underset{\sim}{x}) \equiv |\alpha\rangle \equiv |j_\alpha m_\alpha\rangle \equiv a_\alpha^\dagger|0\rangle$$

with the state label α representing the angular momentum j_α, and the magnetic projection m_α, plus any other quantum numbers (such as the principal, n_α, orbital, ℓ_α, and spin, s_α, quantum numbers) necessary to complete the specification. The label also includes the isospin quantum numbers t_α and $m_{t\alpha}$ which satisfy the same angular momentum algebra as j_α and m_α; but for economy of notation t_α and $m_{t\alpha}$ are not explicitly displayed (as discussed in the appendix A.2).

The many-body functions $\phi(\underset{\sim}{x}_1\ldots\underset{\sim}{x}_A)$, constructed from antisymmetric combinations of product functions, do not in general have definite total angular momentum. For example, the two-particle state

$$|\alpha\beta\rangle_A = a^\dagger_\beta\, a^\dagger_\alpha|0\rangle \tag{1.35}$$

is not an eigenfunction of $\underset{\sim}{J}^2$, but a linear combination of such states weighted by Clebsch-Gordan coefficients is. Such a state is referred to as a coupled two-particle state and is defined as

$$|j_\alpha j_\beta;JM\rangle = \sum_{m_\alpha m_\beta} \langle j_\alpha m_\alpha j_\beta m_\beta|JM\rangle\, a^\dagger_\beta a^\dagger_\alpha|0\rangle. \tag{1.36}$$

The single-particle wavefunctions $\psi_\alpha(\underset{\sim}{x})$ comprise an orbital part and a spin part coupled to total angular momentum j_α: see eqn (A.1) in the appendix. In forming the two-particle state eqn (1.36) we have chosen to couple $\underset{\sim}{j}_\alpha$ to $\underset{\sim}{j}_\beta$ — this being the so-called j-j coupling scheme. An alternative procedure is to couple the orbital functions of the two particles together, $\underset{\sim}{\ell}_\alpha + \underset{\sim}{\ell}_\beta = \underset{\sim}{L}$; similarly to couple the two spin functions, $\underset{\sim}{s}_\alpha + \underset{\sim}{s}_\beta = \underset{\sim}{S}$; and finally to couple $\underset{\sim}{L} + \underset{\sim}{S} = \underset{\sim}{J}$ to form a state of good total angular momentum. This is the so-called L-S coupling scheme. Generalizations for more than two particles are straightforward.

The choice between the two coupling schemes is a matter of convenience they both lead to a complete set of many-body functions $\Phi(\underset{\sim}{x}_1 \ldots \underset{\sim}{x}_A)$, the dimension of the set being the same in both cases. Furthermore the functions constructed through L-S coupling can always be expressed as a linear combination of the j-j coupled functions and vice versa. For two-particle states the transformation coefficients are just the normalized 9j-coefficients (see eqn (A.17)); for more than two particles the transformation involves more complicated recoupling coefficients.

The choice between j-j and L-S coupling is best decided by the particular calculation at hand. For example, if one is using a spin-independent Hamiltonian, then it will save in computation time to use L-S coupling. On the other hand, j-j coupling is preferred in the presence of strong spin-orbit forces. In light nuclei, the spin-orbit force is not particularly strong and either coupling scheme can be used with equal advantage. We choose to work exclusively with j-j coupling in this book and present our formulae in this scheme.

The two-particle state (eqn (1.36)) is not normalized, but the normalization constant is simply found:

$$N^2 = \sum_{m_\alpha m_\beta m'_\alpha m'_\beta} \langle j_\alpha m_\alpha j_\beta m_\beta | JM\rangle \langle j_\alpha m'_\alpha j_\beta m'_\beta | JM\rangle \langle 0 | a_{\alpha'} a_{\beta'} a^\dagger_\beta a^\dagger_\alpha | 0\rangle$$

$$= \sum_{m_\alpha m_\beta} \langle j_\alpha m_\alpha j_\beta m_\beta | JM\rangle \langle j_\alpha m_\alpha j_\beta m_\beta | JM\rangle \, [1-(-)^{j_\alpha + j_\beta - J} \delta_{\alpha\beta}]$$

$$= 1 - (-)^{j_\alpha + j_\beta - J} \delta_{\alpha\beta}, \qquad (1.37)$$

where the orthonormal properties of the Clebsch-Gordan coefficients have been used. Note that if the two particles are occupying the same orbital, i.e. $n_\alpha \ell_\alpha j_\alpha = n_\beta \ell_\beta j_\beta$, then a restriction is placed on the possible J values, namely J = even (or J+T = odd in the isospin formalism). Thus a normalized two-particle state is written

$$|j_\alpha j_\beta; JM\rangle_A = (1+\delta_{\alpha\beta})^{-1/2} \sum_{m_\alpha m_\beta} \langle j_\alpha m_\alpha j_\beta m_\beta | JM\rangle \, a^\dagger_\beta a^\dagger_\alpha |0\rangle. \qquad (1.38)$$

We shall want to work with antisymmetric two-body matrix elements evaluated between coupled two-particle states. These can be expressed in terms of uncoupled antisymmetric matrix elements, viz.

$$\langle j_1 j_2; JM | G | j_3 j_4; J'M'\rangle_A$$

$$= [(1+\delta_{12})(1+\delta_{34})]^{-\frac{1}{2}} \frac{1}{4} \sum_{\substack{m_1 m_2 m_3 m_4 \\ \alpha\beta\gamma\delta}} \langle j_1 m_1 j_2 m_2 | JM\rangle \langle j_3 m_3 j_4 m_4 | J'M'\rangle \langle \alpha\beta | G | \gamma\delta\rangle_A \times$$

$$\times \langle 0 | a_1 a_2 \, a^\dagger_\alpha a^\dagger_\beta \, a_\delta a_\gamma \, a^\dagger_4 a^\dagger_3 | 0\rangle$$

$$= [(1+\delta_{12})(1+\delta_{34})]^{-\frac{1}{2}} \frac{1}{4} \sum_{m_1 m_2 m_3 m_4} (\text{Clebschs}) \{ \langle 1\ 2 | G | 3\ 4\rangle_A - \langle 2\ 1 | G | 3\ 4\rangle_A -$$

$$- \langle 1\ 2 | G | 4\ 3\rangle_A + \langle 2\ 1 | G | 4\ 3\rangle_A \}$$

$$= [(1+\delta_{12})(1+\delta_{34})]^{-\frac{1}{2}} \sum_{m_1 m_2 m_3 m_4} (\text{Clebschs}) \langle 1\ 2|G|3\ 4\rangle_A \tag{1.39}$$

where (Clebschs) represents the two Clebsch-Gordan coefficients written out in the first line. Finally the anti-symmetric coupled two-body matrix elements can be expressed in terms of coupled unnormalized, unsymmetrized matrix elements. Starting from eqn (1.39) and re-ordering the arguments of the Clebsch-Gordan coefficients for the exchange matrix element gives the result

$$\langle j_1 j_2;JM|G|j_3 j_4;J'M'\rangle_A$$
$$= [(1+\delta_{12})(1+\delta_{34})]^{-\frac{1}{2}} \{\langle j_1 j_2;JM|G|j_3 j_4;J'M'\rangle - \tag{1.40}$$
$$- (-)^{j_3+j_4-J'} \langle j_1 j_2;JM|G|j_4 j_3;J'M'\rangle\}.$$

This expression is the starting point in a calculation of two-body matrix elements. Some examples are to be found in appendix A, section (A.8) et seq.

1.5 CLOSED-SHELL NUCLEI

A closed-shell state is defined as one in which all possible magnetic substates allowed by the Pauli exclusion principle are populated. For example, a single closed-shell wave-function is written

$$|C\rangle = \prod_{m_i=-j_i \ldots +j_i} a_i^\dagger\ |0\rangle \tag{1.41}$$

where the product runs over all the distinct magnetic sub-states available in the one orbital j_i. Up to this point, we have been using Greek letters to designate the quantum state of the single-particle function. Very soon we shall want to differentiate between states which are normally occupied in a closed shell configuration $|C\rangle$ and states normally unoccupied in $|C\rangle$. We shall use Roman letters i, j, k, and ℓ for occupied states, letters m, n, p and q for unoccupied states, and as before Greek letters when not differentiating between the two categories.

We have already mentioned that a Slater determinant state constructed as in eqn (1.41) is not in general an eigenfunction of the angular momentum operator J^2. However, a closed-shell state $|C\rangle$, by its particular construction, turns out to be spherically symmetric with definite angular momentum J=0. We show this by calculating the expectation value in $|C\rangle$ of a one-body operator $F_M^{(L)}$, a spherical tensor of rank L and magnetic projection M, and show that the only non-vanishing matrix element is for the case L=0. Explicitly using eqn (1.24), we find that

$$\langle C|F_M^{(L)}|C\rangle = \sum_{m_i=-j_i\ldots+j_i} \langle j_i m_i|F_M^{(L)}|j_i m_i\rangle. \tag{1.42}$$

The dependence on the magnetic substates can be separated using the Wigner-Eckhart theorem,

$$\langle C|F_M^{(L)}|C\rangle = \langle j_i\|F^{(L)}\|j_i\rangle \sum_{m_i=-j_i\ldots j_i} \langle j_i m_i LM|j_i m_i\rangle \tag{1.43}$$

where the double bar indicates a reduced matrix element as defined in Appendix A.5. Manipulation of the Clebsch-Gordan coefficients gives

$$\begin{aligned}\langle C|F_M^{(L)}|C\rangle &= \langle j_i\|F^{(L)}\|j_i\rangle \left[\frac{2j_i+1}{2L+1}\right]^{1/2} \sum_{m_i} (-)^{j_i-m_i} \langle j_i m_i j_i -m_i|LM\rangle \\ &= \frac{\langle j_i\|F^{(L)}\|j_i\rangle}{(2L+1)^{1/2}} (2j_i+1)\delta_{L,0}\,\delta_{M,0}\end{aligned} \tag{1.44}$$

where we have used the identity

$$(-)^{j-m} = (2j+1)^{1/2} \langle jm\; j-m|00\rangle, \tag{1.45}$$

and the orthogonality of Clebsch-Gordan coefficients.

Thus we prove that the only non-zero diagonal matrix elements of the state $|C\rangle$ occur for scalar operators. In particular, the angular momentum operator $\underset{\sim}{J}$, being a spherical tensor of rank one, has zero expectation value, thereby demonstrating that the closed shell state $|C\rangle$ is spherically symmetric.

The definition, eqn (1.41), of the state $|C\rangle$ can be extended to one comprising several closed single shells $j_1, j_2 \ldots j_F$. Again we are envisaging the use of an independent particle model, in which the individual single-particle states $\psi_i(\underset{\sim}{x})$ are ordered according to their eigenenergies ε_i, eqn (1.3). Then the ground state for an assembly of A nucleons is defined as one in which the single-particle states are filled in order, starting with the lowest-energy eigensolution and successively filling the next-lowest energy shell, until all A nucleons have been placed in the lowest available orbits. For a core state the number of nucleons A is such that there are no partially-filled shells. The last completely-filled shell we denote by j_F, and its associated energy ε_F is called the Fermi energy. All single-particle states with energy $\varepsilon > \varepsilon_F$ will be unoccupied.

The wavefunction for the closed-shell core is now written

$$|C\rangle = \prod_{\substack{i=1,2\ldots,F \\ m_i=-j_i,\ldots,+j_i}} a_i^\dagger |0\rangle \tag{1.46}$$

and the expectation value of a scalar one-body operator follows as a simple extension of eqn (1.44) and is

$$\langle C|F^{(0)}|C\rangle = \sum_{i=1}^{F} (2j_i+1)\langle j_i \| F^{(0)} \| j_i\rangle. \tag{1.47}$$

The factor $(2j_i+1)$ merely represents the number of particles that can be placed in orbital j_i.

Similarly the only two-body operators having non-zero expectation values are scalar operators, and in this case we obtain, using eqn (1.32):

$$\langle C|G|C\rangle = \frac{1}{2} \sum_{ij=1}^{F} \langle j_i m_i, j_j m_j |G| j_i m_i, j_j m_j\rangle_A. \tag{1.48}$$

Rewriting in terms of coupled two-body matrix elements, gives

$$\langle C|G|C\rangle = \frac{1}{2} \sum_{\substack{j_i j_j \\ JM}} \sum_{m_i m_j} \langle j_i m_i j_j m_j | JM\rangle^2 \times$$

$$\times \{\langle j_i j_j ; JM|G|j_i j_j ; JM\rangle - (-)^{j_i + j_j - J} \langle j_i j_j ; JM|G|j_j j_i ; JM\rangle\}$$

$$= \frac{1}{2} \sum_{\substack{j_i j_j \\ JM}} (1+\delta_{ij}) \langle j_i j_j ; J|G|j_i j_j ; J\rangle_A \tag{1.49}$$

where we have used the definition, eqn (1.40), of an anti-symmetric coupled two-body matrix element. For a scalar operator, the matrix element does not depend on the magnetic projection M (the Clebsch-Gordan coefficient in the Wigner-Eckhart theorem has value unity), so the sum over M merely counts the number of magnetic substates. The final result is

$$\langle C|G|C\rangle = \sum_{j_i J} (2J+1) \langle j_i{}^2 ; J|G|j_i{}^2 ; J\rangle_A +$$

$$+ \sum_{\substack{j_i < j_j \\ J}} (2J+1) \langle j_i j_j ; J|G|j_i j_j ; J\rangle_A \tag{1.50}$$

where $j_i < j_j$ represents a sum over all the distinct pairs of orbits $j_i \neq j_j$ that make up the core state $|C\rangle$.

1.6 CLOSED-SHELL-PLUS-ONE NUCLEI

Next the definition of a single-particle state is extended to the case in which a particle is coupled to a closed shell core:

$$|m\rangle = a_m^\dagger |C\rangle \tag{1.51}$$

where the core wavefunction is as defined in eqn (1.46). By implication the orbital j_m is *not* one of the closed shell orbitals $j_1, j_2 \ldots j_F$ comprising the core. Since the core is spherically symmetric J=0, the angular momentum of the state, eqn (1.51), is necesarily that of the single particle, namely j_m.

First, we calculate the diagonal matrix element of a one-body operator $F_M^{(L)}$, a spherical tensor of rank L. Using eqn (1.24) we obtain:

$$\langle m|F_M^{(L)}|m\rangle = \langle j_m m_m|F_M^{(L)}|j_m m_m\rangle + \langle C|F_M^{(L)}|C\rangle \tag{1.52}$$

where the second term on the right-hand side represents a contribution from the core. This term was evaluated in eqn (1.44) and is only non-zero for scalar one-body operators, viz.

$$\langle m|F_M^L|m\rangle = \langle j_m m_m|F_M^{(L)}|j_m m_m\rangle + \sum_{i=1}^{F} (2j_i+1)\langle j_i\|F^{(L)}\|j_i\rangle\,\delta_{L,0}. \tag{1.53}$$

For off-diagonal one-body matrix elements, there is no contribution from the core (see eqn (1.25)) and the result in this case is simply

$$\langle m|F_M^{(L)}|m'\rangle = \langle j_m m_m|F_M^{(L)}|j_{m'} m_{m'}\rangle. \tag{1.54}$$

Next let us evaluate the matrix element of a scalar two-body operator. For a diagonal matrix element, eqn (1.32) is used, and the result is divided into two terms:

$$\langle m|G|m\rangle = \sum_{i=1}^{F} \langle im|G|im\rangle_A + \sum_{i,j=1}^{F} \langle ij|G|ij\rangle_A. \tag{1.55}$$

The second term on the right-hand side is again just the contribution from the core and was evaluated in eqn (1.49). The first term can be expressed in terms of coupled matrix elements:

$$\sum_{i=1}^{F} \langle im|G|im\rangle_A = \sum_{i=1}^{F} \sum_{m_i} \langle j_i m_i, j_m m_m|G|j_i m_i, j_m m_m\rangle$$

$$= \sum_{i=1}^{F} \sum_{\substack{m_i \\ JM}} \langle j_i m_i j_m m_m|JM\rangle^2 \langle j_i j_m; J|G|j_i j_m; J\rangle_A$$

$$= \sum_{i=1}^{F} \sum_{\substack{m_i \\ JM}} \frac{(2J+1)}{(2j_m+1)} \langle JM\, j_i\, {-m_i} | j_m m_m \rangle^2 \, \langle j_i j_m; J|G|j_i j_m; J\rangle_A$$

$$= \sum_{i=1}^{F} \sum_{J} \frac{(2J+1)}{(2j_m+1)} \langle j_i j_m; J|G|j_i j_m; J\rangle_A . \tag{1.56}$$

When G is the potential energy operator V, eqn (1.56) can be interpreted as the energy of a single-particle interacting via two-body forces with a closed-shell core. This is an important quantity, the so-called single-particle energy, which will be discussed further in connection with shell model calculations in Chapters 5 and 6.

Before leaving closed-shell-plus-one nuclei altogether, it is instructive to evaluate a specific example. Let us calculate the quadrupole moment of ^{17}F assuming a ground-state configuration for this nucleus of a closed ^{16}O core plus a single proton in the $d_{5/2}$ orbital. The moment, Q, is defined as the expectation value of the operator

$$\left(\frac{16\pi}{5}\right)^{\frac{1}{2}} \sum_i r_i^{\,2}\, Y_{20}(\hat{\underset{\sim}{r}}_i)$$

evaluated in the m=j magnetic substate. The sum i runs over all the protons in the nucleus, but there is no contribution from the core, see eqn (1.53), so that finding Q just reduces to evaluating a single-particle matrix element:

$$Q = \left(\frac{16\pi}{5}\right)^{\frac{1}{2}} \langle jj|r^2\, Y_{20}|jj\rangle .$$

By first employing the Wigner-Eckhart theorem and then using eqn (A.39) in the appendix to evaluate the reduced matrix element of a spherical harmonic, we obtain

$$\begin{aligned} Q &= \left(\frac{16\pi}{5}\right)^{\frac{1}{2}} \langle jj20|jj\rangle \langle j\| r^2\, Y_2 \| j\rangle \\ &= \left(\frac{16\pi}{5}\right)^{\frac{1}{2}} \left[\frac{j(2j-1)}{(j+1)(2j+3)}\right]^{\frac{1}{2}} (-)^{j-\frac{1}{2}} \left(\frac{2j+1}{4\pi}\right)^{\frac{1}{2}} \langle j\tfrac{1}{2} j\, {-\tfrac{1}{2}}|20\rangle \, \langle r^2\rangle \\ &= -\,\frac{4}{7}\,\langle r^2\rangle , \end{aligned}$$

where the value j = 5/2 was inserted to obtain the last line. The radial integral $\langle r^2 \rangle$ is

$$\langle r^2 \rangle = \int_0^\infty |R_d(r)|^2 \, r^4 \, dr$$

with $R_d(r)$ being the normalized radial function for a $d_{5/2}$ proton.

Let us repeat the calculation in the isospin formalism. The one-body operator is given by the expression

$$\left(\frac{16\pi}{5}\right)^{\frac{1}{2}} \sum_i r_i^{\,2} \, Y_{20}(\hat{\underset{\sim}{r}}_i) \, \frac{1}{2} \, [1 - \underset{\sim}{\tau}_z(i)]$$

with $\underset{\sim}{\tau}_z$ having an expectation value +1 for neutron states and -1 for proton states. The sum i now runs over all nucleons, but as before the core gives no contribution and Q is once again given by a single-particle matrix element

$$\begin{aligned} Q = \left(\frac{16\pi}{5}\right)^{\frac{1}{2}} & \{\tfrac{1}{2} \langle jj,\tfrac{1}{2}\, m_t | r^2 \, Y_{20} | jj,\tfrac{1}{2}\, m_t \rangle - \\ & - \tfrac{1}{2} \langle jj,\tfrac{1}{2}\, m_t | r^2 \, Y_{20} \, \underset{\sim}{\tau} | jj,\tfrac{1}{2}\, m_t \rangle\} \\ = \left(\frac{16\pi}{5}\right)^{\frac{1}{2}} & \{\tfrac{1}{2} \, \langle \tfrac{1}{2} \| 1 \| \tfrac{1}{2} \rangle - \tfrac{1}{2} \, \langle \tfrac{1}{2}\, m_t 1 0 | \tfrac{1}{2}\, m_t \rangle \langle \tfrac{1}{2} \| \tau \| \tfrac{1}{2} \rangle\} \times \\ & \langle jj20|jj\rangle \langle j \| r^2 \, Y_2 \| j \rangle . \end{aligned}$$

We have denoted the isospin quantum numbers of a single nucleon by $t=\frac{1}{2}$ with magnetic projection m_t. Note there are two terms; the first is called the isoscalar and the second the isovector term. For ^{17}F with $m_t=-\frac{1}{2}$ these two terms add equally (the quantity in braces summing to unity), and the resulting expression for Q reduces to that obtained pre-iously, as of course it should.

Had we asked for the quadrupole moment of ^{17}O, the mirror nucleus, the same expression could be used, but this ime with $m_t=+\frac{1}{2}$. Now the isoscalar and isovector terms can-el each other, giving zero quadrupole moment.

Experimentally the quadrupole moment of ^{17}O is far from ero, in fact its value $|Q| = 2.65 \pm 0.3$ e fm^2 (Ajzenberg-

Selove 1971) is not far short of the single proton value of -5 e fm^2. This latter figure derives from the formula $-4/7\ \langle r^2\rangle$ with the radial integral $\langle r^2\rangle$ estimated at 9 fm^2. Experiment therefore guides us to the inescapable conclusion that the description of ^{17}O as a closed shell plus a single neutron is over simplistic. Nevertheless we are reluctant to give up these simplistic ideas. Instead we take the closed-shell-plus-one description as a starting point and use perturbation theory to calculate a correction for not only the quadrupole moment but also for the expectation value of other operators as well. The details of this approach are spelt out in Chapter 5.

1.7 CLOSED-SHELL-PLUS-TWO NUCLEI

The definition of a coupled two-particle state (eqn (1.38)) is only slightly modified by the presence of a closed-shell core, and a two-particle wavefunction is now written

$$|j_m j_n; JM\rangle_A = (1+\delta_{mn})^{-1/2} \sum_{m_m m_n} \langle j_m m_m j_n m_n | JM\rangle\, a_n^\dagger\, a_m^\dagger |C\rangle. \quad (1.58)$$

The orbitals j_m and j_n are unoccupied in the core. Matrix elements of one-body and two-body operators can be found in a straightforward manner, a contribution from the core only arising in the case of diagonal matrix elements of scalar operators.

As an illustration, let us reconsider the quadrupole moment example of the last section and this time calculate Q for ^{18}F. We assume the ground state has a shell model configuration of a $d_{5/2}$ proton coupled to a $d_{5/2}$ neutron to produce a state of spin J=1. Then the quadrupole moment is given by the matrix element

$$Q = \left(\frac{16\pi}{5}\right)^{\frac{1}{2}} \langle [j_p, j_n]; J\ M{=}J | r^2\ Y_2 | [j_p, j_n]; J\ M{=}J\rangle$$

where j_p represents the $d_{5/2}$ proton, j_n the $d_{5/2}$ neutron. In the proton-neutron formalism these two particles are distinct, so there is no antisymmetrization required. To evaluate Q, first the Wigner-Eckhart theorem is applied, then since the operator r^2Y_2 acts only on protons, the neutron dependence can be factored out of the matrix element (see eqn (A.38a) in the Appendix) to produce the result.

$$Q = \left(\frac{16\pi}{5}\right)^{\frac{1}{2}} \langle JJ20|JJ\rangle\, U(J2j_nj_p;J\,j_p)\langle j_p\| r^2Y_2\| j_p\rangle$$

$$= +\frac{32}{175}\langle r^2\rangle \tag{1.59}$$

when values $j_p = j_n = 5/2$ and J=1 are inserted.

Let us repeat the calculation in the isospin formalism. This time the matrix element is

$$Q = \left(\frac{16\pi}{5}\right)^{\frac{1}{2}} \frac{1}{2}\langle [j,j];JMTN|\sum_i r_i^2\, Y_{20}(\hat{r}_i)\tfrac{1}{2}[1-\tau_z(i)]|[j,j];JMTN\rangle$$

where the factor 1/2 comes from the normalization $(1+\delta_{mn})^{-1/2}$ in eqn (1.58) for two equivalent nucleons. First note that ^{18}F ground state has isospin T=0, so that only the isoscalar part of the operator contributes to the matrix element. Second, the sum i runs over both extra-core nucleons, and as illustrated in eqn (1.23), the manipulation of the creation and annihilation operators for two-particle states leads to four terms, each term giving an identical contribution. Evaluating just the first term and multiplying the result by four gives

$$Q = \left(\frac{16\pi}{5}\right)^{\frac{1}{2}}\cdot\frac{1}{2}\cdot\frac{1}{2}\cdot 4\,\langle [jj];JMTN|r^2Y_2|[jj];JMTN\rangle$$

here the operator r^2Y_2 now acts only on the first nucleon. hus proceeding as before, using the Wigner-Eckhart theorem nd the factorization theorem, we get

$$Q = \left(\frac{16\pi}{5}\right)^{\frac{1}{2}} \langle JJ20|JJ\rangle\langle TN00|TN\rangle\, U(J2j_nj_p;Jj_p)\,\times$$

$$\times\, U(T0\tfrac{1}{2}\tfrac{1}{2};T\tfrac{1}{2})\,\langle j\|r^2Y_2\|j\rangle\,\langle\tfrac{1}{2}\|1\|\tfrac{1}{2}\rangle \tag{1.60}$$

hich reduces to the previous result, since all the isospin actors give unity.

1.8 CLOSED-SHELL-MINUS-ONE

Consider for the moment, a core state comprising just a single closed shell (eqn (1.41)),

$$|C\rangle = \prod_i a_i^\dagger|0\rangle$$

and the state of angular momentum j and magnetic projection m formed by annihilating the particle with projection -m in that closed-shell wavefunction. Such a state is referred to as a single-hole state and is defined as

$$|j^{-1}m\rangle = S_m\, a_{-m}|C\rangle \qquad (1.61)$$

where S_m is a simple sign factor to be determined. The requirement is that this single-hole state, eqn (1.61), be formally identical with the state made from 2j-particles acting on a vacuum. That is

$$S_m\, a_{-m}|C\rangle \equiv \prod_i{}' a_i^\dagger|0\rangle \qquad (1.62)$$

where the product Π' is over 2j-creation operators; the one operator missing is $a_{-m}^\dagger$, being just the one required to make up the closed shell. Examining the left-hand side of eqn (1.62), the sign factor is determined by the number of commutations that a_{-m} has to make with the operators in $|C\rangle$ before $a_{-m}^\dagger$ is reached. Thus S_m is +1 if that number is even and -1 if that number is odd. To evaluate S_m, the particular ordering of creation operators that form the closed shell $|C\rangle$ has to be specified. This ordering is arbitrary, so we make the choice:

$$|C\rangle = a_j^\dagger\, a_{(j-1)}^\dagger \ldots\ldots a_{-(j-1)}^\dagger a_{-j}^\dagger|0\rangle \qquad (1.63)$$

for which

$$S_m = (-)^{j+m}.$$

This choise is particularly convenient since a single-hole

state is now related to a single-particle state by the time-reversal operation. Let $b_m^\dagger$ be an operator creating a single-hole state of magnetic projection +m, then

$$
\begin{aligned}
|j^{-1}\ m\rangle &\equiv b_m^\dagger|C\rangle \\
&= (-)^{j+m}\ a_{-m}|C\rangle \qquad (1.65) \\
&= a_{\tilde{m}}|C\rangle
\end{aligned}
$$

where the state $|\tilde{m}\rangle$ is the time-reversed state $S_m|-m\rangle$.

The calculation of a one-body matrix element between single-hole states follows in a straightforward way:

$$
\begin{aligned}
\langle k^{-1}|F_M^{(L)}|\ell^{-1}\rangle &= \sum_{ij} \langle C|a_{\tilde{k}}^\dagger\ a_i^\dagger\ a_j^\dagger\ a_{\tilde{\ell}}|C\rangle\langle i|F_M^{(L)}|j\rangle \\
&= -\langle\tilde{\ell}|F_M^{(L)}|\tilde{k}\rangle + \langle C|F_M^{(L)}|C\rangle\delta_{k,\ell}\ \delta_{L,0},
\end{aligned}
\qquad (1.66)
$$

there being a contribution from the core only for scalar one-body operators. Eqn (1.66) illustrates how a matrix element between hole states is related to a particle matrix element in the time-reversed states.

The calculation of a two-body operator follows similarly:

$$
\begin{aligned}
\langle k^{-1}|G|\ell^{-1}\rangle &= \frac{1}{4} \sum_{iji'j'} \langle C|a_{\tilde{k}}^\dagger\ a_i^\dagger\ a_j^\dagger\ a_{i'}\ a_{j'}\ a_{\tilde{\ell}}|C\rangle\langle ij|G|i'j'\rangle_A \\
&= -\sum_i \langle i\tilde{\ell}|G|i\tilde{k}\rangle_A + \langle C|G|C\rangle\delta_{k,\ell},
\end{aligned}
\qquad (1.67)
$$

where the sum i is over all states in the core including k and ℓ. Again all matrix elements on the right of eqn (1.67) are particle matrix elements.

At the risk of belabouring a point, let us calculate the quadrupole moment of ^{27}Aℓ on the assumption that this nucleus can be described as a single proton $d_{5/2}$-hole coupled to a closed-shell ^{28}Si core. The quadrupole moment, then, is easily evaluated using eqn (1.66):

$$Q = \left(\frac{16\pi}{5}\right)^{\frac{1}{2}} \langle j^{-1}\ m{=}j \,|\, r^2 Y_2 \,|\, j^{-1}\ m{=}j \rangle$$

$$= -\left(\frac{16\pi}{5}\right)^{\frac{1}{2}} \langle j\ m{=}{-j} \,|\, r^2 Y_2 \,|\, j\ m{=}{-j} \rangle$$

$$= -\left(\frac{16\pi}{5}\right)^{\frac{1}{2}} \langle j\, {-j}\ 20 \,|\, j\, {-j} \rangle \langle j \| r^2 Y_2 \| j \rangle$$

$$= +\frac{4}{7} \langle r^2 \rangle,$$

where the value $j = 5/2$ has been inserted to obtain the last line. Notice the sign reversal with respect to the single-particle calculation presented in section (1.6).

Experimental measurements of the quadrupole moment of ^{27}Aℓ average at +15.1 ± 0.3 e fm^2 (Endt and Van der Leun 1973), which is three times larger than our single-hole estimate and indicates the need for a more complicated description.

1.9 WICK'S THEOREM

In the last section, matrix elements involving hole states were converted into equivalent matrix elements in the time-reversed particle states. This is an extremely useful device, which will now be generalized for more complicated wavefunctions involving mixtures of hole and particle states. A core is specified, for which all orbitals $j_1, j_2 \ldots, j_F$ up to some chosen Fermi level are filled. States occupied in this core will be denoted with subscript labels i, j, k, and ℓ, and states unoccupied (i.e., above the Fermi level) will be denoted by m, n, p, and q. Greek subscripts α, β, γ, and δ will label states which may be above or below the Fermi surface. The single-particle and single-hole creation and annihilation operators have the property

$$a_m^\dagger |C\rangle = |m\rangle \qquad\qquad a_m|C\rangle = \text{zero}$$

$$b_i^\dagger |C\rangle = a_{\tilde{i}} |C\rangle = S_i\ a_{-i} |C\rangle = |i^{-1}\rangle \qquad\qquad b_i|C\rangle = \text{zero}.$$

The minus sign in the subscript a_{-i} implies a minus sign on the magnetic quantum numbers. The operators b and $b^\dagger$ satisfy the same anticommutation properties as $a^\dagger$ and a.

It is convenient to introduce another set of creation and

annihilation operators $c_\alpha^\dagger$ and c_α which are defined as

$$\begin{aligned} c_\alpha^\dagger &= u_\alpha\, a_\alpha^\dagger - v_\alpha\, b_\alpha^\dagger \\ c_\alpha &= u_\alpha\, a_\alpha - v_\alpha\, b_\alpha . \end{aligned} \tag{1.68}$$

u_α and v_α are real coefficients having the value zero or one depending whether α is an occupied or unoccupied state, namely

$$\begin{aligned} u_i &= 0 \qquad v_i = 1 \\ u_m &= 1 \qquad v_m = 0. \end{aligned} \tag{1.69}$$

Thus $v_i^{\;2}$ can be interpreted as an occupation probability for the state $|i\rangle$. The requirement that the operators $c_\alpha^\dagger$ and c_α should satisfy the anticommutation relation $[c_\alpha^\dagger, c_\alpha]_+ = 1$ leads to the relation $u_\alpha^{\;2} + v_\alpha^{\;2} = 1$. Furthermore, the requirement that $[c_\alpha, c_{-\alpha}]_+ = 0$ shows that v_α and $u_{-\alpha}$ are not independent but satisfy

$$\frac{u_\alpha}{u_{-\alpha}} = -\frac{S_\alpha v_\alpha}{S_{-\alpha} v_{-\alpha}}.$$

We choose a convention that u_α and v_α are always positive numbers for $\alpha > 0$. If the division between occupied and unoccupied orbits is taken at a shell closure, so that when a state $|\alpha\rangle$ is occupied (unoccupied) its time-reversed state $|\tilde{\alpha}\rangle$ is also occupied (unoccupied), then we can choose $u_{-\alpha} = u_\alpha$ and hence find $v_{-\alpha}$ to be

$$v_{-\alpha} = -S_\alpha S_{-\alpha} v_\alpha . \tag{1.70}$$

The inverse transformation of eqns (1.68) is

$$\begin{aligned} a_\alpha^\dagger &= u_\alpha\, c_\alpha^\dagger + v_\alpha\, S_\alpha\, c_{-\alpha} \\ a_\alpha &= u_\alpha\, c_\alpha + v_\alpha\, S_\alpha\, c_{-\alpha}^\dagger . \end{aligned} \tag{1.71}$$

The main purpose in introducing these new operators is

that they have the property

$$c_\alpha |C\rangle = \text{zero} \tag{1.72}$$

for *all* α. This property is extremely useful for simplifying expressions involving strings of creation and annihilation operators which are acting on the core state $|C\rangle$. The technique is to apply the standard commutation relations to reorder the string of operators so that all the creation operators $c_\alpha^\dagger$ are on the left of the annihilation operators, c_α. This is called the *normal order* and the string of operators in such an order is referred to as a *normal product*. Examples of normal products are $c_\alpha^\dagger c_\beta$, $c_\alpha^\dagger c_\beta^\dagger$, $c_\alpha c_\beta$ but the product $c_\alpha c_\beta^\dagger$ is not in normal order. Wick's theorem provides a prescription for arranging a string of operators in normal order.

Suppose that A is an operator comprising two parts

$$A = A^{(+)} + A^{(-)}, \tag{1.73}$$

where $A^{(+)}$ is a part containing only creation operators and $A^{(-)}$ contains only annihilation operators. For example, operators $a_\alpha^\dagger$ and a_α in eqn (1.71) are of this form. Consider a second operator, B, of the same type; then the product AB written out is

$$\begin{aligned} AB &= (A^{(+)} + A^{(-)})(B^{(+)} + B^{(-)}) \\ &= A^{(+)}B^{(+)} + A^{(+)}B^{(-)} + A^{(-)}B^{(+)} + A^{(-)}B^{(-)}. \end{aligned} \tag{1.74}$$

Only the third term on the right-hand side is not in normal order. The normal product of operators A and B is defined as

$$N(AB) = A^{(+)}B^{(+)} + A^{(+)}B^{(-)} - B^{(+)}A^{(-)} + A^{(-)}B^{(-)} \tag{1.75}$$

such that

$$AB = N(AB) + \overline{AB} \tag{1.76}$$

where $\overline{AB}$ is just a pure number known as the *contraction* of the operators A and B. The contraction is expressible in terms of the anticommutator

$$\overline{AB} = [A^{(-)}, B^{(+)}]_+. \tag{1.77}$$

Three properties of normal products follow immediately from these definitions.

$$\begin{aligned} &\text{(i)} && N(BA) = -N(AB) \\ &\text{(ii)} && \langle C|N(AB)|C\rangle = \text{zero} \\ &\text{(iii)} && \overline{AB} = \langle C|AB|C\rangle. \end{aligned} \tag{1.78}$$

Property (i) follows from the anticommutator relations, property (ii) uses eqn (1.72), and property (iii) combines property (ii) and the definition eqn (1.76). It is clear that the definitions of the normal product and the contraction of two operators A and B depend upon the choice of core.

The idea of a normal product can be extended to an arbitrary number of operators. Let $A_1A_2...A_N$ be a mixed group of creation and annihilation operators. The normal product of these operators is defined as

$$N(A_1A_2...A_N) = \lambda_p(A_iA_jA_k....) \tag{1.79}$$

where $A_iA_jA_k...$ are the same operators as the original sequence, but ordered so that the creation operators are on the left of the destruction operators. The factor λ_p is just a sign, $(-)^p$, where p represents the number of permutations of the operators required to go from the original to the normal ordering sequence. If the operator sequence contains sums of products, as would be obtained with operators of type eqn (1.73), each term in the sum must be treated in the same way. For example,

$$\begin{aligned} N[(A^{(+)}+A^{(-)})(B^{(+)}+B^{(-)})....] &= N(A^{(+)}B^{(+)}...)+N(A^{(-)}B^{(+)}...) + \\ &+ N(A^{(+)}B^{(-)}...)+N(A^{(-)}B^{(-)}...)+.... \end{aligned} \tag{1.80}$$

With these preliminaries, we are now ready to state Wick's theorem.

For N operators $A_1A_2...A_N$ the following expansion holds

$$\begin{aligned} A_1A_2...A_N &= N(A_1A_2...A_N) + \\ &+ \sum_{i<j} \lambda_{ij}\, A_iA_j\, N_{ij}(A_1A_2...A_N) + \\ &+ \sum_{\substack{i<j\\k<\ell}} \lambda_{ij}\lambda_{k\ell}\, A_iA_j\, A_kA_\ell\, N_{ijk\ell}(A_1A_2...A_N) + \\ &+ \ldots \end{aligned} \tag{1.81}$$

where $N_{ij}(A_1A_2...A_N)$ is the normal product with operators A_i and A_j removed and λ_{ij} is a sign, $(-)^{p_{ij}}$, with p_{ij} being the number of permutations required to bring A_i and A_j to the front of the sequence of operators. The proof of this theorem will not be given here (see, for example, section 8.4(b) of Muirhead (1965)), but follows from an inductive argument which assumes the theorem if true for N-2 operators and proves the theorem for N operators. The theorem is trivially true for one and two operators.

We shall require this theorem for two and four operators. For two operators

$$AB = N(AB) + \overbrace{AB}$$

which just repeats eqn (1.76). For four operators

$$\begin{aligned} ABCD = N(ABCD) &+ \overbrace{AB}\, N(CD) - \overbrace{AC}\, N(BD) + \overbrace{AD}\, N(BC) + \\ &+ \overbrace{BC}\, N(AD) - \overbrace{BD}\, N(AC) + \overbrace{CD}\, N(AB) + \\ &+ \overbrace{AB}\, \overbrace{CD} - \overbrace{AC}\, \overbrace{BD} + \overbrace{AD}\, \overbrace{BC} \end{aligned} \tag{1.82}$$

which is a sum of ten terms.

1.10 ONE-BODY AND TWO-BODY OPERATORS IN PARTICLE-HOLE FORMALISM

Earlier a one-body operator was defined in eqn (1.20) as

$$F = \sum_{\alpha\beta} \langle\alpha|F|\beta\rangle \ a^{\dagger}_{\alpha}a_{\beta}. \tag{1.83}$$

This formula is rather cumbersome to use for systems containing a large number of particles. Since most of these particles contribute to closed-shell configurations and only a few particles occupy partially-filled shells, it is more convenient to take the closed-shell configuration as a new reference and to compute the expectation value of F relative to its value in the reference closed-shell configuration.

To achieve this, the operators $a^{\dagger}_{\alpha}$ and a_{β} are first expressed in terms of the new set of operators $c^{\dagger}_{\alpha}$, eqn (1.71). Using Wick's theorem, viz.

$$a^{\dagger}_{\alpha} a_{\beta} = N(a^{\dagger}_{\alpha} a_{\beta}) + \overbrace{a^{\dagger}_{\alpha}a_{\beta}}$$

and expanding out the normal product, four terms are obtained

$$N(a^{\dagger}_{\alpha}a_{\beta}) = u_{\alpha}u_{\beta} \ c^{\dagger}_{\alpha}c_{\beta} + S_{\alpha}v_{\alpha}u_{\beta}c_{-\alpha}c_{\beta} + S_{\beta}u_{\alpha}v_{\beta}c^{\dagger}_{\alpha}c^{\dagger}_{-\beta} - \\ - S_{\alpha}S_{\beta}v_{\alpha}v_{\beta}c^{\dagger}_{-\beta}c_{-\alpha}. \tag{1.84}$$

Similarly expanding out the contraction, again four terms are obtained

$$\begin{aligned} \overbrace{a^{\dagger}_{\alpha}a_{\beta}} &= \langle C|a^{\dagger}_{\alpha}a_{\beta}|C\rangle \\ &= u_{\alpha}u_{\beta}\langle C|c^{\dagger}_{\alpha}c_{\beta}|C\rangle + S_{\alpha}v_{\alpha}u_{\beta}\langle C|c_{-\alpha}c_{\beta}|C\rangle + \\ &\quad + S_{\beta}u_{\alpha}v_{\beta}\langle C|c^{\dagger}_{\alpha}c^{\dagger}_{-\beta}|C\rangle + S_{\alpha}S_{\beta}v_{\alpha}v_{\beta}\langle C|c_{-\alpha}c^{\dagger}_{-\beta}|C\rangle \qquad (1.85) \\ &= S_{\alpha}S_{\beta}v_{\alpha}v_{\beta}\langle C|c_{-\alpha}c^{\dagger}_{-\beta}|C\rangle \\ &= v_{\alpha}{}^{2} \ \delta_{\alpha\beta} \end{aligned}$$

from which only one term survives, the other three terms vanishing on account of the property $c_\alpha|C\rangle$ = zero (eqn (1.72)). Thus the expansion of a one-body operator in a particle-hole formalism becomes

$$F = \sum_{\alpha\beta} \langle\alpha|F|\beta\rangle \{u_\alpha u_\beta\, c_\alpha^\dagger c_\beta + S_\alpha v_\alpha u_\beta c_{-\alpha} c_\beta + S_\beta u_\alpha v_\beta c_\alpha^\dagger c_{-\beta}^\dagger - S_\alpha S_\beta v_\alpha v_\beta c_{-\beta}^\dagger c_{-\alpha} + v_\alpha^2\, \delta_{\alpha\beta}\} \quad (1.86)$$

where the coefficients u and v are either zero or one as enumerated in eqn (1.70). If we use subscripts i,j,k,ℓ for occupied states and m,n,p,q for unoccupied states, and write the expressions in terms of the original a and b operators, then we get our final result

$$F = \sum_{mn} \langle m|F|n\rangle\, a_m^\dagger\, a_n + \quad (1.87a)$$

$$+ \sum_{in} \langle \tilde{i}|F|n\rangle\, b_i\, a_n + \quad (1.87b)$$

$$+ \sum_{mj} \langle m|F|\tilde{j}\rangle\, a_m^\dagger\, b_j^\dagger - \quad (1.87c)$$

$$- \sum_{ij} \langle \tilde{i}|F|\tilde{j}\rangle\, b_j^\dagger\, b_i + \quad (1.87d)$$

$$+ \sum_{ij} \langle i|F|j\rangle\, \delta_{ij} \quad (1.87e)$$

where $|\tilde{j}\rangle$ is the time-reversed state, i.e. $|\tilde{j}\rangle = S_j|-j\rangle$. Here the first (fourth) term describes a transition between two single-particle (hole) states, and the second (third) term describes the creation (annihilation) of a particle-hole pair during the transition. The last term represents the expectation value of the core. This term only gives a contribution when a diagonal matrix element of a scalar operator is being evaluated.

What has been achieved here is that the general definition of a one-body operator, eqn (1.83), involving a sum over *all* states, has been replaced by a series of terms in which the summations over occupied states have been differentiated from the summations over unoccupied states. This is particularly useful when one is working with wavefunctions which are mixtures of particle and hole states. For example, the expectation value of a one-body operator between a particle-hole state $|j^{-1}\ m\rangle$ and the core is

$$\langle C|F|j^{-1}\ m\rangle = \langle C|F\ a_m^\dagger\ b_j^\dagger|C\rangle.$$

The expansion for F (eqn (1.87)) is now inserted, however, we can see immediately that only the term with one b-operator and one a-operator is required; all other terms give zero expectation value. Thus

$$\begin{aligned}\langle C|F|j^{-1}m\rangle &= \sum_{in} \langle \tilde{i}|F|n\rangle\langle C|b_i a_n a_m^\dagger b_j^\dagger|C\rangle \\ &= \langle \tilde{j}|F|m\rangle. \end{aligned} \tag{1.88}$$

This, of course, is a trivial example in which a particle-hole matrix element has been expressed in terms of a particle-particle matrix element, but it does illustrate the two simplifying features. First in the expansion of F, one can very easily pick out which terms are going to contribute to the evaluation of a particular matrix element, and secondly with the operators arranged in normal order, the contraction of a string of creation and annihilation operators is achieved rapidly with a minimum of rearrangement.

Using the same procedure, we seek a particle-hole representation for a two-body operator, eqn (1.30):

$$G = \frac{1}{4}\sum_{\alpha\beta\gamma\delta} \langle\alpha\beta|G|\gamma\delta\rangle_A\ a_\alpha^\dagger\ a_\beta^\dagger\ a_\delta\ a_\gamma.$$

As before the aim is to replace this expression by one in which the sums over occupied states have been differentiated from the sums over unoccupied states. The various steps are: (i) rewrite $a_\alpha^\dagger\ a_\beta^\dagger\ a_\delta\ a_\gamma$ as a sum of terms involving normal products and contractions, viz.

$$a^\dagger_\alpha a^\dagger_\beta a_\delta a_\gamma = N(a^\dagger_\alpha a^\dagger_\beta a_\delta a_\gamma) - \overbrace{a^\dagger_\alpha a_\delta}N(a^\dagger_\beta a_\gamma) + \overbrace{a^\dagger_\alpha a_\gamma} N(a^\dagger_\beta a_\delta) + \overbrace{a^\dagger_\beta a_\delta} N(a^\dagger_\alpha a_\gamma) - \overbrace{a^\dagger_\beta a_\gamma} N(a^\dagger_\alpha a_\delta) - \overbrace{a^\dagger_\alpha a_\delta}\,\overbrace{a^\dagger_\beta a_\delta} + \overbrace{a^\dagger_\alpha a_\gamma}\,\overbrace{a^\dagger_\beta a_\delta}; \tag{1.89}$$

(ii) expresses the a-operators in terms of the new operators c_α and $c^\dagger_\alpha$ using eqn (1.71); (iii) expand out the normal products and contractions as in eqns (1.84) and (1.85); (iv) gather together all terms which are simply related to each other by an interchange of labels and (v) express the result in terms of the original a and b operators. The conclusion of this tedious but straightforward exercise is the following:

$$G = \frac{1}{4} \sum_{mnpq} \langle mn|G|pq\rangle_A\, a^\dagger_m a^\dagger_n a_q a_p + \tag{1.90a}$$

$$+ \frac{1}{2} \sum_{\ell mnp} \langle mn|G|p\tilde{\ell}\rangle_A\, a^\dagger_m a^\dagger_n b^\dagger_\ell a_p + \tag{1.90b}$$

$$+ \frac{1}{2} \sum_{jmpq} \langle m\tilde{j}|G|pq\rangle_A\, a^\dagger_m b_j a_q a_p + \tag{1.90c}$$

$$+ \frac{1}{4} \sum_{k\ell mn} \langle mn|G|\tilde{k}\tilde{\ell}\rangle_A\, a^\dagger_m a^\dagger_n b^\dagger_\ell b^\dagger_k + \tag{1.90d}$$

$$+ \frac{1}{4} \sum_{ijpq} \langle \tilde{i}\tilde{j}|G|pq\rangle_A\, b_i b_j a_q a_p + \tag{1.90e}$$

$$+ \sum_{jkmq} \langle m\tilde{j}|G|\tilde{k}q\rangle_A\, a^\dagger_m b^\dagger_k b_j a_q + \tag{1.90f}$$

$$+ \frac{1}{2} \sum_{jk\ell m} \langle m\tilde{j}|G|\tilde{k}\tilde{\ell}\rangle_A\, a^\dagger_m b^\dagger_\ell b^\dagger_k b_j + \tag{1.90g}$$

$$+ \frac{1}{2} \sum_{ij\ell p} \langle \tilde{i}\tilde{j}|G|p\tilde{\ell}\rangle_A\, b^\dagger_\ell b_i b_j a_p + \tag{1.90h}$$

$$+ \frac{1}{4} \sum_{ijk\ell} \langle \tilde{i}\tilde{j}|G|\tilde{k}\tilde{\ell}\rangle_A\, b^\dagger_\ell b^\dagger_k b_i b_j + \tag{1.90i}$$

$$+ \sum_{iknq} \langle in|G|kq\rangle_A\, \delta_{ik}\, a^\dagger_n a_q + \tag{1.90j}$$

$$+ \sum_{ik\ell n} \langle in|G|k\tilde{\ell}\rangle_A\, \delta_{ik}\, a^\dagger_n b^\dagger_\ell + \tag{1.90k}$$

$$+ \sum_{ijkq} \langle i\tilde{j}|G|kq\rangle_A \, \delta_{ik} \, b_j \, a_q \, - \tag{1.90ℓ}$$

$$- \sum_{ijk\ell} \langle i\tilde{j}|G|k\tilde{\ell}\rangle_A \, \delta_{ik} \, b_\ell^\dagger \, b_j \, + \tag{1.90m}$$

$$+ \frac{1}{2} \sum_{ijk\ell} \langle ij|G|k\ell\rangle_A \, \delta_{ik} \, \delta_{j\ell}. \tag{1.90n}$$

Terms (1.90a), (1.90f), (1.90i) describe transitions between two-particle, particle-hole, and two-hole states respectively. Terms (1.90b), (1.90c), (1.90g), (1.90h) correspond to the creation or annihilation of a particle-hole pair while one particle or one hole undergoes a transition. Terms (1.90d) and (1.90e) create or annihilate two pairs. Terms (1.90j), (1.90k), (1.90ℓ) and (1.90m) all involve one contraction and consequently have the appearance of a one-body operator. Finally, term (1.90n) contains two contractions and appears as a zero-body operator. This term represents the expectation value of the core. Notice that as soon as we define a 'model' space in which we are interested only in the movement of a few particles and/or holes defined relative to some chosen core, then a true two-body operator becomes an effective operator comprising zero-body, one-body and two-body terms in the model space.

The expansion of a two-body operator into fourteen terms appears rather cumbersome at first sight; however, using this formula can save a considerable amount of work. Suppose one wished to calculate the expectation value of a two-body operator between a two-particle one-hole state, $|a^{-1}\,rs\rangle$ and a single-particle state $|t\rangle$, i.e.

$$\langle t|G|a^{-1}rs\rangle = \langle C|a_t \; G \; a_s^\dagger \; a_r^\dagger \; b_a^\dagger \; |C\rangle.$$

To evaluate this matrix element, one only needs to insert terms (1.90c) and (1.90ℓ) for G, the other twelve terms giving zero contribution. Thus

$$\langle t|G|a^{-1}rs\rangle = \frac{1}{2} \sum_{jmpq} \langle m\tilde{j}|G|pq\rangle_A \langle C|a_t \; a_m^\dagger \; b_j \; a_q \; a_p \; a_s^\dagger \; a_r^\dagger \; b_a^\dagger|C\rangle \, +$$

$$+ \sum_{ijkq} \langle i\tilde{j}|G|kq\rangle_A \, \delta_{ik} \langle C|a_t \, b_j \, a_q \, a_s^\dagger \, a_r^\dagger \, b_a^\dagger|C\rangle$$

$$= \frac{1}{2} \sum_{jmpq} \langle m\tilde{j}|G|pq\rangle_A \, \delta_{ja} \, \delta_{tm} \, [\delta_{ps} \, \delta_{rq} - \delta_{pr} \, \delta_{sq}] +$$

$$+ \sum_{ijkq} \langle i\tilde{j}|G|kq\rangle_A \, \delta_{ik} \, \delta_{ja} \, [\delta_{qs} \, \delta_{rt} - \delta_{qr} \, \delta_{st}]$$

$$= \langle t\tilde{a}|G|sr\rangle_A + \sum_i \{\langle i\tilde{a}|G|is\rangle_A \, \delta_{rt} - \langle i\tilde{a}|G|ir\rangle_A \, \delta_{st}\}.$$

This will be the last time we write out an example in such detail; however, it does illustrate the usefulness of eqn (1.90) in saving labour.

1.11 DIAGRAMS

Eqns (1.87) and (1.90) describe the expansion of a one-body and two-body operator into products of single-particle and single-hole creation and annihilation operators arranged in normal order. Each term can be pictured in an elegant manner by the use of Feynman-Goldstone diagrams. A diagram consists of a number of vertices with lines joining them. Each vertex represents a point in coordinate space at which an interaction takes place. For example, the terms in eqn (1.87) will be represented by graphs containing one vertex, since $F(\underset{\sim}{x})$ is a one-body operator, and as such depends only on one point in coordinate space. Two-body operators $G(\underset{\sim}{x}_1,\underset{\sim}{x}_2)$ depend on two points $\underset{\sim}{x}_1$ and $\underset{\sim}{x}_2$ in coordinate space and are represented by graphs containing two vertices.

The convention is adopted in which time increases up the page, so that particles present in the initial state are found at the bottom of the diagram and particles present in the final state at the top of the diagram. The lines drawn to and between vertices are given a physical interpretation according to the rules:

(1) A line with one end entering a vertex and the other end free is used to represent a single-particle or single-hole, creation or annihilation operator. In Fig. 1.1, graph (a) represents a single-particle creation operator $a_m^\dagger$, graph (b) a single-particle annihilation operator a_n, graph (c) a single-hole creation operator $b_j^\dagger$, and graph (d) a single-hole

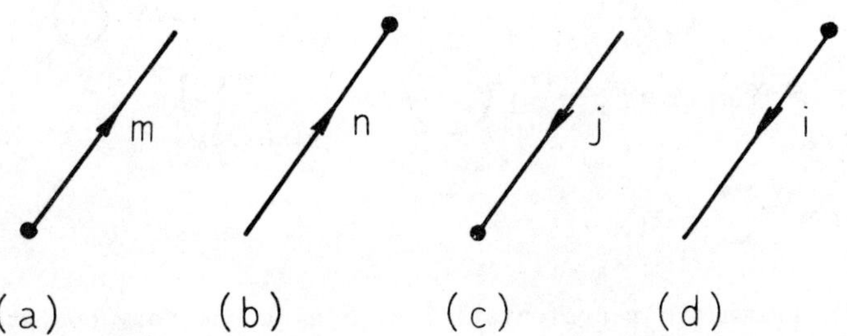

FIG. 1.1. Feynman-Goldstone diagrammatic representation of: (a) a simple-particle creation operator $a_m^\dagger$, (b) a single-particle annihilation operator a_n, (c) a single-hole creation operator b_j, and (d) a single-hole annihilation operator $b_i^\dagger$.

annihilation operator b_i. The solid dot represents the vertex. Notice the particle operators are represented by lines with arrows directed up the page, and hole operators by lines with arrows directed down the page.

(2) The matrix element of a one-body operator will be represented by a dashed line with one end entering a vertex and a cross at the other end, as in Fig. 1.2(a).

(3) The antisymmetrized matrix element of a two-body operator will be represented by a dashed line joining two vertices, as in Fig. 1.2(b).

(4) Each contraction is represented by a closed loop attached to a vertex, as in Fig. 1.2(c). Notice the arrow on the closed loop is directed downwards indicating that it represents a hole line. Remember the contraction only has a non-zero value for occupied states; one never has a closed loop of this type with an upgoing particle line.

(5) Solid lines joining two vertices, where the two vertices occur at different times (i.e. are displaced vertically from one another in the diagram, see for example Fig. 1.2(d)) represent the formation and subsequent annihilation of an intermediate state. The quantum mechanical description of such a process is furnished in perturbation theory. Associated with the formation of each intermediate state, perturbation theory introduces an energy denominator; thus in

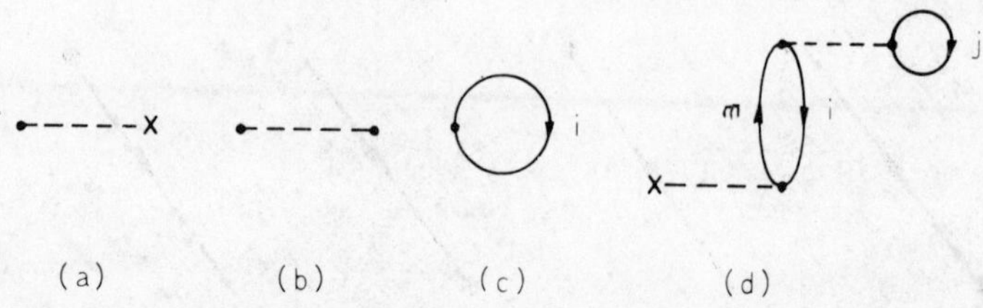

FIG. 1.2. Diagrammatic representation of: (a) a one-body operator, (b) a two-body operator, and (c) a contraction. Diagram (d) is a typical situation where an intermediate state has been created by the action of a one-body operator and subsequently annihilated.

the interpretation of a diagram such as Fig. 1.2(d), an energy denominator has to be included for each intermediate state. More on this later.

Thus the five ingredients of a Feynman-Goldstone diagram are (i) external lines, (ii) one-body matrix elements, (iii) two-body antisymmetrized matrix elements, (iv) closed loops, and (v) energy denominators with a summation occurring over all intermediate states. By way of illustration let us draw the diagrams corresponding to the terms in eqn (1.87) for the expansion of a one-body operator. These are given in Fig. 1.3 with graph (a) corresponding to term (1.87a), graph (b) corresponding to term (1.87b) and so on. Notice that particles present in the initial state are found at the bottom of the diagram and particles present in the final state

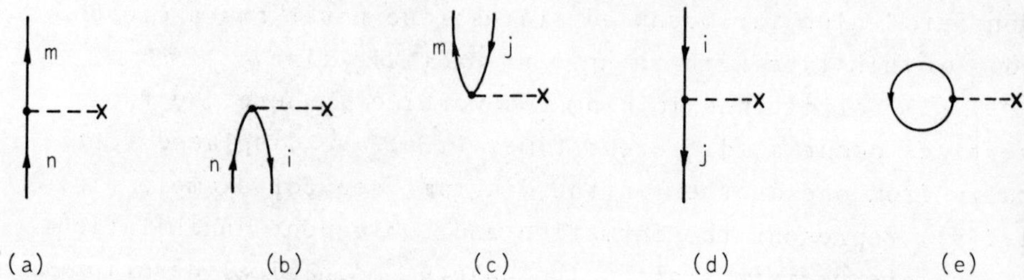

FIG. 1.3. Diagrammatic representation of the five terms in the expansion of a one-body operator in a particle-hole formalism, eqn (1.87). Diagram (a) corresponds to term (1.87a), and so on.

at the top. Furthermore each line with an arrow pointing upwards represents a particle (i.e. unoccupied) state and a line with an arrow pointing downwards represents a hole (i.e. occupied) state. Thus diagram (b) in Fig. 1.3 clearly describes the process in which an initial particle-hole state is annihilated by the action of a one-body operator. Similarly in Fig. 1.4 we have drawn the graphs corresponding to the fourteen terms in eqn (1.90) for the expansion of a two-body operator. Again the physical interpretation of the diagrams

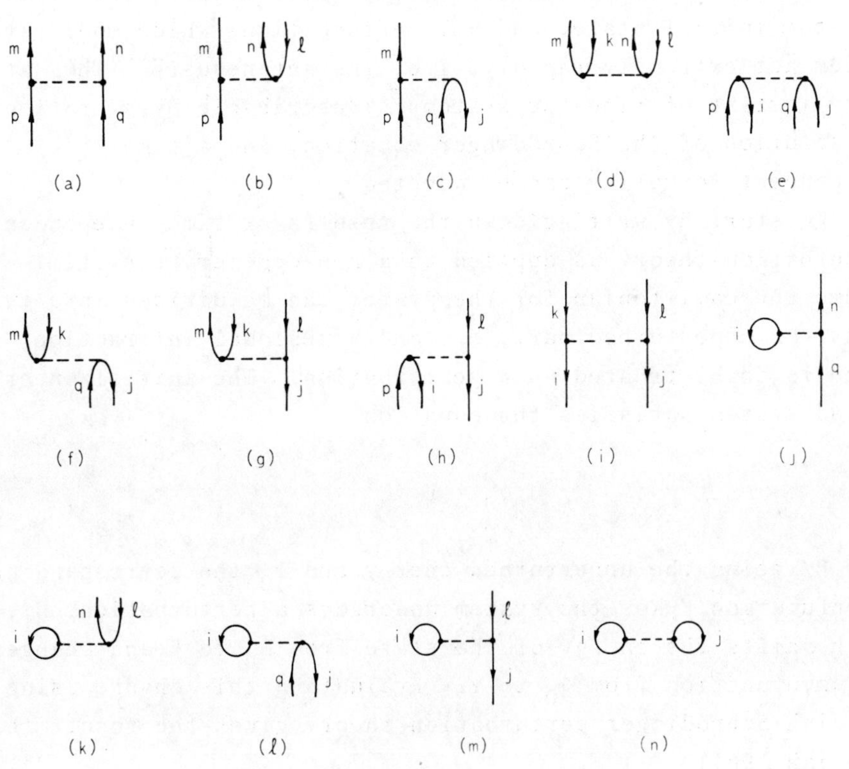

FIG. 1.4. Diagrammatic representation of the fourteen terms in the expansion of a two-body operator in a particle-hole formalism, eqn (1.90). Diagram (a) corresponds to term (1.90a), and so on.

is straightforward. Less straightforward, however, is the precise mathematical interpretation; the main problem is to obtain the correct overall sign factor. In appendix B we write down a set of rules for interpreting diagrams but before doing this we must look a little closer at perturbation theory, and decide precisely what energy denominator is being introduced with each intermediate state.

1.12 ENERGY DENOMINATORS

As mentioned in the last section, a Feynman-Goldstone diagram describes the time evolution of a quantum system as it undergoes a series of small perturbations. In nuclear physics it is frequently assumed that the final state evolves very slowly from the initial state, and all perturbations which modify the system act extremely rapidly, i.e. instantaneously. Thus at every instant of time the system is describable by a stationary solution of the Schrödinger equation, and a time-independent formalism can be adopted.

We start by writing down the results of time-independent perturbation theory as applied to a non-degenerate system. Assume the Hamiltonian for the system can be divided into two parts, an unperturbed part, H_0, and a residual interaction H_1 which is to be treated as a perturbation. The initial unperturbed system satisfies the equation

$$H_0\Phi_0 = E_0\Phi_0$$

with E_0 being the unperturbed energy and Φ_0 the corresponding eigenfunction. Now the system undergoes a perturbation, H_1, which shifts the energy of the state from E_0 to E and changes the wavefunction from Φ_0 to Ψ. Evaluating this change using Rayleigh-Schrödinger perturbation theory gives the result that (Messiah 1964)

$$E = E_0 + \langle \Phi_0|H_1|\Phi_0\rangle + \langle \Phi_0|H_1(E_0-H_0)^{-1}\, Q\, H_1|\Phi_0\rangle + \ldots \tag{1.91}$$

$$|\Psi\rangle = |\Phi_0\rangle + (E_0-H_0)^{-1}\, Q\, H_1|\Phi_0\rangle + \ldots \tag{1.92}$$

$$Q = 1 - |\Phi_0\rangle\langle\Phi_0|.$$

Here the operator Q is a projection operator which ensures

that Φ_0 does not occur as an intermediate state in any of the matrix elements. Inserting a complete set of states which are eigenfunctions of H_0, viz.

$$\frac{Q}{E_0-H_0} \rightarrow \sum_{\alpha}{}' \frac{|\Phi_\alpha\rangle\langle\Phi_\alpha|}{E_0-E_\alpha},$$

into eqns (1.91) and (1.92) gives the expansions

$$E = E_0 + \langle\Phi_0|H_1|\Phi_0\rangle + \sum_{\alpha}{}' \langle\Phi_0|H_1|\Phi_\alpha\rangle(E_0-E_\alpha)^{-1}\langle\Phi_\alpha|H_1|\Phi_0\rangle + \dots \tag{1.93}$$

$$|\Psi\rangle = |\Phi_0\rangle + \sum_{\alpha}{}' |\Phi_\alpha\rangle(E_0-E_\alpha)^{-1}\langle\Phi_\alpha|H_1|\Phi_0\rangle. \tag{1.94}$$

The prime on the summation symbol implies that Φ_0 is excluded as an intermediate state. It is expansions such as these which are being represented by Feynman-Goldstone diagrams. We illustrate this using the following two examples.

Example (1). Suppose the unperturbed wavefunction $|\Phi_0\rangle$ was a closed-shell Slater determinant, $|C\rangle$. The residual interaction H_1 is turned on, and to first order in H_1 it produces an energy shift of $\langle C|H_1|C\rangle$ and to second order in H_1 it produces an energy shift $\Sigma'_\alpha|\langle C|H_1|\Phi_\alpha\rangle|^2/(E_0-E_\alpha)$. This second contribution describes the process of scattering to some intermediate state Φ_α and subsequent scattering back to the closed shell. Notice the presence of the energy denominator (E_0-E_α). If H_1 is an operator comprising one-body and two-body parts but no higher-body parts, then the intermediate states Φ_α can be one-particle one-hole states or two-particle two-hole states. We consider each possibility in turn. Substituting the expansion for a two-body operator (eqn (1.90)) into eqn (1.93) and simplifying where possible, the following expression for the energy shift is obtained:

$$E-E_0 = \frac{1}{2}\sum_{ij}\langle ij|H_1|ij\rangle_A + \sum_{ijkm}\frac{\langle kj|H_1|mj\rangle_A\langle mi|H_1|ki\rangle_A}{(\varepsilon_k-\varepsilon_m)} +$$

$$+ \frac{1}{4}\sum_{k\ell mn}\frac{\langle k\ell|H_1|mn\rangle_A\langle mn|H_1|k\ell\rangle_A}{(\varepsilon_k+\varepsilon_\ell-\varepsilon_m-\varepsilon_n)}. \tag{1.95}$$

Sums $ijk\ell$ run over states occupied in the Slater determinant; and sums mn over unoccupied states. Each term in eqn (1.95) can be represented by a Feynman-Goldstone diagram and these are illustrated in Fig. 1.5. Diagram (a) represents the first order contribution to the energy shift, $E-E_0$, and is just the expectation value of the perturbation H_1 on the closed shell state, viz. $\langle C|H_1|C\rangle$. This matrix element has been evaluated previously in eqn (1.48). Diagrams (b) and (c) describe the second order contribution to the energy shift with the intermediate states being one-particle one-hole and two-particle two-hole states respectively. Rules for interpreting diagrams are given in appendix B; for the present we just want to consider the energy denominators.

The perturbation theory expansions, eqns (1.93) and (1.94) contain the factor $(E_0-E_\alpha)^{-1}$ where E_0 is the unperturbed energy of the closed shell, and E_α is the eigenenergy associated with the intermediate state Φ_α. These intermediate states are eigenfunctions of the unperturbed Hamiltonian H_0. Let us assume that H_0 is a one-body Hamiltonian; then Φ_0 is a Slater determinant of single particle wavefunctions, and the set of states Φ_α are formed by lifting one or two particles from the occupied states to the unoccupied states. The unperturbed energy E_α is then the energy of the core plus the sum of the single-particle energies of the newly occupied states minus the sum of the energies of the newly vacated states. Thus for Fig. 1.5(b) the energy denominator becomes

$$E_0-E_\alpha = E_0 - (E_0 + \varepsilon_m - \varepsilon_k) = \varepsilon_k - \varepsilon_m \, ,$$

and for Fig. 1.5(c)

$$E_0-E_\alpha = E_0 - (E_0+\varepsilon_m+\varepsilon_n-\varepsilon_k-\varepsilon_\ell) = \varepsilon_k+\varepsilon_\ell-\varepsilon_m-\varepsilon_n,$$

where ε_m is the single-particle energy for the state m. Notice that the energy denominators are negative quantities.

Example (2). Consider two single-particle states in a closed-shell-plus-one nucleus with unperturbed energies ε_m and ε_n

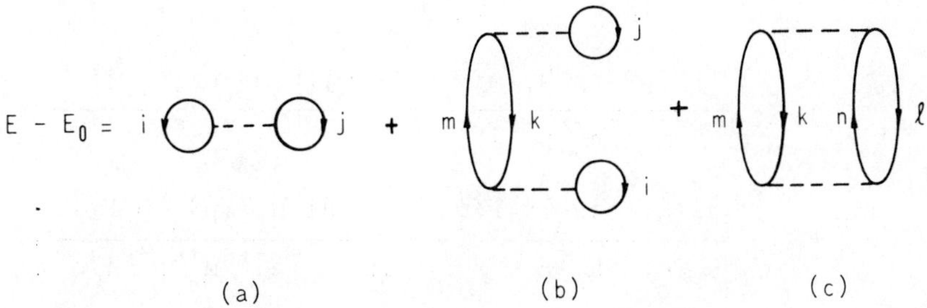

FIG. 1.5. First- and second-order corrections to the energy of a closed-shell state from non-degenerate time-independent perturbation theory. Diagram (a) represents $\langle C|H_1|C\rangle$, the expectation value of the perturbation in the closed-shell state, and diagrams (b) and (c) represent corrections due to scattering to intermediate one-particle one-hole and two-particle two-hole states respectively, and their subsequent annihilation. The three diagrams represent the three terms in eqn (1.95).

(expressed relative to the energy of the closed shell) and unperturbed wavefunctions $|m\rangle$ and $|n\rangle$. Suppose now a residual interaction H_1 is switched on, then the single-particle wavefunctions $|m\rangle$ and $|n\rangle$ are modified to $|\Psi_m\rangle$ and $|\Psi_n\rangle$. Let us now calculate the transition matrix element of a one-body operator, F, between these modified single-particle states $\langle\Psi_n|F|\Psi_m\rangle$ to first order in the perturbation H_1. Using eqn (1.94), we obtain the result

$$\langle\Psi_n|F|\Psi_m\rangle = \langle n|F|m\rangle + \sum_{\alpha}{}' \frac{\langle n|F|\Phi_\alpha\rangle\langle\Phi_\alpha|H_1|m\rangle}{(E_0+\varepsilon_m-E_\alpha)} + \sum_{\alpha}{}' \frac{\langle n|H_1|\Phi_\alpha\rangle\langle\Phi_\alpha|F|m\rangle}{(E_0+\varepsilon_n-E_\alpha)} . \tag{1.96}$$

In this case the intermediate state Φ_α can be a single-particle state (but not m or n), or a two-particle one-hole state. Considering each possibility in turn, and inserting the expansions for a one-body and a two-body operator, eqns (1.87) and (1.90), into eqn (1.96) and simplifying where possible, we obtain the result

$$\langle \Psi_n | F | \Psi_m \rangle = \langle n|F|m\rangle + \sum_{ip} \frac{\langle n|F|p\rangle \langle pi|H_1|mi\rangle_A}{\varepsilon_m - \varepsilon_p} +$$

$$+ \sum_{ip} \frac{\langle i|F|p\rangle \langle np|H_1|mi\rangle_A}{\varepsilon_m - (\varepsilon_n + \varepsilon_p - \varepsilon_i)} - \sum_{ij} \frac{\langle j|F|m\rangle \langle ni|H_1|ji\rangle_A}{\varepsilon_j - \varepsilon_n} +$$

$$+ \sum_{ip} \frac{\langle ni|H_1|pi\rangle_A \langle p|F|m\rangle}{\varepsilon_n - \varepsilon_p} + \sum_{ip} \frac{\langle ni|H_1|mp\rangle_A \langle p|F|i\rangle}{\varepsilon_n - (\varepsilon_m + \varepsilon_p - \varepsilon_i)} -$$

$$- \sum_{ij} \frac{\langle ji|H_1|mi\rangle_A \langle n|F|j\rangle}{\varepsilon_j - \varepsilon_m} . \tag{1.97}$$

A distinct Feynman-Goldstone diagram can be drawn for each of the seven terms in eqn (1.97), as shown in Fig. 1.6. Again we just want to concentrate on the energy denominators involved. Referring back to eqn (1.96), notice that the energy denominator in the second term on the right-hand side of the equation is written as the energy of the initial state

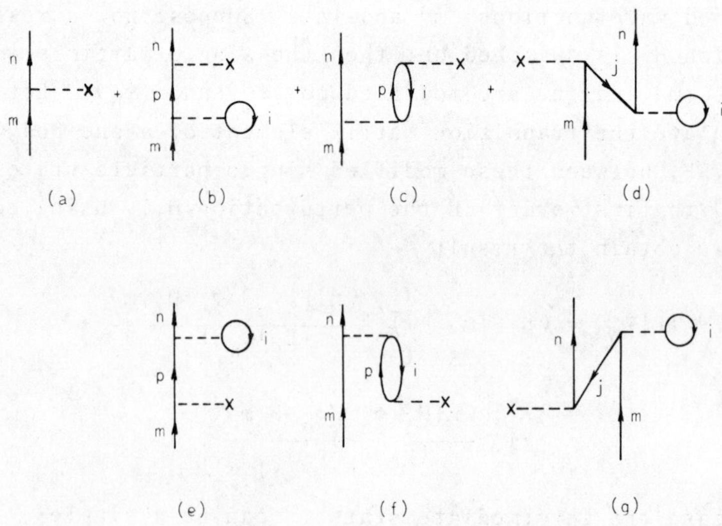

FIG. 1.6. Zeroth-order (diagram (a)) and first-order (diagrams (b) to (g)) perturbation corrections to the expectation value of a one-body operator in a closed-shell-plus-one state. The seven diagrams represent the seven terms in eqn (1.97).

minus the energy of the intermediate state, whereas in the third term the denominator is written as the energy of the final state minus that of the intermediate state. Thus when faced with interpreting diagrams such as those in Fig. 1.6 a question has to be settled as to which of these two options should be used. An examination of eqn (1.96) shows that when H_1 operates on the initial state one uses the second option. Thus in diagrams (b), (c), and (d) of Fig. 1.6 in which the H_1 interaction line lies below the F line one uses $E_0+\varepsilon_m-E_\alpha$, while in diagrams (e), (f), and (g) one uses $E_0+\varepsilon_n-E_\alpha$. Again notice that all the energy denominators are negative quantities.

2
HARTREE-FOCK THEORY

2.1 INTRODUCTION

In the last chapter, we briefly mentioned the problem of calculating the ground-state energy of a many-body system. In this and the next chapter we look more closely at this problem by first introducing Hartree-Fock theory and then the Brueckner-Goldstone expansion. To keep the discussion simple we shall for the present confine ourselves to closed-shell nuclei.

For a system of A nucleons, the Hamiltonian is written as a sum of single-particle kinetic energies and two-body interactions:

$$H = \sum_{i=1}^{A} T_i + \sum_{ij}^{A} V_{ij}. \tag{2.1}$$

The two-body potential V_{ij} represents the realistic nucleon-nucleon force, which is strongly repulsive at short range and attractive at long range. The first question then is how does one handle such a strong short-range potential in a calculation of the binding energy for finite nuclei? The answer is not simple, and for the moment the question will be avoided. Instead, we invoke the standard shell model assumption, as borrowed from atomic physics, and assert that the main effect of the two-body interaction is to produce an average potential, U, in which the particles can move almost independently. The justification for this assumption is to be found in Brueckner theory and this will be discussed at some length in the next chapter.

For the present, we divide the Hamiltonian into two parts: an unperturbed Hamiltonian

$$H_0 = \sum_{i=1}^{A} (T_i + U_i) \tag{2.2}$$

being the sum of the kinetic energy T and the average one-body potential U, and a remainder

$$H_1 = \sum_{i<j=1}^{A} V_{ij} - \sum_{i=1}^{A} U_i. \tag{2.3}$$

The hope is that the remainder will be weak enough that a perturbation calculation, in powers of H_1, will be of some practical use.

In the absence of the residual interaction, H_1, the ground-state wavefunction for the system of A nucleons, is simply a single Slater determinant constructed from the A lowest-energy eigenfunctions of $T_i + U_i$. For closed-shell nuclei the number of nucleons A is such that all the degenerate single-particle orbitals are either completely full or completely empty. This eigensolution of H_0 is spherically symmetric and non-degenerate.

Next the residual interaction is switched on, and the energy of the system evaluated as a perturbation expansion in H_1. The trick is to try and choose U in such a way that the expansion converges rapidly. In the last chapter an expression, eqn (1.95). for the energy E (to second order in H_1) was written as a sum of three diagrams (see Fig. 1.5). We repeat the expression here; however, since H_1 = V-U we will draw separate diagrams for V-interactions (represented by a dashed line joining two vertices) and for U-interactions (a dashed line with a vertex at one end and a cross at the other end). The result is plotted in Fig. 2.1.

The Hartree-Fock way of choosing U is to make the sum of contributions from graphs (c), (d), (e), and (f) vanish. The contribution from these four graphs is simply expressed in the formula

$$\sum_{km} \frac{\{\sum_i \langle ki|V|mi\rangle_A - \langle k|U|m\rangle\}\{\sum_j \langle mj|V|kj\rangle_A - \langle m|U|k\rangle\}}{(\varepsilon_k - \varepsilon_m)}; \tag{2.4}$$

thus if the potential U is chosen such that

$$\langle k|U|m\rangle = \sum_i \langle ki|V|mi\rangle_A , \tag{2.5}$$

which is the Hartree-Fock choice, then we obtain the desired cancellation in the perturbation expansion for the energy E.

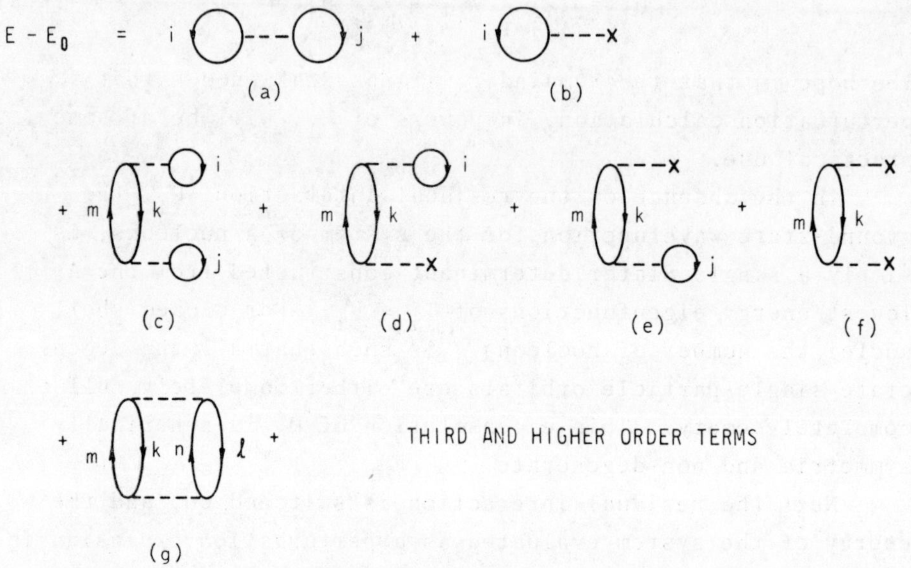

$E - E_0$ = i j (a) + i X (b)

+ m k i j (c) + m k i X (d) + m k X j (e) + m k X X (f)

+ m k n ℓ (g) + THIRD AND HIGHER ORDER TERMS

FIG. 2.1. Diagram expansion for the energy of a closed-shell system. Two-body vertices, such as those contained in the diagrams (a), (c), and (g), represent the V part of the residual interaction, H_1, whereas the one-body vertices, such as those contained in diagrams (b), (d), (e), and (f), represent the -U part of H_1.

Note that no effort is made to define U so that the first-order diagrams (a) and (b) cancel. These diagrams are easy to calculate and hopefully they constitute the major correction to the unperturbed energy. In standard Hartree-Fock calculations these are the only terms evaluated, and second-order corrections, such as diagram (g), are assumed to be small. As a result of this definition of U, eqn (2.5), we do not need to include any diagrams in the perturbation expansion involving one or more 'bubble' interactions, as these are cancelled by corresponding diagrams involving U. Thus, a tremendous number of diagrams is eliminated in this way. This choice of U is, of course, not unique (in fact we shall discuss another choice later on); however, it is extremely popular since this choice of U also leads to the familiar Hartree-

Fock self-consistency condition. In the next few sections we shall discuss the Hartree-Fock theory and return to the problem of calculating the ground-state energy of a many-body system later.

2.2 HARTREE-FOCK EQUATIONS

In Hartree-Fock theory it is assumed that the wavefunction of the ground state of a many-particle system can be represented by a single Slater determinant, Φ_0. The Hartree-Fock problem then is to find the 'best' single Slater determinant Φ_0, where 'best' is defined in this context to mean that the energy expectation value of the determinant Φ_0 should be a minimum. This statement is expressed as

$$\frac{\langle \Phi_0 | H | \Phi_0 \rangle}{\langle \Phi_0 | \Phi_0 \rangle} = \text{minimum}. \qquad (2.6)$$

If Φ_0 is normalized, then the minimization condition can be written†

$$\delta \langle \Phi_0 | H | \Phi_0 \rangle = 0$$

or

$$\langle \delta\Phi_0 | H | \Phi_0 \rangle + \langle \Phi_0 | H | \delta\Phi_0 \rangle = 0. \qquad (2.7)$$

What, then, is a small variation in a single Slater determinant, $\delta\Phi_0$? For a system of A-nucleons, one single Slater determinant is

$$|\Phi_0\rangle = \prod_{i=1}^{A} a_i^{\dagger} |0\rangle \qquad i=1,2...A \qquad (2.8)$$

The $|0\rangle$ is the vacuum state, and $a_i^{\dagger}$ a single-particle creation operator. Thouless (1960) proves that any other Slater determinant, Φ, which is not orthogonal to Φ_0 can be

---

†Strictly speaking the conditions discussed only provide an extremum. It is necessary that $\delta^2\langle H\rangle$ be greater than zero for the energy to be a minimum. We shall return to this when discussing the stability of the Hartree-Fock solution in section (4.8).

written in the form

$$|\Phi\rangle = \exp(\sum_{im} C_{mi}\, a_m^\dagger\, b_i^\dagger)|\Phi_0\rangle \qquad (2.9)$$

where $a_m^\dagger\, b_i^\dagger\, |\Phi_0\rangle$ creates a particle-hole state with respect to the first Slater determinant, Φ_0. We denote states occupied in Φ_0 with subscripts i, j, k, and ℓ, states empty in Φ_0 with m, n, p, and q, and $b_i^\dagger$ is the hole-creation operator. If Φ is to describe a small variation in Φ_0, then it is sufficient to expand the exponential to first order in the coefficient, C_{mi}, and take $\delta\Phi_0$ to be the difference between Φ and Φ_0. Thus

$$|\delta\Phi_0\rangle = \sum_{im} C_{mi}\, a_m^\dagger\, b_i^\dagger|\Phi_0\rangle. \qquad (2.10)$$

In order that $\delta\langle\Phi_0|H|\Phi_0\rangle = 0$ should give the minimum energy, Φ_0 must be normalized and the variation $|\delta\Phi_0\rangle$ be orthogonal to $|\Phi_0\rangle$. This condition

$$\begin{aligned}\langle\Phi_0|\delta\Phi_0\rangle &= \sum_{im} C_{mi}\, \langle\Phi_0|a_m^\dagger\, b_i^\dagger|\Phi_0\rangle \\ &= 0\end{aligned} \qquad (2.11)$$

is satisfied, because $\langle\Phi_0|a_m^\dagger$ is zero, since m is not an occupied state in $|\Phi_0\rangle$. Inserting the variation $|\delta\Phi_0\rangle$ into eqn (2.7) leads to

$$\sum_{im} C_{mi}{}^*\, \langle\Phi_0|b_i\, a_m\, H|\Phi_0\rangle + \sum_{im} C_{mi}\, \langle\Phi_0|H\, a_m^\dagger\, b_i^\dagger|\Phi_0\rangle = 0. \qquad (2.12)$$

As the first term is just the complex conjugate of the second term, each term individually can be equated to zero. Furthermore the coefficients C_{mi} are arbitrary, thus we obtain the condition that

$$\langle\Phi_0|b_i\, a_m\, H|\Phi_0\rangle = 0 \quad \text{for all i and m.} \qquad (2.13)$$

This is the basic defining equation of the Hartree-Fock Slater determinant Φ_0 and implies that all matrix elements of the Hamiltonian H between the Hartree-Fock state Φ_0 and one-particle one-hole states built on Φ_0 are zero.

The Hamiltonian H consists of a one-body and a two-body part and can be written as an expansion in particle and hole operators as shown in eqns (1.87) and (1.90). When these equations are substituted in the basic Hartree-Fock eqn (2.13), the only non-zero contributions to the matrix element come from terms (1.87c) and (1.90k), and eqn (2.13) reduces to

$$\langle m|T|i\rangle + \sum_j \langle jm|V|ji\rangle_A = 0. \tag{2.14}$$

The sum j is over all states occupied in the determinant Φ_0. Next we introduce a one-body Hartree-Fock Hamiltonian, h, by writing eqn (2.14) as

$$\langle m|h|i\rangle = 0 \tag{2.15}$$

for i occupied and m unoccupied in Φ_0. The Hartree-Fock Hamiltonian is defined by its matrix elements between *any* two single-particle states as

$$\langle\alpha|h|\beta\rangle = \langle\alpha|T|\beta\rangle + \sum_j \langle j\alpha|V|j\beta\rangle_A. \tag{2.16}$$

Note that h is a one-body operator whose matrix elements can only be evaluated when it is known which states are occupied in the determinant Φ_0. Furthermore the Hartree-Fock condition, eqn (2.15), requires that this operator h has no non-zero matrix elements connecting the occupied states with the unoccupied states. If the original nuclear Hamiltonian H is Hermitian, then the Hartree-Fock one-body Hamiltonian h is Hermitian.

It is useful to write h as a sum of a kinetic energy and a one-body potential energy operator

$$h = T + U \tag{2.17}$$

where the matrix elements of U are given by the second term on the right in eqn (2.16), namely

$$\langle\alpha|U|\beta\rangle = \sum_j \langle j\alpha|V|j\beta\rangle_A. \tag{2.18}$$

The potential U is called the Hartree-Fock potential. Notice that this definition of a one-body potential, which comes rather naturally from the Hartree-Fock condition, is the same as that required to make the 'one-bubble' contribution to the ground-state energy of an A particle system vanish, as shown in eqn (2.4).

The procedure for solving the Hartree-Fock equation, eqn (2.14), has been discussed by many authors (Villars 1963, Baranger 1963, Ripka 1968, Bouten 1970). We shall follow the method of Ripka.

It is an advantage to expand the Hartree-Fock single-particle states in a basis of known wavefunctions

$$|i\rangle = \sum_\alpha C_{i\alpha}|\alpha\rangle. \tag{2.19}$$

We denote the single-particle states occupied in the Hartree-Fock Slater determinant Φ_0 by i with i = 1,2...,A, and the basis states α with α = 1,2...,A,A+1,...,B. The basis states form a complete set of infinite dimension, so for any practical calculation the set has to be truncated to dimension B with B $\gtrsim$ A. The choice of basis functions is arbitrary, but a popular choice is to use harmonic oscillator functions since matrix elements of the interaction V are generally available in this representation. The convergence of the expansion, eqn (2.19), has been studied for spherical closed-shell nuclei by Davies, Krieger, and Baranger (1966) and harmonic oscillator orbits were found to give excellent convergence for the bound states making up the Hartree-Fock determinantal state Φ_0.

The Hartree-Fock single-particle states $|i\rangle$ are assumed to form an orthonormal set, viz.

$$\sum_\alpha C^*_{i\alpha} C_{j\alpha} = \delta_{ij} \qquad \sum_i C^*_{i\alpha} C_{i\beta} = \delta_{\alpha\beta}. \tag{2.20}$$

The Slater determinant Φ_0 is now specified by the coefficient $C_{i\alpha}$ which become the variational parameters, and the variational parameters, and the variational principle $\delta\langle\Phi_0|H|\Phi_0\rangle$ expressed in eqn (2.7) is generalized to read

$$\frac{\partial}{\partial C^*_{i\alpha}}[\langle\Phi_0|H|\Phi_0\rangle - \varepsilon_i \sum_\alpha C^*_{i\alpha} C_{i\alpha}] = 0. \tag{2.21}$$

Notice that a Lagrange multiplier ε_i is introduced to maintain the normalization condition. Again writing H as a sum of a one-body and a two-body operator, T and V respectively, and expanding these operators as in eqns (1.87) and (1.90), we obtain

$$\langle\Phi_0|H|\Phi_0\rangle = \sum_i \langle i|T|i\rangle + \sum_{i<j} \langle ij|V|ij\rangle_A \tag{2.22}$$

which are the closed-shell expressions, eqn (1.42) and (1.48), previously derived. Next the Hartree-Fock states $|i\rangle$ are expanded in the basis, $|\alpha\rangle$, eqn (2.19), and the expression differentiated with respect to $C^*_{i\alpha}$, so that the variational condition eqn (2.21) reduces to the series of equations

$$\sum_\beta [\langle\alpha|T|\beta\rangle + \sum_j \langle\alpha j|V|\beta j\rangle_A]C_{i\beta} = \varepsilon_i C_{i\alpha}$$

or

$$\sum_\beta \langle\alpha|h|\beta\rangle C_{i\beta} = \varepsilon_i C_{i\alpha}. \tag{2.23}$$

These equations take the form of an eigenvalue problem, with the Lagrange multipliers, ε_i, introduced in eqn (2.21) to maintain the normalization condition, being interpreted as the energy eigenvalues of the Hartree-Fock Hamiltonian, h.

The set of equations, (2.23) cannot immediately be solved since the matrix elements $\langle\alpha|h|\beta\rangle$ cannot be constructed without knowledge of which states are the occupied ones. The procedure, however, is to initiate an iteration scheme which proceeds as follows.

Step (1): Choose some trial initial Slater determinant Φ_0 containing A occupied states $|i\rangle$, i = 1...A, and make some initial guess to the expansion coefficients $C^{(0)}_{i\alpha}$. We use the superscript to denote the sequence of iterations.

This initial selection requires some care. Since the iteration procedure is non-linear, more than one solution is possible and the lowest-energy solution may be missed by improper choice of the starting values. The simplest choice is to take $C_{i\alpha}^{(0)} = \delta_{i\alpha}$, i.e. the initial states are simply the first A basis states.

This initialization works quite adequately for spherical solutions in closed-shell nuclei. However Hartree-Fock theory can also be applied to nuclei away from closed-shells, in which case spherical symmetry is no longer preserved by the iteration procedure and deformed solutions are generated. In this case it may be better to initialize the calculation with a deformed solution. For example, one might select for the lowest occupied states those Nilsson orbits which have lowest energy and for the expansion coefficients use the Nilsson model wavefunction expansion in a spherical basis (Nilsson 1955). However, we shall not discuss deformed Hartree-Fock theory in this book.

Step (2): Calculate the Hartree-Fock single-particle Hamiltonian

$$\langle\alpha|h|\beta\rangle^{(0)} = \langle\alpha|T|\beta\rangle + \sum_{i\gamma\delta} C_{i\gamma}^{*(0)} C_{i\delta}^{(0)} \langle\alpha\gamma|V|\beta\delta\rangle_A . \quad (2.24)$$

Step (3): Solve the eigenvalue problem

$$\sum_{\beta} \langle\alpha|h|\beta\rangle^{(0)} C_{i\beta}^{(1)} = \varepsilon_i^{(1)} C_{i\alpha}^{(1)} . \quad (2.25)$$

The lowest A eigenvalues now form the new set of occupied orbitals and the corresponding eigenvectors the new set of expansion coefficients.

Step (4): Go back to step (2) and recalculate the Hartree-Fock single-particle Hamiltonian. Steps (2) and (3) are then iterated until convergence.

At the conclusion of the calculation one has the single-particle energies and the wavefunction coefficients $C_{i\alpha}$. From these one can calculate the total energy of the Hartree-

Fock state

$$\begin{aligned} E &= \langle \Phi_0|H|\Phi_0\rangle \\ &= \sum_i \langle i|T|i\rangle + \frac{1}{2}\sum_{ij}\langle ij|V|ij\rangle_A \\ &= \frac{1}{2}\sum_i \langle i|T|i\rangle + \frac{1}{2}\sum_i\{\langle i|T|i\rangle + \sum_j \langle ij|V|ij\rangle_A\} \\ &= \frac{1}{2}\sum_i\{\langle i|T|i\rangle + \varepsilon_i\}. \end{aligned} \tag{2.26}$$

This is the Hartree-Fock approximation to the ground-state energy. Other quantities can also be calculated such as density distributions, r.m.s. radii, form factors for elastic electron scattering etc. These are all bulk properties of the nucleus and depend on knowing the wavefunction of all the occupied states. A complete set of calculations of this type for light nuclei has been carried out by Lee and Cusson (1972).

Another interesting property to calculate is the energy of the (A+1)-particle state, $a_m^\dagger|\Phi_0\rangle$, relative to that of the Hartree-Fock A-particle state, $|\Phi_0\rangle$:

$$\begin{aligned} E_{A+1} - E_A &= \langle \Phi_0|a_m\, H\, a_m^\dagger|\Phi_0\rangle - \langle \Phi_0|H|\Phi_0\rangle \\ &= \langle m|T|m\rangle + \sum_i \langle mi|V|mi\rangle_A \\ &= \langle m|h|m\rangle \\ &= \varepsilon_m. \end{aligned}$$

where eqns (1.52) and (1.55) were used to evaluate $\langle\Phi_0|a_m\, H\, a_m^\dagger|\Phi_0\rangle$. Similarly the energy of an (A-1)-particle state $b_i^\dagger|\Phi_0\rangle$ can be evaluated from eqns (1.66) and (1.67):

$$\begin{aligned} E_{A-1} - E_A &= \langle \Phi_0|b_i\, H\, b_i^\dagger|\Phi_0\rangle - \langle \Phi_0|H|\Phi_0\rangle \\ &= -\langle i|T|i\rangle - \sum_j \langle ij|V|ij\rangle_A \end{aligned}$$

$$= -\langle i|h|i\rangle$$

$$= -\varepsilon_i.$$

This latter result is known as Koopmans' theorem and states that ε_i, the Hartree-Fock single-particle energy, is the energy required to remove a nucleon from the state i in the Hartree-Fock state Φ_0. It is not legitimate, however, to apply Koopmans' theorem successively to the system with (A-1) particles, (A-2) particles etc. If the Ath particle has already been removed, then in removing the $(A-1)^{th}$ particle, its interactions with the Ath nucleon should no longer be counted. This expresses itself by the fact that $E_A \neq \sum_i \varepsilon_i$ but rather is given by eqn (2.26). One should redo the Hartree-Fock minimization calculation for each new system.

If on repeating a Hartree-Fock calculation for the (A-1) system it so happened that the Hartree-Fock orbits coincided with those of the A-particle system, then Koopmans' theorem would apply. More generally, however, we have that

$$E_{A-1} - E_A = -\varepsilon_i - \Delta$$

where Δ is the Hartree-Fock rearrangement energy, and is non-negative if both Hartree-Fock calculations for the A and (A-1) systems are absolute minima (Lee and Cusson 1972).

2.3 SYMMETRIES IN HARTREE-FOCK THEORY

From an examination of the iteration equations (2.24) and (2.25), it is clear that most of the labour involved is in evaluating the matrix elements of the Hartree-Fock Hamiltonian, $\langle\alpha|h|\beta\rangle$. The number of such matrix elements depends on the dimension of the basis. For Hartree-Fock calculations in light nuclei the basis is truncated to the lowest few oscillator orbits; a typical example might be a restriction to the 0s, 0p, 0d-1s and 0f-1p orbitals. The number of single-particle states, α (i.e. number of different values for $n\ell jm\tau$), would be 80 in this case, and at each step in the iteration there would be $\frac{1}{2}(80 \times 81) = 3240$ matrix elements to be evaluated. Furthermore each matrix element is calculated from eqn (2.24) which involves a sum over i, γ, and δ containing A × 80 × 80 terms. This is clearly a sizeable numerical problem which has to be met at each stage of the iteration. One way of reducing the amount of labour is to impose certain restrictions on the solution by making the assumption that it has a certain symmetry. However, at the same time one is restricting the variations allowed to the trial wavefunction, thus if one was particularly unlucky, one might generate a solution which was not the lowest-energy Hartree-Fock solution. With this proviso in mind let us see how imposing symmetry restrictions reduces the amount of computation.

Suppose that the initial trial solution has a definite symmetry, viz.

$$L\;|i\rangle^{(0)} = \lambda_i^{(0)}\;|i\rangle^{(0)} \tag{2.27}$$

with

$$|\lambda_i^{(0)}|^2 = 1$$

Here L is a unitary operator with eigenvalues $\lambda_i^{(0)} = \pm 1$ when acting on the trial single-particle wavefunction $|i\rangle^{(0)}$. Suppose further that L commutes with the nuclear Hamiltonian H, then the Hartree-Fock Hamiltonian $h^{(0)}$ constructed from the $|i\rangle^{(0)}$ will commute with L and the symmetry will be preserved at each stage in the iteration. The proof (Bouten 1970, Ripka 1968) is as follows: we have that

$$\langle\alpha|h|\beta\rangle^{(0)} = \langle\alpha|T|\beta\rangle + \sum_i \langle\alpha i|V|\beta i\rangle_A^{(0)}$$

and that

$$\begin{aligned}\langle\alpha|L^{-1}hL|\beta\rangle^{(0)} &= \langle L\alpha|h|L\beta\rangle^{(0)} \\ &= \langle\alpha|L^{-1}TL|\beta\rangle + \sum_i \langle L\alpha,i|V|L\beta,i\rangle_A^{(0)}.\end{aligned}$$

Remembering that L commutes with T and V,

$$\begin{aligned}\langle\alpha|L^{-1}hL|\beta\rangle^{(0)} &= \langle\alpha|T|\beta\rangle + \sum_i \langle\alpha, Li|V|\beta, Li\rangle_A^{(0)} \\ &= \langle\alpha|T|\beta\rangle + \sum_i |\lambda_i^{(0)}|^2 \langle\alpha i|V|\beta i\rangle_A^{(0)} \\ &= \langle\alpha|T|\beta\rangle + \sum_i \langle\alpha i|V|\beta i\rangle_A^{(0)} \\ &= \langle\alpha|h|\beta\rangle^{(0)}. \end{aligned} \tag{2.28}$$

Thus we have that L commutes with $h^{(0)}$. The eigenvalue problem

$$h^{(0)}\ |i\rangle^{(1)} = \varepsilon_i^{(1)}\ |i\rangle^{(1)}$$

is next solved and it has to be shown that $L|i\rangle^{(1)}$ has the eigenvalue property possessed by the initial guess eqn (2.27). Since L is unitary and the $|i\rangle^{(1)}$ being solutions of an eigenvalue problem form a complete set, then

$$L\ |i\rangle^{(1)} = \sum_j \lambda_{ij}^{(1)}\ |j\rangle^{(1)}. \tag{2.29}$$

Thus

$$\langle k|h^{(0)}L|i\rangle^{(1)} = \sum_j \lambda_{ij}^{(1)} \langle k|h^{(0)}|j\rangle^{(1)} = \lambda_{ik}^{(1)}\varepsilon_k^{(1)} \tag{2.30}$$

$$\langle k|Lh^{(0)}|i\rangle^{(1)} = \varepsilon_i^{(1)}\langle k|L|i\rangle = \varepsilon_i^{(1)}\lambda_{ik}^{(1)}. \tag{2.31}$$

These two equations (2.30) and (2.31) are equal since L commutes with $h^{(0)}$, which implies that $\lambda_{ik}^{(1)} = \lambda_i^{(1)}\delta_{ik}$ and therefore the iterated solution $|i\rangle^{(1)}$ has the same symmetry property, eqn (2.27), as the trial solution $|i\rangle^{(0)}$. A symmetry L of this type which commutes with the Hartree-Fock Hamiltonian, h, and is conserved at each step in the iteration is called a 'consistent symmetry'.

Examples of consistent symmetries are:

(i) parity: $P|n\ell jm\tau\rangle = (-)^{\ell}|n\ell jm\tau\rangle$

(ii) axial symmetry: $e^{i\phi J_z}|n\ell jm\tau\rangle = e^{im\phi}|n\ell jm\rangle$

(iii) axial symmetry in isospin space:

$$e^{i\phi T_z}|n\ell jm\tau\rangle = e^{i\tau\phi}|n\ell jm\tau\rangle$$

(iv) time reversal for even-even nuclei.

Consider the example (Bouten 1970) of the 80-dimensional space involving the first four major oscillator shells and let us impose parity as a consistent symmetry. There are 28 positive parity states and 52 negative parity states. Thus the expansion eqn (2.23) can be split into

$$|i\rangle = \sum_{\alpha=1}^{28} C_{i\alpha}\,|\alpha\rangle \qquad \text{for a positive parity state}$$

and

$$|i\rangle = \sum_{\alpha=29}^{80} C_{i\alpha}\,|\alpha\rangle \qquad \text{for a negative parity state,}$$

and the matrix $\langle\alpha|h|\beta\rangle$ reduces to the block diagonal form (see Fig. 2.2). Thus the simplification in the computation is threefold:

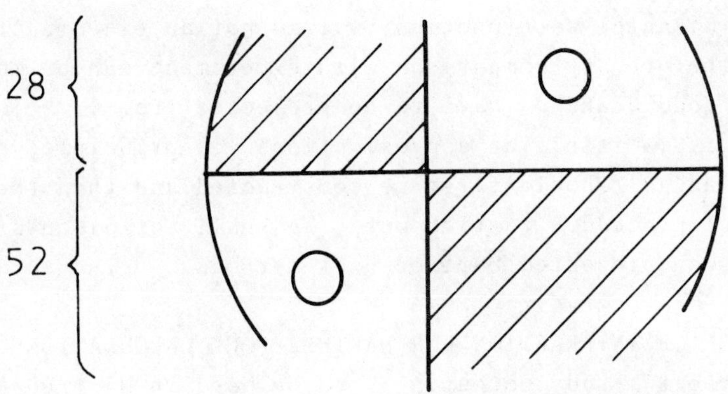

FIG. 2.2. For the example discussed in the text, the matrix $\langle\alpha|h|\beta\rangle$ can be written in block diagonal form with $\alpha = 1,\ldots,28$ representing positive parity states, and $\alpha = 29,\ldots,80$ representing negative parity states (Bouten 1970).

(i) there are fewer non-zero matrix elements $\langle\alpha|h|\beta\rangle$;

(ii) the evaluation of each matrix element involves smaller sums of two-body matrix elements;

(iii) smaller matrices are to be diagonalized.

Clearly imposing a consistent symmetry on the Hartree-Fock solution reduces the amount of computational labour, but there is always a risk that the solution of lowest energy may be lost.

The Hartree-Fock Hamiltonian does not in general commute with the total angular momentum operator $\underset{\sim}{J} \equiv (J_x, J_y, J_z)$, for, if one assumes as an initial guess that the orbits i are eigenstates of the angular momentum operator

$$|i\rangle^{(0)} = |j_i\ m_i\rangle$$

then J_x and J_y, which raise and lower the magnetic quantum number m_i, will leave the set of occupied orbits invariant only if for each j_i all the magnetic quantum numbers m_i are filled or empty. Hence spherical symmetry can exist only in nuclei having the appropriate number of particles to form closed shells (Ripka 1968).

For other than closed-shell configurations, the Hartree-Fock determinantal wavefunction, Φ_0, is not an eigenfunction of $\underset{\sim}{J}$, and before any comparison with experiment can be made, states of good J and J_z have to be projected from Φ_0. This procedure is in itself an approximation. In principle, states of good J and J_z should first be constructed and then the variational procedure carried out. We shall not discuss at all here such projected Hartree-Fock methods.

2.4 CHOICE OF INTERACTION FOR HARTREE-FOCK CALCULATIONS

Ideally, the two-body potential V to be used in Hartree-Fock calculations should be one which fits nucleon-nucleon scattering data. Such potentials frequently exhibit strongly repulsive short-range behaviour, and this makes their use inappropriate if not impossible in a first-order-in-V calculation, such as a Hartree-Fock calculation. Instead the short-range repulsion has to be explicitly considered and the Brueckner-

Hartree-Fock procedure was developed just to cope with this situation. We shall discuss such calculations in the next chapter.

One realistic potential, which more or less fits the available scattering data and yet is soft enough to be used directly in a Hartree-Fock calculation is the non-local potential of Tabakin (1964). But even for this potential, Kerman and his co-workers (Kerman, Svenne, and Villars 1966, Bassichis, Kerman, and Svenne 1967, Kerman and Pal 1967) find that second-order and even third-order corrections are sizeable. Furthermore the calculations have difficulties in explaining binding energies and nuclear radii simultaneously. The Brueckner-Hartree-Fock calculations also suffer from this difficulty.

With realistic potentials generating such enormous complexities, a second class of Hartree-Fock calculations has come into existence. These calculations leave out completely the problem of higher-order corrections and instead try to construct an effective potential, which will reproduce in lowest order as many nuclear properties as possible. There are two approaches. The first uses a realistic two-body force in lowest order and parameterizes the higher-order corrections. The second approach parameterizes the effective interaction directly as a whole.

Typical of the first approach has been the work of Negele (1970). He starts with the Reid potential, and by using a local density approximation constructs an effective potential with significant density dependence. When used in a Hartree-Fock calculation, a rather satisfactory fit is obtained for nuclear radii, binding energies and single-particle energies of doubly-closed-shell nuclei. This is typical of all density dependent effective interactions.

In the second approach are the potentials of Brink and Boeker (1967) and Volkov (1965), which are constructed from linear combinations of two or more gaussians, with one gaussian being repulsive and short-range and another attractive and long-range. These very simple interactions nevertheless have difficulties in fitting binding energies and radii simultaneously. Other examples of effective interactions are the potentials of Saunier and Pearson (1970), which contain a short-range momentum-dependent part, and the Skyrme (1959) interaction which contains a zero-range three-body part which in turn is equivalent to a two-body density dependent interaction. The former potential has been used extensively by Lee and Cusson (1972) and the latter by the Saclay group (e.g., Vautherin and Brink 1972).

The results obtained with these effective interactions are in impressive agreement with experiment. Although less fundamental, the use of effective interactions has two distinct advantages. First it enables one to calculate in regions where realistic calculations become impracticable, and second it allows one to make systematic studies with the least amount of numerical work.

3
BRUECKNER-HARTREE-FOCK THEORY

3.1 GOLDSTONE EXPANSION

We return to the problem of calculating the ground-state energy of a many-particle system. The Hamiltonian is divided as before into two pieces $H = H_0+H_1$. The unperturbed ground state is represented by a single Slater determinant, Φ_0, and the unperturbed energy E_0 is the sum of the single particle energies of the occupied states, i.e.

$$H_0 \Phi_0 = E_0 \Phi_0$$
$$E_0 = \sum_{i=1}^{A} \langle i|T+U|i\rangle . \tag{3.1}$$

The exact ground state Ψ satisfies

$$H\Psi = E\Psi \tag{3.2}$$

and it is the eigenvalue E that we want to calculate. Perturbation theory written out to third order in H_1 gives the expansion (Messiah 1964):

$$E = E_0 + \langle \Phi_0|H_1|\Phi_0\rangle + \langle \Phi_0|H_1(E_0-H_0)^{-1} QH_1|\Phi_0\rangle +$$
$$+ \langle \Phi_0|H_1(E_0-H_0)^{-1} QH_1(E_0-H_0)^{-1} QH_1|\Phi_0\rangle -$$
$$- \langle \Phi_0|H_1|\Phi_0\rangle \langle \Phi_0|H_1(E_0-H_0)^{-2} QH_1|\Phi_0\rangle + \ldots \tag{3.3}$$

The operator Q is a projection operator which ensures that the initial state Φ_0 does not occur as an intermediate state in any of the matrix elements. Remember that $H_1 = V-U$ (eqn (2.3)), and let us take the Hartree-Fock potential for U. Then to first order in H_1, the energy (summing diagrams (a) and (b) of Fig. 2.1) is

$$E = E_0 + \langle \Phi_0 | H_1 | \Phi_0 \rangle$$

$$= \sum_i \langle i|T+U|i\rangle + \frac{1}{2}\sum_{ij} \langle ij|V|ij\rangle_A - \sum_i \langle i|U|i\rangle$$

$$= \sum_i \langle i|T|i\rangle + \frac{1}{2}\sum_{ij} \langle ij|V|ij\rangle_A . \tag{3.4}$$

Note that the terms involving U drop out. However, the expression for E is not independent of U, since the sums involved are over the occupied states which have been determined from a Hartree-Fock procedure, and the states $|i\rangle$ are therefore eigenfunctions of the Hartree-Fock Hamiltonian $h = T+U$.

The second-order correction is given by Fig. 3.1, and was evaluated in eqn (1.97), namely

$$\frac{1}{4}\sum_{ijmn} \frac{\langle ij|V|mn\rangle_A \langle mn|V|ij\rangle_A}{(\varepsilon_i+\varepsilon_j-\varepsilon_m-\varepsilon_n)} .$$

Again there are no terms involving matrix elements of U as discussed earlier, but the single-particle energies ε are eigenvalues of $h = T+U$.

The perturbation series gets more complicated in third order. There are two terms in eqn (3.3) instead of one. Some typical diagrams which arise from the first of these

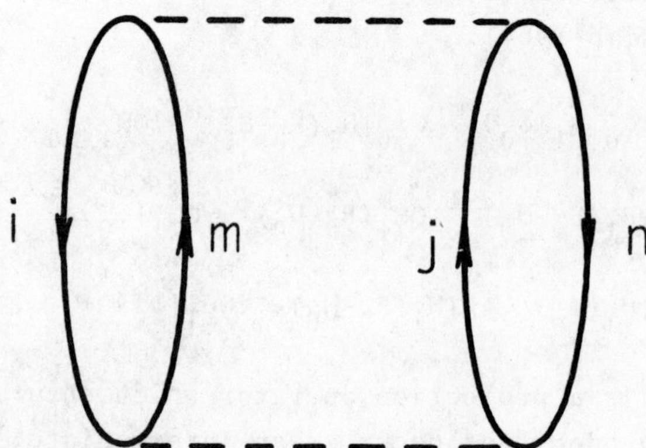

FIG. 3.1. The only second-order correction to the energy of a closed-shell nucleus, in a case where a Hartree-Fock Hamiltonian describes the unperturbed system.

terms are shown in Fig. 3.2. Consider diagram (c) which consists of two separated pieces. That such a diagram should appear is a catastrophe (Day 1967). The reason is that in

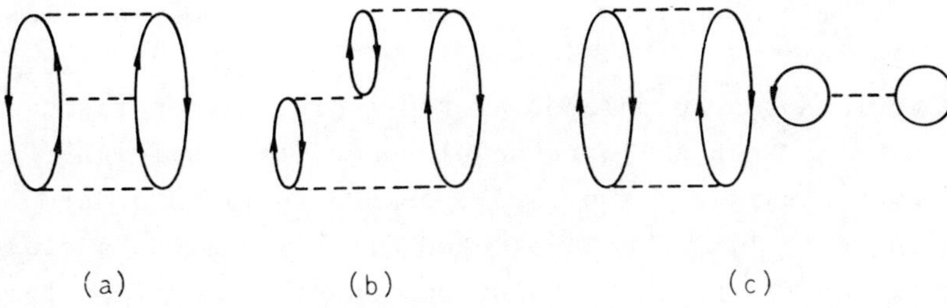

Fig. 3.2. Some typical third-order corrections to the energy of a closed-shell nucleus. Diagram (c) is a disconnected diagram and gives no contribution to the energy, since it is exactly cancelled by another term in the perturbation expansion.

the limit of a large number of particles, the contribution of a connected diagram is proportional to the number of particles, and so the contribution of a diagram with two disconnected pieces is proportional to the square of the number of particles (Brandow 1967). On the other hand the energy of a heavy nucleus is known to be proportional to the number of particles. This leads to a dilemma which is resolved by considering the second third-order term in the perturbation series (eqn (3.3)). One can show that this second term exactly cancels all the disconnected diagrams that contribute to the first term, and thereby leaves only connected diagrams for the third-order contribution. A similar result holds for all higher orders. Goldstone (1957) showed that this cancellation is exact in every order of perturbation theory and leads to the 'Goldstone Expansion' which states that the energy of a many-body system is given by the sum of all connected (linked) diagrams. This is certainly a very simple prescription and leads to an energy which is proportional to the number of particles. However, there are two technical difficulties.

First, it has been pointed out by Mavromatis (1973) that

the cancellation is complete only when no restrictions are placed on the sums over the intermediate states. However, a typical calculations using harmonic oscillator basis states might as a practical measure retain only intermediate states of energy $2\hbar\omega$ above E_0, the unperturbed energy of the closed shell. In this case the cancellation in fourth and higher orders would not be complete. Furthermore, Mavromatis has shown that for such a finite model space, the Rayleigh-Schrödinger expansion (eqn (3.3)) summed to fifth order (including both linked and unlinked diagrams) gives a closer approximation to the exact solution than the summation to the same order with linked terms only.

The second difficulty concerns our discussion on diagrams. We have adopted the convention that only the diagram corresponding to the direct part of a two-body matrix element should be drawn; the exchange part was always understood to be included. Furthermore in sums over intermediate states we did not have to worry about questions of antisymmetry providing we worked with antisymmetrized two-body matrix elements; all the Pauli violating terms were cancelled. However, consider the diagrams in Fig. 3.3. These are fourth-order diagrams in the perturbation expansion. Both these diagrams violate the exclusion principle in that the state $|j\rangle$ is occupied by two particles, but because diagram (b) is just the exchange of diagram (a), the contribution from the sum of

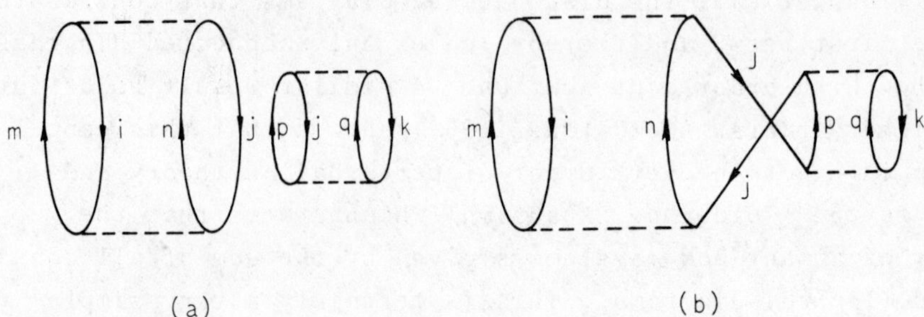

FIG. 3.3. Diagram (b) is an exchange version of diagram (a). However, diagram (b), being fully linked, is included in the Goldstone expansion, whereas diagram (a), being disconnected, is not.

these two diagrams to the energy exactly cancels. However, diagram (a) is disconnected and is therefore not included in the Goldstone expansion. Nevertheless diagram (b) must still be included in the Goldstone expansion and gives a non-zero contribution which is not cancelled by any other term. Thus when using the Goldstone expansion, diagrams must be drawn for both the direct and exchange parts of two-body matrix elements, and all connected diagrams are retained even if they violate the Pauli exclusion principle.

3.2 THE REACTION MATRIX, G

The Goldstone expansion cannot be used in its present form for nuclear calculations because the short-range repulsion in the nucleon-nucleon potential, V, makes all the matrix elements very large and the series is unlikely to converge. This is obvious for the case of an infinite hard core potential for which all the matrix elements of V themselves are infinite. The next step is to introduce Brueckner's reaction matrix and convert the expansion to one in which the potential, V, is eliminated in favour of the reaction matrix, G. This means we are going to re-order the perturbation series and perform a summation of a certain class of diagrams to obtain a reaction matrix that is well behaved even for a singular two-body force.

It was suggested by Hugenholtz (1957), that a series in powers of the nuclear density might converge. In any given diagram, each hole line introduces a summation over all the occupied states, and therefore each hole line gives a contribution proportional to the density. Since there are at least two hole lines in any closed diagram, the lowest order in an expansion in the nuclear density should comprise all diagrams with only two hole lines. We illustrate the first few of these on the right-hand side in Fig. 3.4. This selected class of diagrams, called ladder diagrams, is summed to all orders to form the G-matrix or reaction matrix. In the diagram a matrix element of G is indicated by a wiggly line as shown on the left of Fig. 3.4. Evaluating these diagrams leads to the equation

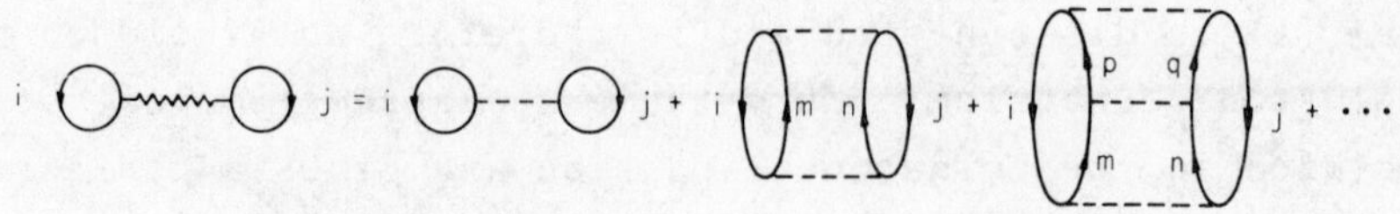

FIG. 3.4. A diagrammatic definition of the reaction matrix, in which all 'ladder' diagrams are summed to all orders.

$$\frac{1}{2}\sum_{ij}\langle ij|G|ij\rangle_A = \frac{1}{2}\sum_{ij}\langle ij|V|ij\rangle_A +$$

$$+\ \frac{1}{4}\sum_{ijmn}\frac{\langle ij|V|mn\rangle_A\langle mn|V|ij\rangle_A}{(W-\varepsilon_m-\varepsilon_n)} +$$

$$+\ \frac{1}{8}\sum_{ijmnpq}\frac{\langle ij|V|mn\rangle_A\langle mn|V|pq\rangle_A\langle pq|V|ij\rangle_A}{(W-\varepsilon_m+\varepsilon_n)(W-\varepsilon_p+\varepsilon_q)} + \tag{3.6}$$

$$+\ \ldots.$$

with

$$W = \varepsilon_i + \varepsilon_j.$$

Note that we are using antisymmetrized matrix elements both for the G-matrix and the V-matrix, and the sums are unrestricted. Hence the presence of the factor $\left(\frac{1}{2}\right)^n$ in the nth-order ladder diagram. We have also introduced an energy W which corresponds to the energy of the two-hole lines in the diagram. This energy is frequently referred to as the starting energy in the calculation of the G-matrix. In this example, the starting energy W is equal to $\varepsilon_i+\varepsilon_j$ which is just the energy of the initial two-particle state. When the starting energy is equal either to the energy of the initial two-particle state or to the energy of the final two-particle state, the G-matrix is said to be calculated *on the energy shell*. In all other cases G is said to be calculated *off the energy shell*. More about this later. Notice that in constructing the ladder diagrams, each successive V-interaction is inserted in the diagram between two upgoing particle lines, never between two hole-lines.

It is convenient to define the two-particle operators Q and e by the equations

$$Q|rs\rangle = \begin{cases} |rs\rangle \text{ if r and s unoccupied in } \Phi_0 \\ 0 \quad \text{otherwise} \end{cases} \tag{3.7}$$

$$e|rs\rangle = (W-\varepsilon_r-\varepsilon_s).$$

The operator Q is often called the 'Pauli operator' since it restricts the two nucleons in intermediate states to lie above the Fermi surface. By using these definitions we can write a formal expansion for the two-body operator G as

$$G(W) = V + V\frac{Q}{e}V + V\frac{Q}{e}V\frac{Q}{e}V + \ldots, \tag{3.8}$$

where G is written as a function of the starting energy. This in fact is the definition of G and the treatment in terms of sums of ladders can be seen to be equivalent. The expansion can be rewritten as

$$G(W) = V + V\frac{Q}{e}G(W). \tag{3.9}$$

For a fixed starting energy W, the reaction matrix G is a well-defined Hermitian two-body operator (Day 1967). It is Hermitian because Q, e and V are all Hermitian and therefore every term in the expansion eqn (3.8) is Hermitian.

The procedure now is to convert each diagram in the Goldstone expansion to one containing only G-matrices. Starting with an arbitrary V-matrix diagram, one successively converts each V to a G by summing the proper sequence of ladder diagrams. Some clarifying examples of this procedure is given in Appendix B of the review article by Day (1967). The resulting expansion is called the Brueckner-Goldstone expansion. It tells us simply to replace V by G in the Goldstone expansion, and to omit ladder diagrams. This avoids double counting. In this new expansion, every term is finite and well behaved, even when the potential contains strong short-range repulsion. The Brueckner-Goldstone expansion is a frequent starting point for nuclear calculations using realistic forces.

However, this procedure of replacing V-interactions by G-interactions is not without its difficulties. Consider the set of diagrams shown in Fig. 3.5. The first and last interactions have already been replaced by G-interactions and the

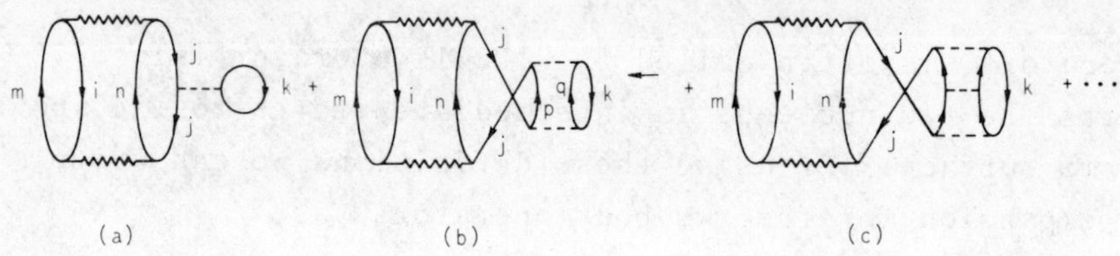

FIG. 3.5. Sequence of diagrams that have to be summed if the middle V-intersection in diagram (a) is to be converted to a G-interaction. Note, however, that this G-interaction is 'off-the-energy-shell', as can be seen by evaluating the energy of the intermediate state formed at the position of the arrow in diagram (b).

sequence of diagrams shown is those which have to be summed if the middle V-interaction (connecting a hole-line with a bubble) is to be replaced by a G-interaction. At first sight it might appear that these diagrams are redundant ladder diagrams since we have two G-interactions in the particle lines m and n. However, this is not the case. The rule is that two *successive* G-matrices should never appear between the same pair of particle lines. In our case the two G-interactions are separated by an interaction in the hole line j. In diagram (a), the V-interaction gives a contribution $\langle jk|V|jk\rangle_A$ to the energy. In diagram (b), the contribution becomes

$$+ \frac{1}{2} \sum_{pq} \langle jk|V|pq\rangle_A \frac{Q}{e} \langle pq|V|jk\rangle_A$$

with the energy denominator e being given by $W-\varepsilon_p-\varepsilon_q$. The problem is that the starting energy W depends upon the particular diagram in question. This can be seen in the case of diagram (b) where the energy denominator at the level indicated by the arrow can also be written as the sum of the particle energies minus the sum of the hole energies. That is

$$e = W - \varepsilon_p - \varepsilon_q = \varepsilon_i + 2\varepsilon_j + \varepsilon_k - \varepsilon_p - \varepsilon_q - \varepsilon_m - \varepsilon_n$$

and

$$W = \varepsilon_j + \varepsilon_k - (\varepsilon_m + \varepsilon_n - \varepsilon_i - \varepsilon_j).$$

Therefore the G-matrix $\langle jk|G(W)|jk\rangle_A$ for Fig. 3.5 has to be calculated off-the-energy shell by an amount $(\varepsilon_m+\varepsilon_n-\varepsilon_i-\varepsilon_j)$. This can have serious effects since W can be off the energy shell by as much as 100 MeV in some cases. It is also undesirable since we should like to calculate the G-matrix by solving eqn (3.9) without reference to the diagram into which the G-matrix is to be inserted. Thus we should always like to evaluate the G-matrix on the energy shell.

Fortunately it was noted by Brueckner and Goldman (1960) that for every diagram of type 3.5(b) there exists a corresponding diagram (Fig. 3.6) which is of the same order as Fig. 3.5(b) when all the interactions are V-interactions, but

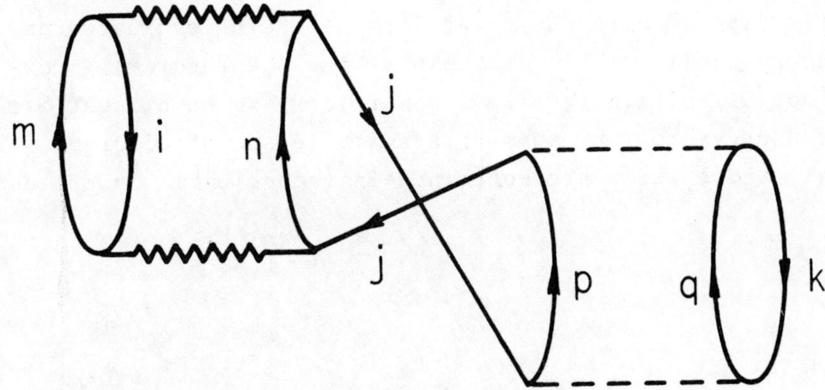

FIG. 3.6. A compensating diagram, which when added to diagram 3.5(b), gives the same result as that obtained from diagram 3.5(b) alone, but with the G-matrix evaluated on the energy shell.

is of one higher order when all the interactions have been replaced by G-interactions. These two diagrams have the same number of particle lines and hole lines, the same interactions and only differ from each other in the energy denominators. Let us denote $e_A = \varepsilon_i + \varepsilon_j - \varepsilon_m - \varepsilon_n$ and $e_B = \varepsilon_j + \varepsilon_k - \varepsilon_p - \varepsilon_q$, then the sum of diagrams Fig. 3.5(b) and Fig. 3.6 can be written symbolically as

$$(\text{matrix elements})\left\{\frac{1}{e_A}\cdot\frac{1}{e_A+e_B}\cdot\frac{1}{e_A}+\frac{1}{e_A}\cdot\frac{1}{e_A+e_B}\cdot\frac{1}{e_B}\right\}$$

$$=(\text{matrix elements})\ \frac{1}{e_A}\cdot\frac{1}{e_B}\cdot\frac{1}{e_A}\,.$$

That is, the sum of both diagrams is as if we evaluated only Fig. 3.5(b), treating the hole-line bubble interactions as if it were uncoupled from the rest of the diagram and evaluated on the energy shell. Such a piece of a diagram which can be uncoupled in this way is called a 'factorizable insertion' (more on this in the next section). Bethe, Brandow, and Petschek (1963) generalized this example to cases involving all orders of perturbation theory to show that a similar sequence of diagrams can be summed together to produce one G-matrix diagram with the G-matrices all being evaluated on the energy shell. (This is known as the BBP theorem). However, not all off-energy-shell G-matrices can be brought back on the energy shell by summing a suitable set of diagrams, but only those which are contained in factorizable insertions.

3.3 INSERTIONS

In order to illustrate how the BBP theorem is used it is necessary to write down a few definitions of insertions. We have taken these definitions from Kirson (1968, 1969) where extensive discussion is given to the following statements.

The diagrams of a reaction matrix perturbation expansion (G-matrices) can be divided into two groups – diagrams without and diagrams with insertions. An insertion may be defined as a subdiagram connected to the main part of the diagram by two fermion lines. If these two lines are cut, the diagram separates into two parts. The gap in the main part of the diagram may be closed by bringing the cut ends together, the resulting diagram being a lower term of the perturbation series considered. The insertion has two external lines, which may be joined together to produce one of the Brueckner-Goldstone diagrams for the ground-state energy.

Insertions may be further classified as 'essential' or 'factorizable' insertions. The latter are insertions which can be put on the energy shell relative to the main part of the diagram by an application of the BBP theorem. The former cannot be so treated.

A 'dangling' insertion is defined as a subdiagram which can be separated from the main part by cutting one hole line and one particle line at the *same* horizontal level. Dangling insertions are factorizable.

Since an insertion is formed by breaking a fermion line in a ground-state energy diagram there are essentially two kinds – those formed by breaking a particle line and those formed by breaking a hole line. Any insertion, of either kind, may be inserted into a particle line or into a hole line, or as a dangling insertion. Of the diagrams so formed some contain essential insertions while others contain factorizable insertions. The factorizable insertions are all the dangling insertions and all the insertions formed by breaking a hole (particle) line in a ground-state diagram and inserting it in a hole (particle) line in the main part of the diagram. The essential insertions are those formed by breaking a hole (particle) line in a ground-state diagram and inserting it in a particle (hole) line in the main part of the diagram. Consider Fig. 3.7. The boxed part of Fig. 3.7(a) contains an essential insertion, whereas the boxed part of Fig. 3.7(b) contains a factorizable insertion.

We conclude this section by stating the BBP theorem in a slightly more generalized form (due to Kirson 1967, 1968, 1969). The purpose of the theorem, which Kirson calls the generalized factorization theorem, is to separate Brueckner-Goldstone diagrams into independent parts, in which G-matrices and energy denominators in one part are entirely independent of particle or hole energies in the other part. It is necessary to consider the Goldstone diagrams (V-interactions) which

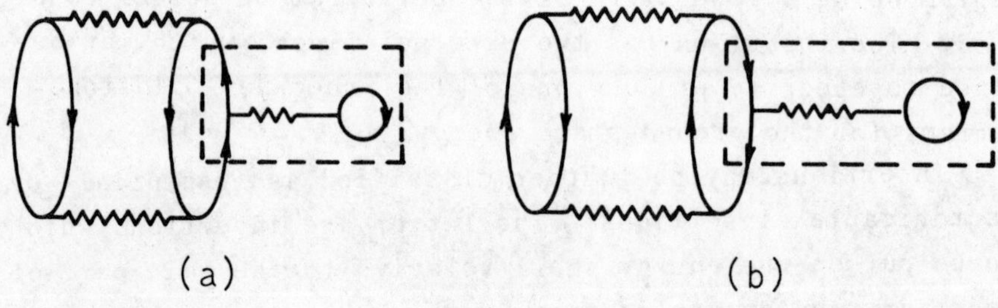

FIG. 3.7. The boxed parts of the diagrams are examples of insertions. Diagram (a) illustrates an 'essential' insertion and diagram (b) a 'factorizable' insertion.

are summed to produce a given Brueckner-Goldstone diagram, and the theorem is first stated for Goldstone diagrams.

Consider a class of Goldstone diagrams, to be called a gto class (generalized-time-ordering class), having the follow-following properties:

(i) there exists a horizontal level in each diagram above which all members of the class are identical and have at least one interaction;

(ii) below this separation level, every diagram may be divided into a left-hand part and a right-hand part such that no interaction lines connect the left- and right-hand parts of the diagram;

(iii) the left-hand parts of all diagrams are identical and the right-hand parts of all diagrams are identical;

(iv) the first interaction below the separation level in every diagram is a right-interaction.

All members of a gto class thus differ only in the ordering of left-interactions relative to right-interactions below the separation level, the left-interactions having a fixed order among themselves and the right-interactions having a fixed order among themselves. Thus the contributions of all the diagrams in a gto class will have the same matrix elements of V, the same intermediate-state sums and the same overall sign, and will differ only in their energy demoninators, and only below the separation level. The generalized factorization theorem now states that the sum of all diagrams in a gto class is equal to a *single* diagram of identical structure (to be called the primitive element of the gto class) with all right-interactions occurring above all left-interactions, but with the energy denominators on the right independent of the fermion lines on the left.

It is clear that the generalized factorization theorem can be applied above a separation level, as well as below, by a suitable change in the wording (i) to (iv), and that it could factor off the left-interactions, rather than the right-interactions, by similar appropriate changes of wording. An example is shown in Fig. 3.8 in which three Goldstone diagrams are summed to give a result:

$$\frac{1}{8}\sum \langle ij|V|rs\rangle_A \frac{1}{(\varepsilon_i+\varepsilon_j-\varepsilon_r-\varepsilon_s)} \langle rs|V|mn\rangle_A \frac{1}{(\varepsilon_i+\varepsilon_j-\varepsilon_m-\varepsilon_n)} \langle jk|V|pq\rangle_A \times$$

$$\times \frac{1}{(\varepsilon_j+\varepsilon_k-\varepsilon_p-\varepsilon_q)} \langle pq|V|jk\rangle_A \frac{1}{(\varepsilon_i+\varepsilon_j-\varepsilon_m-\varepsilon_n)} \langle mn|V|ij\rangle_A .$$

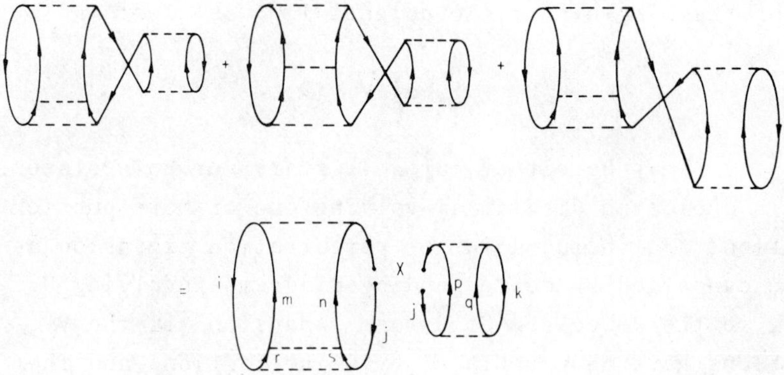

FIG. 3.8. An example of the factorization theorem. The sum of the three diagrams in the upper half of the figure is equal to the single diagram in the lower half with the energy denominators in its two parts being evaluated independently of each other.

One may now consider the primitive element of the gto class as a representative of a Brueckner-Goldstone diagram.

This diagram may be written as a sum of Goldstone diagrams, each of which belongs to a gto class to which the theorem can be applied. The set of gto classes generated by these Goldstone diagrams, when resummed, gives rise to a set of Brueckner-Goldstone diagrams called the Brueckner-Goldstone gto class, with a primitive element which is just the Brueckner-Goldstone diagram containing the primitive elements of the original gto classes. The strength of the generalized factorization theorem then lies in the fact that the sum of all the diagrams in the Brueckner-Goldstone gto class is equal to the primitive element of that class, with all G-matrices and energy denominators on the left (right) side independent of the energies of all fermion lines on right (left) side.

This is the theorem that was used in the last section and enabled the G-interaction between a hole line and a bubble to be evaluated on the energy shell.

3.4 SINGLE-PARTICLE POTENTIAL IN BRUECKNER THEORY

In our discussion of the Goldstone expansion, the Hartree-Fock choice of the single-particle potential U was used:

$$\langle\alpha|U|\beta\rangle = \sum_j \langle\alpha j|V|\beta j\rangle_A,$$

where α and β may be either particle states or hole states. With this choice no diagrams involving one or more bubble interactions are included in the perturbation expansion as they are cancelled by corresponding diagrams involving U. However, in the Brueckner-Goldstone expansion all the V-interactions have been replaced by G-interactions and the question of choice of single-particle potential has to be considered again.

The first few terms of the Brueckner-Goldstone expansion are shown in Fig. 3.9. Only a few of the third-order diagram have been drawn. Compare this with Fig. 2.1 in which the first few terms of the Goldstone expansion were drawn. Notice that Fig. 2.1(g) forms part of the ladder sequence of diagrams that makes up the G-interaction in Fig. 3.9(a), so that

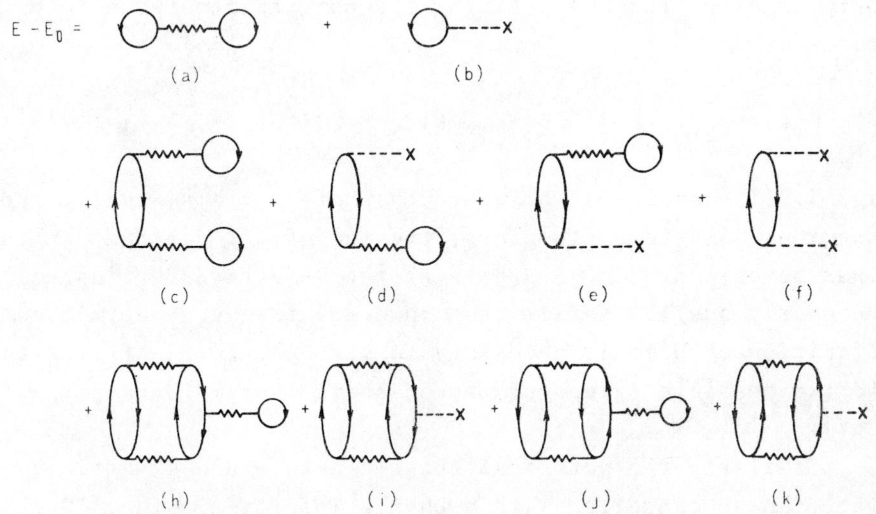

FIG. 3.9. First-, second-, and some third-order diagrams in the Brueckner-Goldstone expansion for the energy of a closed-shell nucleus.

diagram (g) is excluded from the Brueckner-Goldstone expansion.

Perhaps the obvious first choice of U is to follow the Hartree-Fock result and attempt to cancel G-interaction bubble insertions by defining

$$\langle \alpha | U | \beta \rangle = \sum_j \langle \alpha j | G(W) | \beta j \rangle_A .$$

This definition is ambiguous, however, since the G-matrix depends on the starting energy W and the Gs entering in different diagrams have different values of W. The potential U has no such dependence, so that the cancellation can take place only for one particular value of W. It is convenient to work with 'on-energy-shell' G-matrices so only diagrams with bubble insertions which can be made on the energy shell by application of the BBP theorem can be cancelled. Consider Fig. 3.9(h). This bubble insertion is a factorizable insertion, as discussed in the last section, and so can be evaluated on the energy shell. Thus by defining the potential between hole states as

$$\langle i|U|k\rangle = \frac{1}{2}\sum_j \{\langle ij|G(\varepsilon_i+\varepsilon_j)|kj\rangle_A + \langle ij|G(\varepsilon_k+\varepsilon_j)|kj\rangle_A\}, \quad (3.10)$$

Fig. 3.9(i) cancels a whole sequence of diagrams, namely the Brueckner-Goldstone gto class, the sum of all diagrams in this class being just Fig. 3.9(h) with the G-matrices evaluated on the energy shell. Notice that in order to have a Hermitian definition of U it is necessary to average two G-matrices for the two possible values of the on-shell energy (Baranger 1967).

Similarly the potential between a hole and a particle state can be cancelled with a bubble insertion, since in this case the bubble insertion is a dangling insertion and is therefore factorizable. Thus with

$$\langle i|U|m\rangle = \sum_j \langle ij|G(\varepsilon_i+\varepsilon_j)|mj\rangle_A \quad (3.11)$$

Figs. 3.9(c), (d), (e), and (f) mutually cancel each other.

However, the potential between particle states causes many problems and is still the subject of much discussion. One might try to define U so that Figs. 3.9(j) and (k) cancel, but the bubble insertion in the particle line of diagram (j) is an essential insertion and cannot be put on the energy shell. This means that U must be defined in terms of off-energy-shell G-matrices and therefore all particle-bubble insertions cannot be cancelled exactly, but only in some average sense which is not well defined.

However, not all third-order diagrams have been considered in Fig. 3.9. Rajaraman (1963a, b) showed that there were other third order (in G) diagrams which were as large as the particle-bubble diagram (j) and which could not be cancelled by such a choice of U. Later it was realized that selected diagrams of fourth and higher order in G were just as large and Bethe (1965) showed that if one tried to sum these diagrams order by order in G, then the series diverged.

However, the situation was saved when Rajaraman (1963a, b) pointed out that all diagrams having three and only three hole lines were of the same order of magnitude irrespective of the number of G-interactions.

Each hole line introduces a sum over occupied states and is therefore proportional to the density. All Goldstone diagrams (V-interactions) with only two hole lines were originally summed in our first definition of the G-matrix. Thus we now look upon the first-order terms of the Brueckner-Goldstone expansion as being the sum of all two-hole line diagrams, or two-body cluster diagrams as they have been called. The next order in the series is to sum all three-body cluster diagrams. This has only been attempted for nuclear matter for which a number of simplifications occur but the results can hopefully be carried over to finite nuclei. Bethe (1967) and Rajaraman and Bethe (1967) succeeded in summing all the three-body cluster diagrams by adapting the Faddeev equations and they found that the total contribution to the binding energy of nuclear matter was very much smaller than the contribution of the single-particle bubble diagram (fig. 3.9(j)) which was our original concern. Thus if one requires the matrix elements of U between particle states to cancel the three-body cluster diagrams they will not have to be very large. In fact, it may be a reasonable prescription to set $\langle m|U|n\rangle$ to be zero, viz.

$$\langle m|U|n\rangle = 0. \tag{3.12}$$

This would certainly simplify the computations. Another prescription is to follow the definition of the hole-hole matrix element and set

$$\langle m|U|n\rangle = \frac{1}{2}\sum_j \langle mj|[G(\bar{\varepsilon}_m+\varepsilon_j) + G(\bar{\varepsilon}_n+\varepsilon_j)]|nj\rangle_A \tag{3.13}$$

where $\bar{\varepsilon}_m$ is parameterized in some way to take into account the off-shell behaviour of the particle-bubble diagram, Fig. 3.9(j).

3.5 BRUECKNER-HARTREE-FOCK THEORY

By a judicious choice of the single-particle potential U, such that particle-hole matrix elements $\langle i|U|m\rangle$ cancel diagrams (c), (d), (e), and (f) of Fig. 2.11 (eqn (3.10)), hole-hole matrix elements $\langle i|U|k\rangle$ cancel diagrams (h) and (i) (eqn (3.11)) and particle-particle matrix elements $\langle m|U|n\rangle$ cancel three-body clusters (see various prescriptions in eqns (3.12) and (3.13)), then it might be hoped that the first-order terms of the Brueckner-Goldstone expansion will give a reasonable estimate of the binding energy of a closed-shell nuclear system. This estimate is

$$E = \sum_i \langle i|T|i\rangle + \frac{1}{2}\sum_{ij} \langle ij|G(\varepsilon_i+\varepsilon_j)|ij\rangle_A, \tag{3.14}$$

where the sums run over the occupied states, and the G-matrix elements are evaluated on-the-energy-shell. The matrix elements of G satisfy eqn (3.9)

$$\langle ij|G(\varepsilon_i+\varepsilon_j)|ij\rangle_A = \langle ij|V|ij\rangle_A + \frac{1}{2}\sum_{mn} \frac{\langle ij|V|mn\rangle\langle mn|G(\varepsilon_i+\varepsilon_j)|ij\rangle_A}{\varepsilon_i+\varepsilon_j-\varepsilon_m-\varepsilon_n}.$$

The single-particle and single-hole energies, ε_m and ε_i respectively, are the eigenvalues of T+U, viz.

$$\varepsilon_i = \langle i|T|i\rangle + \langle i|U|i\rangle,$$

and similarly for ε_m. Finding a solution to this sequence of equations poses a formidable problem, complicated by the fact that there are two self-consistency conditions that have to be met (Baranger 1967).

The first self-consistency condition is the familiar Hartree-Fock one discussed in the last chapter. To find the binding energy E of the closed-shell system we need to know which are the occupied states, so that we know which matrix elements of G are to be evaluated and which terms in eqn (3.6) are to be summed. However, the single-particle states are eigenstates of T+U so that U is required; but U is expressed in terms of the occupied states which we do not know *a priori*.

The second self-consistency condition is as follows. The

one-body potential U is expressed in terms of the two-body G-matrix, but before G can be evaluated one needs to know U for two separate reasons: one is that the value of the starting energy W to be used is given in terms of the single-particle energies which are eigenvalues of T+U; the other is that in the sum over intermediate particles states m,n..., the single-particle wavefunctions and energies used must also be eigenfunctions and eigenvalues of T+U. This is called the Brueckner self-consistency problem.

In principle the same iteration procedure can take care of both Hartree-Fock and Brueckner self-consistencies. One starts by choosing a reasonable set of single-particle states and single-particle energies. These are used to calculate G(W) and then U. Next one diagonalizes T+U to yield a new set of single-particle states and energies, and the process is repeated until sufficient convergence has been achieved. The operator G has to be recalculated at every step of the iterative process. Since a single calculation of G can be quite lengthy, the above procedure, although correct in principle, is prohibitively difficult and some short-cuts must be devised.

Most calculations to date use an oscillator basis, that is the single-particle states are expanded in terms of harmonic oscillator functions as is done in most conventional Hartree-Fock theories. The calculation starts by guessing the expansion coefficients and the single-particle energies and continues to successively improve these guesses as the iteration proceeds. However, it is not feasible to solve the Bethe-Goldstone equation for G at each stage of the iteration procedure. Instead one solves the equation beforehand for a few selected values of the starting energy, W, and then interpolates during the iteration process. This procedure does not allow the Pauli operator Q to be handled properly. In the evaluation of the G-matrix, a sum over the unoccupied particle states has to be performed, and which states are the unoccupied ones can vary at each stage of the iteration. Thus a Pauli operator defined in terms of pure oscillator orbitals is frequently used, which implies that states designated as unoccupied in the oscillator model at the start of the calculation remain unoccupied throughout. Another short cut is to fix the single-particle energies of the unoccupied particle states at the start of the calculation and make no attempt to determine them self consistently. This short-cut is motivated by the lack of a definite prescription relating the matrix elements of U between particle states to matrix elements of G.

Computations based on this outline have been performed at Oak Ridge (Davies and McCarthy 1971, Davies and Baranger 1970, McCarthy and Davies 1970, Davies, Baranger, Tarbutton and Kuo 1969, McCarthy 1969, Becker, MacKellar, and Morris 1968). The results were rather disappointing. For a series of closed-shell nuclei it was not possible to fit the experimental binding energy and charge radius simultaneously, when using realistic nucleon-nucleon potentials. It is not clear whether the poor results are due to the deficiencies of the nucleon-nucleon potential or the inadequacy of the many-body theory. Some improvement was obtained by Davies and McCarthy (1971) when a further class of diagrams, called the occupation probability diagrams, were incorporated. On the other hand calculations using effective rather than realistic nucleon-nucleon interactions have obtained impressive agreement with experiment. These effective interactions (Negele 1970, Köhler 1971) usually contain non-local or density-dependent terms which improve the saturation properties. However, these interactions have a number of parameters which are adjusted to fit properties of nuclear matter or selected finite nuclei. It is not clear whether the good fits obtained to binding energies and radii with these effective interactions come because they compensate for neglected higher order terms in the Brueckner-Goldstone expansion, or if the realistic two-body potentials themselves are inadequate.

3.6 CALCULATION OF THE G-MATRIX - REFERENCE SPECTRUM METHOD

So far we have said very little about actually calculating the reaction matrix G. In this and the next section we shall briefly discuss two methods for solving the Bethe-Goldstone equation: (i) the reference spectrum method (Bethe, Brandow, and Petschek 1963); and (ii) the separation method (Moszkowski and Scott 1960). There have been many articles discussing these methods and for a critical discussion we refer the reader to three recent reviews (Baranger 1967, Sprung 1972, Bethe 1971).

Suppose we are at the stage of a Brueckner-Hartree-Fock calculation where we know the single-particle potential U and hence the one-body Hamiltonian $H_0 = T+U$. Furthermore H_0 has been diagonalized giving the single-particle energies and wavefunctions, and it remains to solve the equation

$$G(W) = V + V \frac{Q}{e} G(W). \tag{3.8}$$

Here V is the nucleon-nucleon potential, Q the Pauli operator $Q \equiv \sum_{mn} |mn\rangle\langle mn|$ with m and n being unoccupied single-particle states, and e is an energy denominator defined as

$$\begin{aligned} e|mn\rangle &= (W - \varepsilon_m - \varepsilon_n)\,|mn\rangle \\ &= (W - H_0(m) - H_0(n))\,|mn\rangle. \end{aligned}$$

As discussed in the last section, a formally correct Brueckner-Hartree-Fock calculation is not a practical proposition. Thus, instead of working with the correct one-body Hamiltonian H_0 for which the single-particle potential U has been constructed by summing two-body G-interactions evaluated at the previous iteration, an approximate H_0 is assumed. As an illustration of the techniques used, we will calculate the matrix element of G between two occupied states, viz. $\langle i'j'|G(W)|ij\rangle_A$. For diagonal matrix elements the starting energy, W, is evaluated on-the-energy-shell, i.e. $W = \varepsilon_i+\varepsilon_j$. For off-diagonal matrix elements the prescription

$$\langle i'j'|G|ij\rangle_A = \tfrac{1}{2}[\langle i'j'|G(W=\varepsilon_i+\varepsilon_j)|ij\rangle_A + \langle i'j'|G(W=\varepsilon_{i'}+\varepsilon_{j'})|ij\rangle_A]$$

is used. The state $|ij\rangle$ is called the unperturbed state; by this we mean the wavefunction $|ij\rangle$ is written as a product of single-particle functions, each one being an eigenfunction of

H_0. Thus we introduce a notation

$$\Phi_{ij}(\underset{\sim}{r}_1,\underset{\sim}{r}_2) \equiv \phi_i(\underset{\sim}{r}_1)\phi_j(\underset{\sim}{r}_2) \equiv |ij\rangle \tag{3.15}$$

where the dependence on the radial coordinates $\underset{\sim}{r}_1$ and $\underset{\sim}{r}_2$ is explicitly displayed. Furthermore we have

$$H_0(i)\phi_i = \varepsilon_i\,\phi_i. \tag{3.16}$$

We now define a correlated two-body wavefunction Ψ_{ij} by the equation

$$\Psi_{ij}(\underset{\sim}{r}_1,\underset{\sim}{r}_2;W) = \Phi_{ij}(\underset{\sim}{r}_1,\underset{\sim}{r}_2) + \frac{Q}{e}\,G(W)\,\Phi_{ij}(\underset{\sim}{r}_1,\underset{\sim}{r}_2). \tag{3.17}$$

It then follows that

$$V\,\Psi_{ij} = [V + V\frac{Q}{e}G]\,\Phi_{ij} = G\Phi_{ij}, \tag{3.18}$$

and substituting back in eqn (3.17), the correlated two-particle wavefunction satisfies

$$\Psi_{ij} = \Phi_{ij} + \frac{Q}{e}\,V\,\Psi_{ij}. \tag{3.19}$$

The matrix elements of G are now evaluated using

$$\langle i'j'|G(W)|ij\rangle = \langle \Phi_{i'j'}|V|\Psi_{ij}(W)\rangle \tag{3.20}$$

where the first step in the calculation is to solve eqn (3.19) for the correlated wavefunction Ψ_{ij}.

The approach taken in the reference spectrum method is first to find an approximate G-matrix, G_R, and then to correct G_R in a subsequent calculation. The approximate G-matrix, G_R, is found after making the following simplifications:

(i) the Pauli operator Q is replaced by unity;

(ii) the unperturbed Hamiltonian H_0 is taken to be either the free-particle or harmonic-oscillator Hamiltonian.

The reason for taking only these choices for H_0 is that when operating on the unperturbed two-particle state Φ_{ij}, a convenient separation into centre-of-mass and relative co-ordinates is possible. For example, with a harmonic oscillator Hamiltonian, we have

$$H_0 = -\frac{\hbar^2}{2m}\nabla_1^{\,2} - \frac{\hbar^2}{2m}\nabla_2^{\,2} + \frac{m}{2}\Omega^2 r_1^{\,2} + \frac{m}{2}\Omega^2 r_2^{\,2}$$

$$= -\frac{\hbar^2}{2M}\nabla_R^{\,2} - \frac{\hbar^2}{2\mu}\nabla_r^{\,2} + \frac{1}{2}M\Omega^2 R^2 + \frac{1}{2}\mu\Omega^2 r^2$$

where Ω is the oscillator parameter, $M = 2m$ is the total mass the $\mu = m/2$ is the reduced mass. Thus H_0 is separable, viz.

$$H_0(1) + H_0(2) = H_{cm} + H_{rel} \tag{3.21}$$

with H_{cm} being the centre-of-mass part and H_{rel} the relative part of the unperturbed Hamiltonian.

We require the matrix elements of G between two-particle states which have been coupled to good total angular momentum J and isospin T. In the single-particle representation the Bethe-Goldstone equation reads

$$\langle j_i' j_j'; JT | G_R(W) | j_i j_j; JT\rangle_A$$

$$= \langle j_i' j_j'; JT|V|j_i j_j; JT\rangle_A +$$

$$+ \sum_{j_m \geqslant j_n} \frac{\langle j_i' j_j'; JT|V|j_m j_n, JT\rangle_A \langle j_m j_n; JT|G_R(W)|j_i j_j; JT\rangle_A}{W - \varepsilon_m + \varepsilon_n}. \tag{3.22}$$

Here we use an abbreviated notation with j_i standing for all the quantum numbers of a single-particle state $n_i \ell_i j_i$. We have also inserted the reference spectrum approximation of setting Q = 1. Without this approximation an extra matrix element $\langle j_m j_n; JT|Q|j_m j_n; JT\rangle$ would occur in the last term of eqn (3.22). This matrix element of Q is diagonal in the single-particle representation and has the value unity if

j_m and j_n are unoccupied single-particle states, and zero otherwise. We now use the properties of the harmonic oscillator to transform to a relative and centre-of-mass coordinate representation. Specifically we expand a normalized, antisymmetrized two-particle state as

$$|j_i j_j;JT\rangle_A = (1+\delta_{ij})^{-\frac{1}{2}} \sum_{n\ell S\mathcal{I}\mathcal{N}\mathcal{L}} [1-(-)^{\ell+S+T}] C_{ij}(n\ell S\mathcal{I}\mathcal{N}\mathcal{L};JT)\,|n(\ell S)\mathcal{I}T;\mathcal{N}\mathcal{L};J\rangle \quad (3.23)$$

where the coefficient C_{ij} is given by

$$C_{ij}(n\ell S\mathcal{I}\mathcal{N}\mathcal{L};JT) = \sum_L \begin{bmatrix} \ell_i & \frac{1}{2} & j_i \\ \ell_j & \frac{1}{2} & j_j \\ L & S & J \end{bmatrix} (-)^{L+\mathcal{I}-\ell-J}$$

$$\langle n_i\ell_i, n_j\ell_j;L|n\ell,\mathcal{N}\mathcal{L};L\rangle\; U(\mathcal{L}\ell JS;L\mathcal{I}) \quad (3.24)$$

Here the [ ]-coefficient is the transformation coefficient from j-j to L-S coupling, the bracket $\langle n_i\ell_i, n_j\ell_j;L|n\ell,\mathcal{N}\mathcal{L};L\rangle$ is the Brody-Moshinsky transformation bracket between the single-particle and relative, centre-of-mass coordinate systems; $n\ell$ and $\mathcal{N}\mathcal{L}$ are respectively the relative and centre-of-mass oscillator quantum numbers. The U-coefficient represents a recoupling of the angular momenta, so that the relative ℓ couples with the spin S to form a total relative angular momentum $\mathcal{I}$ which in turn couples to the centre-of-mass $\mathcal{L}$ to form J. Notice that the C_{ij} satisfy an orthogonality relation,

$$\sum_{ij} C_{ij}(n\ell S\mathcal{I}\mathcal{N}\mathcal{L};JT)C_{ij}(n'\ell'S'\mathcal{I}'\mathcal{N}'\mathcal{L}';JT) = \delta(n\ell S\mathcal{I}\mathcal{N}\mathcal{L};n'\ell'S'\mathcal{I}'\mathcal{N}'\mathcal{L}').$$

In the relative, centre-of-mass representation the reference spectrum Bethe-Goldstone equation now reads

$$\langle n'(\ell'S)\mathcal{I}T|G_R(W)|n(\ell S)\mathcal{I}T\rangle$$

$$= \langle n'(\ell'S)\mathcal{I}T|V|n(\ell S)\mathcal{I}T\rangle +$$

$$+ \sum_{n''\ell''} \frac{\langle n'(\ell'S)IT|V|n''(\ell''S)IT\rangle\langle n''(\ell''S)IT|G_R(W)|n(\ell S)IT\rangle}{W - \varepsilon_{n''\ell''} - \varepsilon_{NL}} \qquad (3.25)$$

where we have noted that the two-nucleon potential V is independent of the centre-of-mass variables. Note also that the transformation conserves the oscillator energy for the intermediate two-particle state, i.e. $\varepsilon_m+\varepsilon_n = \varepsilon_{n''\ell''}+\varepsilon_{NL}$. To simplify the writing of these equations we will use one Greek letter α to represent the quantum numbers $n\ell SIT$. Thus

$$\langle\alpha'|G_R(W)|\alpha\rangle = \langle\alpha'|V|\alpha\rangle + \sum_{\alpha''} \frac{\langle\alpha'|V|\alpha''\rangle\langle\alpha''|G_R(W)|\alpha\rangle}{W-\varepsilon_{\alpha''}-\varepsilon_{NL}},$$

or as an operator equation

$$G_R(W) = V + V\frac{1}{W-H_{rel}-H_{cm}}G_R(W).$$

Note that H_{cm} in the denominator commutes with everything and can be considered as just a number, ε_{NL}. This is true as long as Q has been replaced by unity. In general Q does not commute with H_{cm}. The correct version of the Bethe-Goldstone equation would contain an additional matrix element $\langle n''(\ell''S)IT;NL;J|Q|n'''(\ell'''S)IT;N'L';J\rangle$ in the last term of eqn (3.25), indicating that Q is not diagonal in the centre-of-mass coordinates. This provides a major difficulty for an accurate calculation of G in a relative, centre-of-mass representation, and some approximation which makes Q diagonal in this representation has to be found.

Let us define a relative starting energy

$$W_{rel} = W-H_{cm} = W - \varepsilon_{NL}$$

which is state-dependent now, i.e. depends on the centre-of-mass quantum numbers NL, then the reference spectrum equation becomes

$$G_R(W_{rel}) = V + V\frac{1}{W_{rel}-H_{rel}}G_R(W_{rel}), \qquad (3.28)$$

in which all operators act on relative states only. Again we introduce a notation for the unperturbed wavefunction in relative coordinates

$$|\alpha\rangle \equiv |n(\ell S)IT\rangle \equiv \Phi_\alpha(\underset{\sim}{r})$$

and the corresponding correlated wavefunction as $\Psi_\alpha(\underset{\sim}{r},W_{rel})$. The advantage of the reference-spectrum approximations is that the correlated wavefunction Ψ_α can now be found as a solution to a single differential equation in the relative coordinate. This equation reads

$$\Psi_\alpha = \Phi_\alpha + \frac{1}{W_{rel}-H_{rel}} V\Psi_\alpha,$$

that is

$$(H_{rel}+V-W_{rel})\Psi_\alpha = (\varepsilon_\alpha-W_{rel})\Phi_\alpha \tag{3.30}$$

where ε_α is the relative energy of the unperturbed state Φ_α. To solve this equation it is necessary to expand Ψ_α and Φ_α in partial waves. However, for illustrative purposes we will just write the equation for the ^1S partial wave, in which case

$$\begin{aligned}\Phi_\alpha(\underset{\sim}{r}) &= \frac{1}{r} R_\alpha(r)\\ \Psi_\alpha(\underset{\sim}{r};W_{rel}) &= \frac{1}{r} U_\alpha(r;W_{rel})\end{aligned} \tag{3.31}$$

and the differential equation becomes

$$\left[\frac{\partial^2}{\partial r^2} - \frac{\mu^2\Omega^2}{\hbar^2} r^2 - \frac{2\mu}{\hbar^2}(V-W_{rel})\right]U_\alpha = -\frac{2\mu}{\hbar^2}(\varepsilon_\alpha-W_{rel})R_\alpha \tag{3.32}$$

where R_α is a radial harmonic oscillator function. This is a second order linear differential equation which differs from the Schrödinger equation by the appearance of an inhomogeneous term on the right hand side. In practice, however, it is not this equation which is solved; instead a 'defect wave function' ζ_α, being just the difference between the unperturbed and perturbed wavefunctions, is introduced, i.e.

$$\zeta_\alpha = \Phi_\alpha - \Psi_\alpha. \tag{3.33}$$

Substituting eqn (3.33) back into eqn (3.30) and remembering that $H_{rel}\,\Phi_\alpha = \varepsilon_\alpha\,\Phi_\alpha$ a differential equation for the defect function ζ_α is obtained

$$(H_{rel} + V - W_{rel})\zeta_\alpha = V\Phi_\alpha. \tag{3.34}$$

Again to solve this equation we make a partial wave expansion. Writing down only the ^1S wave the differential equation reads

$$\left[\frac{\partial^2}{\partial r^2} - \frac{\mu^2\Omega^2}{\hbar^2}\,r^2 - \frac{2\mu}{\hbar^2}\,(V-W_{rel})\right]\chi_\alpha = -\frac{2\mu}{\hbar^2}\,V(r)R_\alpha \tag{3.35}$$

where

$$\zeta_\alpha(\underset{\sim}{r};W_{rel}) = \frac{1}{r}\,\chi_\alpha(r;W_{rel}). \tag{3.36}$$

Asymptotically the potential V vanishes, $V \to 0$ as $r \to \infty$, and for large r the defect wavefunction satisfies the equation

$$\left[\frac{\partial^2}{\partial r^2} - \frac{\mu^2\Omega^2}{\hbar^2}\,r^2 - \gamma^2\right]\chi_\alpha = 0 \tag{3.37}$$

where

$$\gamma^2 = -\frac{2\mu}{\hbar^2}\,W_{rel}. \tag{3.38}$$

All the above three differential equations, (3.32), (3.35), and (3.37), have been written for a harmonic oscillator unperturbed Hamiltonian H_0; the same equations apply for the plane-wave Hamiltonian except that the term $\mu^2\Omega^2 r^2/\hbar^2$ is dropped. In particular the asymptotic equation for the defect function is just

$$\left[\frac{\partial^2}{\partial r^2} - \gamma^2\right]\chi_\alpha = 0 \tag{3.39}$$

which admits a solution

$$\chi_\alpha \sim e^{-\gamma r}, \tag{3.40}$$

that is, the defect function tends to zero as r tends to infinity. This is said to be the 'healing' property, and the constant $1/\gamma$ is said to be the healing length. The larger the value of γ, the more rapid is the healing. If, however, V(r) has an exponential or Yukawa tail whose range exceeds $1/\gamma$, then the healing length is this range. For the harmonic oscillator choice of H_0, the healing is Gaussian-like for large r; however, at moderate values of r the term γ^2 dominates $\mu^2\Omega^2r^2/\hbar^2$ and so the healing starts out exponential-like and becomes Gaussian only for very large r.

To illustrate the behaviour of the wavefunctions R, U, and χ we plot in Fig. 3.10 the solutions for the ${}^1S$ partial wave for the standard hard-core potential of Moszkowski and Scott (1960)

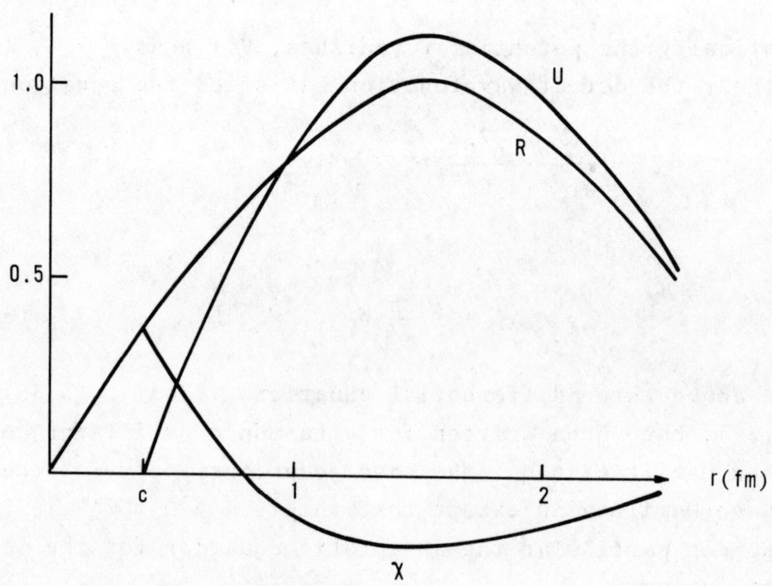

FIG. 3.10. Radial dependence of the uncorrelated function R(r), the correlated function U(r), and the defect function $\chi(r)$ for the ${}^1S$ partial wave, (Day 1967).

$$V(r) = \begin{cases} +\infty & \text{for } r < c \\ -V_0 e^{-\mu(r-c)} & \text{for } r > c. \end{cases}$$

The plane-wave unperturbed Hamiltonian H_0 was used. Inside the hard core the perturbed wavefunction U is zero such that VΨ is finite, and so the defect function just equals the unperturbed function R at $r < c$. At $r = c$ there is a discontinuity in the slope of χ as the perturbed wavefunction U has a non-zero slope at $r = c$. As $r \to \infty$, the unperturbed and perturbed function heal together and $\chi \to 0$.

The differential equation (3.35) for the defect function χ_α is therefore solved in the following way. At large r there is an analytic asymptotic solution. This provides a set of starting values which enables the equation to be numerically integrated inwards to some matching radius. Further, the equation is also integrated outwards from the origin and the solutions are forced to obey the correct boundary conditions at the hard-core edge and at the matching radius. In this way the defect function χ_α and hence U_α are obtained.

The matrix elements of the reference G-matrix (again for the ^1S partial wave) reduce to just a radial integral

$$\langle\alpha'|G_R|\alpha\rangle = \int_0^\infty R_{\alpha'}(r)\ V(r)\ U_\alpha(r;W_{rel})dr. \qquad (3.41)$$

This integral is not defined if V has an infinite hard core and even with a finite core there may be computational difficulties. However, an alternative expression involving the defect function can be obtained from the differential equation (3.37)

$$V(r)U_\alpha(r;W_{rel}) = -\frac{\hbar^2}{2\mu}\left[\frac{d^2}{dr^2} - \frac{\mu^2\Omega^2}{\hbar^2}r^2 - \gamma^2\right]\chi_\alpha(r;W_{rel})$$

to give

$$\langle\alpha'|G_R|\alpha\rangle = -\frac{\hbar^2}{2\mu}\int_0^\infty R_{\alpha'}(r)\left[\frac{\partial^2}{\partial r^2} - \frac{\mu^2\Omega^2}{\hbar^2}r^2 - \gamma^2\right]\chi_\alpha(r;W_{rel})dr. \qquad (3.42)$$

Bethe, Brandow and Petschek (1963) recommend using eqn (3.42) for small r and eqn (3.41) for large r. Thus the range of integration is divided into two parts, the division occurring at some arbitrary radius a. The resultant expression is

$$\langle \alpha' | G_R | \alpha \rangle = -\frac{\hbar^2}{2\mu} \int_0^a R_{\alpha'} \left[\frac{d^2}{dr^2} - \frac{\mu^2 \Omega^2}{\hbar^2} r^2 - \gamma^2\right] \chi_\alpha \, dr + \int_a^\infty R_{\alpha'} V U_\alpha dr \tag{3.43}$$

where the first term is further simplified using Green's theorem and the Schrödinger equation for $R_{\alpha'}$ to give

$$\langle \alpha' | G_R | \alpha \rangle = \int_0^a R_{\alpha'} [-E_{\alpha'} U_\alpha + W_{rel} (U_\alpha - R_\alpha) + E_\alpha R_\alpha] \, dr$$

$$+ \frac{\hbar^2}{2\mu} [R_{\alpha'}(a) \frac{dU_\alpha}{dr}(a) - \frac{dR_{\alpha'}}{dr}(a) U_\alpha(a)] \tag{3.44}$$

$$+ \int_a^\infty R_{\alpha'} V U_\alpha \, dr.$$

In the case of an infinite hard-core potential of radius c, it is convenient to let a = c in the above expression. Noting that U_α is zero for r < c, some simplification occurs:

$$\langle \alpha' | G_R | \alpha \rangle = (E_\alpha - W_{rel}) \int_0^c R_{\alpha'} R_\alpha \, dr$$

$$+ \frac{\hbar^2}{2\mu} R_{\alpha'}(c) \frac{dU_\alpha}{dr}(c) \tag{3.45}$$

$$+ \int_c^\infty R_{\alpha'} V U_\alpha \, dr.$$

The first time on the right is called the 'core volume' term, the middle term is the 'core edge' term, and the third term the 'outer' term. This shows that the dependence of the G-matrix on the hard core has been completely separated from its dependence on the outer attractive part of the potential.

In this way an approximate reference spectrum G-matrix G_R is evaluated. It remains to improve upon this approximation by iterating the exact equation

$$G = G_R - G_R \left(\frac{1}{e_R} - \frac{Q}{e}\right) G. \tag{3.46}$$

Here e_R is the reference energy denominator associated with using an approximate Hamiltonian, H_0. Note, however, that G_R is a first approximation in a systematic expansion of G. This expansion is extremely useful (Day 1967) because the leading term is both simple and accurate. For a central two-body potential the diagonal matrix elements of G may be given to an accuracy of about 5 per cent; for a tensor force the accuracy will be poorer but still may be of order 15 per cent.

To prove the result (3.46), it is necessary to derive some operator relationships (Bethe, Brandow, and Petschek 1963). In analogy to scattering theory a wave matrix Ω is introduced:

$$G = V\,\Omega, \tag{3.47}$$

where Ω satisfies the integral equation

$$\Omega = 1 + \frac{Q}{e}\,G. \tag{3.48}$$

Consider now two problems A and B with different potentials V_A and V_B and different propagators Q_A/e_A and Q_B/e_B. Then

$$\Omega_A = 1 + \frac{Q_A}{e_A}\,G_A \tag{3.49}$$

$$\Omega_B^{\dagger} = 1 + G_B^{\dagger}\,\frac{Q_B^{\dagger}}{e_B^{\dagger}}. \tag{3.50}$$

Pre-multiplying eqn (3.49) by $G_B^{\dagger}$, post-multiplying eqn (3.50) by G_A, and subtracting leads to a general result

$$G_A = G_B^{\dagger} + \Omega_B^{\dagger}(V_A - V_B^{\dagger})\Omega_A - G_B^{\dagger}\left(\frac{Q_B^{\dagger}}{e_B^{\dagger}} - \frac{Q_A}{e_A}\right)G_A. \tag{3.51}$$

We now specialize this for the reference spectrum case. Assume that problem B corresponds to finding an approximate G-matrix, G_R, using an approximate propagator $1/e_R$ and that problem A corresponds to the exact G-matrix. In this case, both problems have the same potential $V_A = V_B$ and the required result follows:

$$G = G_R - G_R\left(\frac{1}{e_R} - \frac{Q}{e}\right)G \tag{3.46}$$

which indicates how an approximate G-matrix, G_R, can be improved iteratively to approach the correct G-matrix.

For finite nuclei, calculations have only been carried out in a harmonic oscillator representation, implying that the distinction between occupied and unoccupied states is determined by a harmonic oscillator spectrum rather than from any self-consistent Brueckner-Hartree-Fock spectrum. Furthermore the energy denominator e is given in terms of oscillator energies. With these restrictions the procedure advocated by Köhler and McCarthy (Köhler and McCarthy 1968, McCarthy 1969) is to consider eqn (3.46) as a system of linear algebraic equations. A matrix P defined as

$$P = 1 - G_R\left(\frac{1}{e_R} - \frac{Q}{e}\right)$$

can be set up and inverted so that the exact G-matrix is given by

$$G = P^{-1}\, G_R.$$

The problem, however, is that the reference G-matrix has been evaluated in relative coordinates and the Pauli operator Q is not diagonal here. McCarthy (1969) therefore approximates Q by an operator which is diagonal. The size of the matrix obtained then depends on the approximation for Q and on the cut-off at high quantum numbers. A fairly low cut-off is desirable otherwise the number of coupled equations gets unmanageably large. An exact treatment has recently been given by Barrett, Hewitt, and McCarthy (1971) in which the reference G-matrix is transformed back to a single particle representation (i.e. coordinates $\underset{\sim}{r}_1$ and $\underset{\sim}{r}_2$, rather than $\underset{\sim}{r}$ and $\underset{\sim}{R}$) where the Pauli operator Q is exactly diagonal. The only approximation is now the truncation of the matrix, P, for high quantum numbers.

3.7 MOSZKOWSKI-SCOTT SEPARATION METHOD

The separation method of Moszkowski and Scott (Moszkowski and Scott 1960, Scott and Moszkowski 1961) is based on a simple physical idea. In the $^1S_0$-scattering state, the nuclear force is strongly repulsive at short distances, whereas at larger distances it becomes attractive. At low enough energies the attraction at large distances more than compensates for the repulsion. Thus the simple physical idea is to separate the nucleon-nucleon potential into two parts, a short-range part, V_S, and a long-range part V_L, viz.

$$V = V_S + V_L. \tag{3.52}$$

Into the short-range part is put the repulsive part of the force together with that part of the attractive force which will exactly cancel the repulsion, i.e. scattering by the potential V_S would produce zero phase shift. The remaining part of the attractive force is put in V_L. The division into V_S and V_L is usually made sharply at some radius called the separation distance, d. The appealing idea, however, is that V_L is a weak attractive potential, the kind of potential that could be used in a Hartree-Fock calculation and for which the Goldstone series converges rapidly without the introduction of the Brueckner G-matrix. Thus one can expand in powers of V_L and keep only low orders. The Pauli operator Q is essential to this expansion; it is only by virtue of Q that the higher order terms in V_L are small. On the other hand the short-range potential V_S which contains the repulsive core does require the introduction of a G-matrix. However, since the presence of V_S is not felt at distances beyond the separation distance, d, it may be sufficiently accurate to evaluate this G-matrix making the Q = 1 approximation. Thus under ideal conditions the separation method achieves a complete separation of the Brueckner problem from the Pauli problem (Baranger 1967). There are, however, disadvantages to the method, but we will discuss these after first giving details of how the separation method works.

In the last section we derived a formula which related two G-matrices, G_A and G_B. The reaction matrix G_A was the solution for problem A which involved a nucleon-nucleon potential V_A and a propagator Q_A/e_A and G_B corresponded to a separate problem B with potential V_B and propagator Q_B/e_B. The relation between them was

$$G_A = G_B{}^\dagger + \Omega_B{}^\dagger (V_A - V_B{}^\dagger)\Omega_A - G_B{}^\dagger \left(\frac{Q_B{}^\dagger}{e_B{}^\dagger} - \frac{Q_A}{e_A} \right) G_A. \tag{3.51}$$

Specializing this formula to the case in which the nucleon-nucleon potential is separated into two parts $V = V_S + V_L$, and identifying problem A with the solution for the full potential, i.e. $V_A = V$, and problem B with the solution for the short-range potential, i.e. $V_B = V_S$, then one obtains (Köhler 1961)

$$G = G_S + \Omega_S{}^\dagger \; V_L \; \Omega, \tag{3.53}$$

where we have assumed the propagators for the two problems are the same. Here G_S is the G-matrix corresponding to the short-range part of the potential and satisfies the integral equation

$$G_S = V_S + V_S \frac{Q}{e} G_S, \tag{3.54}$$

and Ω_S is the corresponding wave matrix. Note that Ω_S is not Hermitian. The wave matrices were defined in eqns (3.49) and (3.50). Specifically they are

$$\Omega = 1 + \frac{Q}{e} G$$

$$\Omega_S{}^\dagger = 1 + G_S{}^\dagger \frac{Q^\dagger}{e^\dagger}.$$

The first-order approximation for the G-matrix from the Moszkowski-Scott separation method is obtained from eqn (3.53) on approximating the wave matrices by unity. This gives

$$G = G_S + V_L \tag{3.55}$$

The next-order approximation maintains all terms involving two interactions, but replaces G by its first-order approximation $G_S + V_L$ to give

$$G \approx G_S + V_L + G_S \frac{Q}{e} V_L + V_L \frac{Q}{e} G_S + V_L \frac{Q}{e} V_L. \tag{3.56}$$

Eqn (3.56) shows that G is given as a sum of five terms, in which the last three correspond to small correction terms. The second and fifth terms are the contributions from the weak attractive long-range part of the force calculated to first order and to second order respectively in the Born approximation. This Born series will converge rapidly if V_L is sufficiently weak. The third and fourth terms represent an interference between short-range and long-range forces, while the first term is the reaction matrix corresponding to the short-range part of the potential. The essence of the calculation is to find a good approximation for G_S.

For this the reference spectrum approximations are made, viz. the Pauli operator Q is replaced by unity and the energy denominator replaced by $e_0 = W - H_0$ where H_0 corresponds to the free-particle Hamiltonian. A first guess to G_S is then given by

$$G_S^{(0)} = V_S + V_S \frac{1}{e_0} G_S^{(0)} \tag{3.57}$$

which can be solved by the method described in the last section. This approximate $G_S^{(0)}$ is then improved to next-order giving

$$\begin{aligned} G_S &= G_S^{(0)} - G_S^{(0)} \left(\frac{1}{e_0} - \frac{Q}{e}\right) G_S \\ &\approx G_S^{(0)} - G_S^{(0)} \left(\frac{1}{e_0} - \frac{Q}{e}\right) G_S^{(0)} \\ &\approx G_S^{(0)} + G_S^{(0)} \left(\frac{Q}{e} - \frac{1}{e}\right) G_S^{(0)} + G_S^{(0)} \left(\frac{1}{e} - \frac{1}{e_0}\right) G_S^{(0)} \end{aligned} \tag{3.58}$$

and the final expression for the Moszkowski-Scott G-matrix now reads

$$G = G_S^{(0)} + V_L + G_S^{(0)}\left(\frac{Q}{e} - \frac{1}{e}\right)G_S^{(0)} + G_S^{(0)}\left(\frac{1}{e} - \frac{1}{e_0}\right)G_S^{(0)} +$$

$$+ G_S^{(0)} \frac{Q}{e} V_L + V_L \frac{Q}{e} G_S^{(0)} + V_L \frac{Q}{e} V_L. \tag{3.59}$$

The third term on the right here is called the Pauli correction term and the fourth term the spectral or dispersion correction term.

It remains to specify exactly how the nucleon-nucleon potential V is divided into a short-range, V_S, and a long-range, V_L, part. The original suggestion of Moszkowski and Scott was that the separation distance d should be chosen such that the diagonal matrix element of $G_S^{(0)}$ would be zero, viz.

$$\langle \Phi_\alpha | G_S^{(0)}(W) | \Phi_\alpha \rangle \sim 0. \tag{3.60}$$

This clearly can never be achieved exactly, since $G_S^{(0)}$ is a function of the starting energy, W, and therefore the matrix element will only be zero for one value of W and not for all starting values. The best that can be hoped for is that $G_S^{(0)}(W)$ is only a weak function of the starting energy. Therefore Moszkowski and Scott had to inject a further assumption. They assumed that the correlated wavefunction Ψ_α was given sufficiently accurately by the solution of the Schrödinger equation for free particle scattering in the short-range potential V_S, namely

$$(H_{rel} + V_S - \varepsilon_\alpha)\Psi_\alpha = 0 \tag{3.61}$$

rather than the correct reference spectrum equation

$$(H_{rel} + V_S - W_{rel})\Psi_\alpha = (\varepsilon_\alpha - W_{rel})\Phi_\alpha. \tag{3.30}$$

Here ε_α is the relative energy of the unperturbed state Φ_α, viz.

$$(H_{rel} - \varepsilon_\alpha)\Phi_\alpha = 0,$$

H_{rel} being that part of the unperturbed free-particle Hamiltonian H_0 depending on the relative coordinate. Comparing eqns (3.30) and (3.61), we see that the Moszkowski and Scott assumption amounts to choosing a starting energy W_{rel} equal to ε_α. That is, the energy denominator e_0 in the Bethe-Goldstone equation is evaluated using just the kinetic energies instead of the true single-particle energies. Thus one can anticipate a sizeable dispersion correction $G_S^{(0)}\left(\frac{1}{e} - \frac{1}{e_0}\right)G_S^{(0)}$ when the correct energies are inserted.

In section (3.6), a general expression, eqn (3.44), was derived for the matrix element of G evaluated by the reference spectrum method for the ^1S partial wave. Repeating the expression here, but choosing the division in the integration range to occur at the separation distance, d, we have

$$\langle\Phi_\alpha|G_S^{(0)}|\Phi_\alpha\rangle = \int_0^d R_\alpha[-\varepsilon_\alpha U_\alpha + W_{rel}(U_\alpha - R_\alpha) + \varepsilon_\alpha R_\alpha]dr +$$

$$+ \frac{\hbar^2}{2\mu}\,[R_\alpha(d)\,\frac{dU_\alpha}{dr}(d) - \frac{dR_\alpha}{dr}(d)\,U_\alpha(d)] +$$

$$+ \int_d^\infty R_\alpha\, V_S\, U_\alpha\, dr. \qquad (3.62)$$

With the assumption $W_{rel} = \varepsilon_\alpha$, the first term vanishes. Furthermore the third term is zero since the short range potential V_S is defined to be zero for $r > d$. Thus the requirement that the separation distance, d, be chosen such that the diagonal matrix element of $G_S^{(0)}$ is zero, reduces to the condition that

$$\frac{1}{U_\alpha}\,\frac{dU_\alpha}{dr} = \frac{1}{R_\alpha}\,\frac{dR_\alpha}{dr} \qquad \text{at } r = d. \qquad (3.63)$$

An equivalent statement is that the defect function $\chi_\alpha = (R_\alpha - U_\alpha)$ and its derivative should vanish at the separation distance. This prescription shows that d depends on the energy ε_α of the unperturbed state and also on the quantum numbers characterizing that state. Note it is only possible to define such a separation distance in any particular partial wave at any particular energy ε_α if the corresponding scattering phase shift due to the entire interaction is positive, i.e. the attraction more than compensates the repulsion.

The attractive feature of the Moszkowski-Scott method is that the zeroth order approximation to the G-matrix, viz.

$$G = G_S^{(0)} + V_L \tag{3.64a}$$

which for the case of diagonal matrix elements reduces to

$$G = V_L \tag{3.64b}$$

is simple to calculate and is moderately accurate. However, the disadvantage is that there are a large number of correction terms, eqn (3.59), each one of which is difficult to calculate and may not be negligible, although some cancellation between corrections does occur. The method does fairly well for singlet S-states but does poorly in the triplet-S, triplet-D states due mainly to the strong tensor interaction. The long-range force V_L is mainly tensor and only begins to contribute in the second order of perturbation theory. Thus, in the ^3S partial wave the term $V_L \frac{Q}{e} V_L$ is almost as large as the first order term V_L. However, in all fairness it should be mentioned that the reference spectrum method also has difficulty handling a strong tensor force.

3.8 G-MATRIX FOR NUCLEAR STRUCTURE CALCULATIONS

Perhaps the one goal that has attracted the most theoretical attention in nuclear structure physics has been the attempt to relate properties of finite nuclei with basic properties of the two-nucleon system. There are three distinct steps in the chain:

(a) First, a nucleon-nucleon potential V has to be constructed which describes the scattering data and certain bound state properties such as the deuteron quadrupole moment.

There are many such potentials in the literature and they all fit the scattering phase shifts (more or less) well. It is not our intention to discuss these potentials; instead we refer the reader to two recent reviews - Signell (1969) and Bethe (1971).

Briefly, the available potentials fall into one of two categories. There are the local phenomenological potentials, such as those of Hamada and Johnston (1962), the Yale Group (Lassila *et al.* 1962) and of Reid (1968), which are purely empirical and which are characterized by a strong short-range core; or there are the super-soft potentials, such as the non-local separable force of Tabakin (1964), which fit the scattering data without invoking repulsive cores. However, nuclear matter calculations (Haftel and Tabakin 1970, 1971) have indicated that these super-soft potentials do not saturate and give too much binding at too high a density. They also require a substantial tensor force.

(b) The second step is to progress from two-nucleon systems to finite nuclei by replacing the nucleon-nucleon potential V with a reaction matrix G. This has been the subject of the current chapter.

(c) The third step is to evaluate selected properties of finite nuclei using these G-matrix elements. This includes both Hartree-Fock calculations of the type discussed in chapter 2, where the aim is to reproduce bulk properties such as total binding energies, r.m.s. radii, etc., as well as shell model calculations where the aim is to calculate excitation energies, transition probabilities etc. — properties which depend on just a few nucleons near the Fermi surface.

For the present we wish to survey briefly the sets of two-body G-matrix elements which have been constructed primarily for use in the calculations outlined in (c). These matrix elements can be categorized as being either 'realistic', 'modelistic' or 'empirical'.

(i) *Realistic*. By realistic we imply that the G-matrix elements have been constructed from a nucleon-nucleon potential which gives a fit to scattering phase shifts. The connection with the two-nucleon system remains clear and unambiguous. Perhaps the most successful calculation of this type and certainly the most quoted has been the G-matrix of Kuo and Brown (Kuo and Brown 1966, Kuo 1967), which is constructed from the Hamada-Johnston nucleon-nucleon potential. The aim of Kuo and Brown was not to calculate bulk properties of nuclei, but rather to find an interaction which was simple to evaluate and useful for shell model calculations near the Fermi surface. Thus a number of simplifying assumptions were made, some of which were removed in the later calculations of Kuo (1967).

A harmonic oscillator basis was used, and relative G-matrix elements were evaluated using the reference spectrum method for $\ell > 0$ partial waves and the separation method for S-states. In the reference spectrum method, plane wave intermediate states were used, and the Q-1 corrections were found to be small and hence ignored. The parameter γ^2 (eqn (3.38)), which is state-dependent since it depends on NL (the centre-of-mass quantum numbers), was found to be insensitive in the calculations and so was replaced by a state-independent value. On the other hand, for S-states γ^2 is sensitive and Q-1 corrections are large, so the separation method was used. In Kuo and Brown (1966), the short-range G-matrix, G_S, was ignored and the G-matrix was taken to be just the long-range part of the potential, V_L. An average state-independent separation distance, d, was used. For the ^{3}S-state the second Born term $V_L(Q/W-H_0)V_L$ has to be included because of the strong tensor force. this is handled by a 'closure approximation' which replaces $Q/W-H_0$ by a pure number $1/e_{eff}$. In the subsequent calculation of Kuo (1967), G_S was also calculated. The resulting set of matrix elements was first used to calculate the spectra of ^{18}F and ^{18}O, assuming these nuclei could be described as two nucleons outside an inert closed-shell core. This severe model truncation is then corrected by first order perturbation theory, and reasonable agreement with experiment obtained.

Some other sets of reaction matrix elements for the same shell model calculation have been derived by McCarthy (1969) and Mercier, Baranger and McCarthy (1969). These authors also work from a Hamada-Johnston potential. They use the reference spectrum method for all partial waves with harmonic oscillator intermediate states and a shell model Q, which they approximate by an average Q, diagonal in the relative centre-of-mass representation. They get reasonable agreement with the matrix elements of Kuo, though some differences are noted. Kahana, Lee, and Scott (1969) work from a non-local separable potential of the Tabakin type into which some repulsive core has been injected. Thus, they need to solve the Bethe-Goldstone equation, but using a separable potential gives some simplification. They also use the reference spectrum method for all partial waves and work with an approximate Q.

Barrett, Hewitt, and McCarthy (1970, 1971) give a simple and exact method for calculating G in a two-particle harmonic-oscillator basis. The method makes use of an expansion of the correlated wavefunction $\Psi_{ij}(\underset{\sim}{r}_1,\underset{\sim}{r}_2;W)$ in terms of solutions of a Schrödinger equation for two interacting particles in a harmonic-oscillator well. Since a two-particle basis is used, the Pauli operator Q is diagonal and can be treated exactly. Their comparisons with calculations using an approximate Q indicate that the approximate-Q results are reasonably accurate for T=1 matrix elements, but less so for T=0 matrix elements for which the treatment of the strong tensor force is particularly crucial.

The use of super-soft potentials such as the Tabakin potential is an attractive proposition, since it is no longer necessary to solve the Bethe-Goldstone equation. The nucleon-nucleon potential itself can be used as the residual shell-model interaction. However, Clement and Baranger (1968) find that the second Born correction must be included before reasonable agreement is obtained with reaction matrix elements of Kuo and Brown. Hartree-Fock calculations of Kerman and Pal (1967) with the Tabakin potential also indicated that satisfactory binding energies could only be obtained when second Born corrections were included.

Finally we mention the Sussex matrix elements (Elliott *et al.* 1968) which are obtained directly from the scattering phase shifts without constructing an explicit potential. The existence of such a potential, which is assumed to be local and non-singular is postulated; then in Born approximation there is a relationship between the phase shift, δ, and the potential, V, given by

$$\tan \delta_\alpha = -\frac{Mk^2}{\hbar^2}\int_0^\infty j_\ell{}^2(kr)\ V_\alpha(r)\ r^2\ dr \tag{3.65}$$

where α denotes the quantum numbers of the scattering channel, k the relative momentum of the two scattering nucleons and M the nucleon mass. On the other hand, a typical matrix element required in a nuclear structure calculation might be

$$\langle n\ell|V|n\ell\rangle = \int_0^\infty R_{n\ell}{}^2(r)\ V_\alpha(r)\ r^2\ dr \tag{3.66}$$

where $R_{n\ell}$ is a relative oscillator wavefunction. The essence of the method then is to note that at a particular value of k there is a very large overlap between the two integrals in eqns (3.65) and (3.66). Thus if one could find a k for which it is sufficiently accurate to assert that

$$j_\ell(kr) \approx A\ R_{n\ell}(r),$$

then the required matrix element $\langle n\ell|V|n\ell\rangle$ is given directly in terms of the scattering phase shift. This idea is then elaborated to deal with the complications of (i) off-diagonal nuclear matrix elements, (ii) mixing of scattering channels, e.g. $^3S_1$-$^3D_1$, and (iii) treating the S-wave phase shifts which are too large for the Born approximation (eqn (3.65)) to be meaningful. The resulting Sussex matrix elements have been compared with the Tabakin matrix elements, and they find a remarkable similarity — the main differences coming in those channels where Tabakin fails to fit the phase shifts well.

(ii) *Modelistic*. For potentials in this category the link with the two-nucleon interaction is lost. Instead one is motivated by the Moszkowski-Scott idea which suggests that the effective interaction in nuclei should be long-ranged, weak and attractive, the short-range repulsion having been cancelled by some part of the attractive force. The resulting effective interaction is then assumed to take some simple, reasonable form. A typical example is

$$V_L(r) = V_0 f(r)[W + B\,P_\sigma + MP_x + HP_H]$$

where r is the relative coordinate, $r_1 - r_2$, between nucleons 1 and 2, and

$$P_\sigma = \tfrac{1}{2}(1 + \underset{\sim}{\sigma}_1 \cdot \underset{\sim}{\sigma}_2)$$

$$P_H = -\tfrac{1}{2}(1 + \underset{\sim}{\tau}_1 \cdot \underset{\sim}{\tau}_2)$$

$$P_x = P_\sigma P_H = -\tfrac{1}{4}(1 + \underset{\sim}{\sigma}_1 \cdot \underset{\sim}{\sigma}_2)(1 + \underset{\sim}{\tau}_1 \cdot \underset{\sim}{\tau}_2),$$

the four terms here representing the general exchange mixtures that can characterize a central force. They are called the Wigner, Bartlett, Majorana, and Heisenberg forces respectively and the coefficients are usually normalized such that W+B+M+H = 1. The functional form of f(r) is commonly chosen to be Yukawa or Gaussian in shape, and the strengths, ranges, and mixture constants are taken as parameters and adjusted until an optimum fit to some selected experimental data is achieved. Such forces are extremely useful if one wants to make a rapid calculation of some nuclear structure property involving only a few nucleons near the Fermi surface. However, the fitted parameters depend critically upon what configurations are allowed in the calculation and upon what data are being fitted.

Another popular parameterization is to make a multipole decomposition of the interaction

$$V(r) = \sum_k v_k(r_1, r_2)\, Y_k(\hat{\underset{\sim}{r}}_1)\, Y_k(\hat{\underset{\sim}{r}}_2)$$

and if the radial function $v_k(r_1, r_2)$ is assumed separable, i.e. $v_k(r_1, r_2) = v_k(r_1)v_k(r_2)$, then a particularly convenient form of interaction is obtained. For example, if only the k = 2 multipole is retained and $v_k(r_1, r_2)$ is chosen as $r_1^{\,2}\, r_2^{\,2}$, then the so-called quadrupole-quadrupole force is obtained. This force can be expressed in terms of generators of the SU_3-group. Thus a problem involving n particles in one major shell of the spherically-symmetric oscillator potential, with the residual interaction being just this quadrupole-quadrupole force, can be solved analytically using the properties of the SU_3-group. A zero range force $V(r) = \delta(r)$ is also separable although all multipoles are present in this case. The advantage of this type of interaction is that it often indicates that a certain nuclear property is sensitive to a certain piece of the nuclear residual interaction. For example, the SU_3 model demonstrated a connection between rotational spectra and quadrupole-quadrupole forces (Harvey 1968).

(iii) *Empirical*. If a calculation for open-shell nuclei is contemplated, using a severely truncated model space, and if the additional assumption is made that only two-body forces are operable, then the number of independent two-body matrix elements required to perform the calculation can be quite small. Therefore, rather than calculate this small set of numbers by proceeding down the lengthy chain of constructing a G-matrix and correcting it for the smallness of the model space, an alternative procedure is to treat these matrix elements as parameters and determine them empirically by least squares fitting to an appropriate set of data. This procedure was pioneered by Talmi (1962) and has been actively pursued since.

The usefulness of this approach is two-fold. First, it enables a large body of experimental data to be correlated, and it is highly successful in this, but more important it singles out those data not fitting in the general scheme - that is states (known as 'intruder' states) can be identified whose basic configuration is not encompassed by the chosen model space. Secondly, there are certain symmetries in the angular momentum algebra which enable the properties of one nucleus to be correlated with those of another. An obvious case is the one-to-one correspondence between configurations j^n, n particles in shell model orbital j, and j^{-n}, n holes in orbital j. We will discuss this type of application in chapter 6.

4
PARTICLE-HOLE MODELS

4.1 INTRODUCTION

In the last two chapters, the ground-state binding energies of closed-shell nuclei were calculated in the Hartree-Fock and the Brueckner-Hartree-Fock approximations. In this chapter we begin our discussion on nuclear spectroscopy by considering the properties of excited states of closed-shell nuclei. The ground-state wavefunction is written (eqn (1.46)) in an occupation number representation as

$$|C\rangle = \prod_{\substack{i=1,2\ldots F \\ m_i=-j_i\ldots+j_i}} a_i^{\dagger}\,|0\rangle \tag{4.1}$$

where the subscript i on the creation operator $a_i^{\dagger}$ stands for all the quantum numbers characterizing the single-particle state $\psi_i(\underset{\sim}{x})$.

We shall assume that the closed-shell wavefunction, eqn (4.1), satisfies the self-consistency conditions of the Hartree-Fock equations. A successful solution to these equations enables the Hartree-Fock one-body Hamiltonian, h, to be constructed such that the lowest A energy eigenvalues of the Schrödinger equation

$$h\psi_i = \varepsilon_i\psi_i \tag{4.2}$$

represent the A occupied single-particle states. The precise radial form of the functions ψ_i is specified in the Hartree-Fock procedure. All other solutions to eqn (4.2) correspond to unoccupied single-particle states. We shall call the eigenfunctions of eqn (4.2) the Hartree-Fock basis states.

Excited states are now constructed by lifting particles out of the occupied orbits and placing them in the unoccupied orbits. Such states are orthogonal to the closed-shell state (eqn (4.1)) since the eigenfunctions of eqn (4.2) form an orthonormal set. In particular, we shall concentrate on the states formed by lifting up just one particle; these are the so-called one-particle one-hole (1p-1h) states. Other excitations such as two-particle two-hole (2p-2h), four-particle four-hole (4p-4h), and n-particle n-hole states can be constructed in an analogous manner, but we will not consider them explicitly here. These many-particle many-hole states are supposed to lie at energies which are substantially higher than the 1p-1h states and therefore do not mix with them. This supposition, which gives a significant truncation to the size of the model calculation, is referred to as the 'Tamm-Dancoff Approximation' (TDA). In practice, however, nuclear interactions are such that some many-particle many-hole states can lie low in the energy spectrum, and in certain cases they are at a lower excitation energy than the 1p-1h states. Later on we will show how the influence of 2p-2h states can be built into the theory using the 'Random Phase Approximation' (RPA).

The attractive feature of the TDA is that in truncating the calculation to include only 1p-1h states, the Hartree-Fock ground-state wavefunction $|C\rangle$ is completely unaffected. There is no mixing between the 1p-1h states and $|C\rangle$; in fact it was precisely this property that was built into the definition of the Hartree-Fock ground state. On the other hand, when the calculation is extended to include 2p-2h via the RPA, a variation is induced into the ground-state wavefunction and we can anticipate an intimate correction between the RPA theory, and the stability of the Hartree-Fock solution. This will be discussed towards the end of the chapter.

Although in principle one should be working with Hartree-Fock basis states ψ_i, in practice one would like to learn something of the properties of these particle-hole states without having to go through a lengthy Hartree-Fock calculation first. Thus it is frequently assumed that harmonic oscillator basis states (or sometimes Saxon-Woods basis states) are a sufficiently accurate approximation to the Hartree-Fock basis; certainly as far as specifying the radial form for ψ_i and enumerating which are the occupied and which are the unoccupied states are concerned. However, one does not use the harmonic oscillator single-particle energies. Instead the low lying states in the closed-shell-plus-one and closed-shell-minus-one nuclei are assumed to be single-particle states such that their measured energies expressed relative to the closed-shell core can be taken for the ε_i.

4.2 TAMM-DANCOFF APPROXIMATION

A particle-hole state is defined in terms of the chosen basis states as

$$|i^{-1}m\rangle = a_m^{\dagger}\, b_i^{\dagger}\, |C\rangle \qquad (4.3)$$

where $b_i^{\dagger}$ is a hole creation operator $b_i^{\dagger} = a_{\tilde{i}} = S_i\, a_{-i}$ with the phase factor $S_i = (-)^{j_i+m_i}$. Throughout this chapter we will be working in an isospin formalism, but for economy in notation we will not write down any isospin quantum numbers explicitly. For example the phase factor, S_i, is to be interpreted as $S_i = (-)^{j_i+m_i+1/2+t_{zi}}$. Thus we are considering closed-shell nuclei like ^{16}O and ^{40}Ca, for which the gap between occupied and unoccupied orbits occurs at the same level for protons and neutrons. In closed-shell nuclei such as ^{208}Pb which have a neutron excess, it is not possible to construct one-particle one-hole states having good total isospin T; instead states of good T are certain linear binations of one-particle one-hole and two-particle two-hole states. Thus an isospin formalism is not very practicable and a proton-neutron formalism is preferred in these cases.

For ^{16}O the occupied orbits are the $1s_{1/2}$, $1p_{3/2}$ and $1p_{1/2}$; the unoccupied orbits are the $1d_{5/2}$, $2s_{1/2}$, $1d_{3/2}$ ..., etc. — an infinite number. Thus there are an infinite number of particle-hole states which can be constructed; in practice, of course, the number is truncated to a manageable size. For example, one might restrict the holes to be from the last filled major oscillator shells and the particles to be placed in the first unfilled major oscillator shells. For ^{16}O this would imply only holes in the 1p-shell and particles in the 2s-1d shell. These particular particle-hole states would all have negative parity and as such they have often been called 'particle-hole vibrations' in the literature.

The particle-hole state as written in eqn (4.3) does not have good total angular momentum, J and T. We prefer to work with coupled particle-hole states defined as

$$|j_i^{-1} j_m; J\rangle = \sum_{m_i m_m} \langle j_i m_i j_m m_m | JM\rangle \, a_m^\dagger \, b_i^\dagger |C\rangle. \tag{4.4}$$

We shall assume that the physical states observed in closed-shell nuclei can be characterized by a wavefunction which is a linear combination of such coupled particle-hole states. Thus we write

$$|q\rangle = \sum_{mi} x_{mi}^{(q)} \; |j_i^{-1} j_m; J\rangle \tag{4.5}$$

where the $x_{mi}^{(q)}$ are coefficients (wavefunction amplitudes) which are determined by solving the Schrödinger equation:

$$H\,|q\rangle = E_q\,|q\rangle \tag{4.6}$$

with E_q being the calculated energy of the state. Note that the sum over m and i in eqn (4.5) is just a sum over the $n\ell j$-values of the orbits but not a sum over the magnetic projections which are explicitly contained in the definition, eqn (4.4). Substituting eqn (4.6) into eqn (4.5) and left multiplying by another member of the complete set leads to the familiar shell model eigenvalue problem

$$\sum_{mi} \langle j_j^{-1} j_n; J | H | j_i^{-1} j_m; J\rangle \, x_{mi}^{(q)} = E_q \, x_{nj}^{(q)}. \tag{4.7}$$

To evaluate the matrix element on the left, it is necessary to return to the uncoupled representation

$$\langle j_j^{-1} j_n; J | H | j_i^{-1} j_m; J\rangle$$

$$= \sum_{\substack{m_i m_m \\ m_j m_n}} \langle j_i m_i j_m m_m | JM\rangle \langle j_j m_j j_n m_n | JM\rangle \langle C | b_j a_n \; H \; a_m^\dagger \; b_i^\dagger | C\rangle. \tag{4.8}$$

Writing H = T+V where T is a one-body kinetic energy operator and V a two-body potential energy operator, and using the expansions (1.87) for T and (1.90) for V, the core expectation value becomes

$$\langle C|b_j a_n\, H\, a_m^\dagger\, b_i^\dagger|C\rangle$$

$$= -\langle n\tilde{i}|V|m\tilde{j}\rangle_A + \delta_{ij}\{\langle n|T|m\rangle + \sum_k \langle kn|V|km\rangle_A\} - \delta_{mn}\{\langle \tilde{i}|T|\tilde{j}\rangle +$$

$$+ \sum_k \langle k\tilde{i}|T|k\tilde{j}\rangle_A\} + \delta_{ij}\delta_{mn}\,\{\sum_k \langle k|T|k\rangle + \frac{1}{2}\sum_{k\ell}\langle k\ell|V|k\ell\rangle_A\}$$

$$= -\langle n\tilde{i}|V|m\tilde{j}\rangle_A + \delta_{ij}\delta_{mn}\,\{\varepsilon_m - \varepsilon_i + E_{core}\}$$

where we are using a Hartree-Fock basis for which

$$\langle \alpha|T|\beta\rangle + \sum_k \langle k\alpha|V|k\beta\rangle_A = \langle \alpha|h|\beta\rangle = \varepsilon_\alpha\,\delta_{\alpha\beta} \qquad (4.10)$$

i.e. the Hartree-Fock single-particle Hamiltonian is diagonal in this basis, and where the energy of the closed shell is given by

$$E_{core} = \sum_k \langle k|T|k\rangle + \frac{1}{2}\sum_{k\ell}\langle k\ell|V|k\ell\rangle_A .$$

Next the two-body matrix element $\langle n\tilde{i}|V|m\tilde{j}\rangle_A$ is expanded in terms of coupled two-body matrix elements, viz.

$$\langle n\tilde{i}|V|m\tilde{j}\rangle_A = \sum_{\overline{JM}} (-)^{j_i+m_i+j_j+m_j}\langle j_n m_n j_i -m_i|\overline{JM}\rangle\langle j_m m_m j_j -m_j|\overline{JM}\rangle \times$$

$$\times \langle j_n j_i;\bar{J}|V|j_m j_j;\bar{J}\rangle_A . \qquad (4.11)$$

Inserting eqns (4.9) and (4.11) back into eqn (4.8) and summing the four Clebsch-Gordan coefficients to give one recoupling coefficient, eqn (A.15), we obtain

$$\langle j_j^{-1} j_n;J|H|j_i^{-1} j_m;J\rangle = \delta_{ij}\delta_{mn}(\varepsilon_m-\varepsilon_i+E_{core})+\langle j_j^{-1} j_n;J|V|j_i^{-1} j_m;J\rangle \qquad (4.12)$$

where

$$\langle j_j^{-1} j_n;J|V|j_i^{-1} j_m;J\rangle = -\sum_{\bar{J}} \frac{\hat{\bar{J}}}{\hat{J}}\, U(j_j j_n j_m j_i;J\bar{J})\langle j_n j_i;\bar{J}|V|j_m j_j;\bar{J}\rangle \qquad (4.13)$$

and $\hat{J} = (2J+1)^{1/2}$. Eqn (4.13) exhibits the relationship between a coupled particle-hole matrix element and a particle-particle matrix element; a result first obtained by Pandya (1956). From here on, we shall drop any explicit mention of the energy E_{core}, and assume that all energies are measured relative to the energy of the closed-shell core.

An approximate solution to the eigenvalue problem, eqn (4.7), can be found when the particle-hole matrix elements of V are small such that a perturbative expansion in V is meaningful. The zeroth-order approximation to the energy spectrum is just the unperturbed energies of the particle-hole states. For example, setting all $\langle V \rangle = 0$ in eqn (4.7), and all $x_{mi} = 0$ except x_{nj}, then the approximate energy, $E^{(0)}$, of the particle-hole state $|j^{-1}n\rangle$ is

$$E^{(0)} = \varepsilon_n - \varepsilon_j .$$

The next order is obtained by setting all $\langle V \rangle = 0$ except those directly influencing the state $|j^{-1}n\rangle$, namely $\langle j^{-1}n|V|i^{-1}m\rangle$. Then the sequence of equations in (4.7) provides one equation of the type

$$(\varepsilon_n - \varepsilon_j)x_{nj} + \langle j^{-1}n|V|j^{-1}n\rangle x_{nj} + \sum_{mi \neq nj} \langle j^{-1}n|V|i^{-1}m\rangle x_{mi} = E\, x_{nj}$$

and a series of equations of the type

$$\langle i^{-1}m|V|j^{-1}n\rangle x_{nj} + (\varepsilon_m - \varepsilon_i)x_{mi} = E\, x_{mi}$$

From the second set we solve for the coefficients x_{mi}:

$$\frac{x_{mi}}{x_{nj}} = \frac{\langle i^{-1}m|V|j^{-1}n\rangle}{E - (\varepsilon_m - \varepsilon_i)}$$

$$\approx \frac{\langle i^{-1}m|V|j^{-1}n\rangle}{(\varepsilon_n - \varepsilon_j) - (\varepsilon_m - \varepsilon_i)} ,$$

where it is recognized that $x_{mi} \ll x_{nj}$, so that it is sufficiently accurate to replace E in the denominator by its zeroth-order approximation. Inserting this in the first equation gives the next order approximation for E, namely,

$$E^{(2)} = (\varepsilon_n-\varepsilon_j)+\langle j^{-1}n|V|j^{-1}n\rangle + \sum_{mi} \frac{\langle j^{-1}n|V|i^{-1}m\rangle\langle i^{-1}m|V|j^{-1}n\rangle}{(\varepsilon_n-\varepsilon_j)-(\varepsilon_m-\varepsilon_i)} \tag{4.14}$$

Continuing in this way the third order will contain terms involving three V-interactions and two energy denominators. This perturbative solution can be represented in terms of the Feynman-Goldstone diagrams discussed in Chapter 1. In Fig. 4.1 the three terms in eqn (4.14) together with the next order are illustrated. This solution is known colloquially as summing the forward going bubbles.

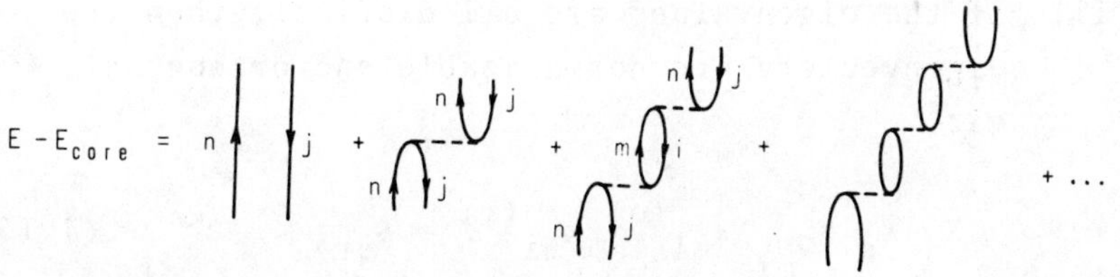

FIG. 4.1. A diagrammatic representation of the perturbation solution to the Tamm-Dancoff particle-hole eigenvalue problem, eqn (4.7).

In most applications in light nuclei, the number of expansion coefficients in eqn (4.5) is quite small, and the eigenvalue problem can easily be solved numerically. All possible matrix elements of the Hamiltonian H between particle-hole states are first evaluated. These constitute the elements of a symmetric d × d matrix, which we will denote by A. The dimension, d, is the number of terms in the expansion, eqn (4.5). The eigenvalue problem, eqn (4.7), is now cast into matrix form

$$A\ X = E\ X$$

where X is a column vector and the energy E is the eigenvalue. The condition for a non-trivial solution is that the determinant det(A-EI) should vanish. Here I is the unit matrix. This condition gives a dth order polynomial equation for the eigenvalues E and there are d solutions, which we write as E_q where q = 1,...d. The method of solution is to find a second matrix S such that the similarity transformation $S^{-1}AS$ gives a diagonal matrix D. Then the eigenvalues of A are the diagonal elements in D, and the qth column vector in the matrix S is the eigenvector X_q of A corresponding to the eigenvalue E_q. There are a number of algorithms for finding the matrix S, for example Jacobi's or Householder's methods (Wilkinson 1969), but these will not be discussed here.

The solutions to the matrix problem have the following properties:

(i) the eigenvalues, E_q, and eigenvectors X_q are real;

(ii) if the eigenvalues are all distinct, then the eigenvectors are normalizable and orthogonal, viz.

$$X_q^\dagger X_r = \sum_{mi} x_{mi}^{(q)} x_{mi}^{(r)} = \delta_{qr}; \tag{4.15a}$$

and (iii) the eigenvectors satisfy the closure property

$$\sum_q x_{mi}^{(q)} x_{nj}^{(q)} = \delta_{mi,nj}. \tag{4.15b}$$

The proofs of these results can be found in most standard textbooks on quantum mechanics.

4.3 TRANSITION MATRIX ELEMENTS AND SUM RULES IN TDA

Suppose a particular particle-hole state $|q\rangle$ is excited from the closed-shell ground state by the action of a one-body operator, $U^{(J)}$. An example is photo-absorption for which the one-body operator $U^{(J)}$ would then correspond to the electromagnetic multipole operator. The transition strength is defined as

$$T_q = |\langle q \| U^{(J)} \| C\rangle|^2$$

and is the square of a reduced matrix element connecting the closed shell with the particle-hole state. An interesting quantity is the total transition strength to all states $|q\rangle$

of a given spin, isospin and parity. In the particle-hole model, this just corresponds to summing the transition strengths over all the solutions of the TDA equations. Thus writing $|q\rangle = \sum x_{nj}^{(q)} |j_j^{-1} j_n;J\rangle$ this sum becomes

$$S_{NEW} = \sum_q |\langle q \| U^{(J)} \| C \rangle|^2$$

$$= \sum_q \sum_{nj,mi} x_{nj}^{(q)} x_{mi}^{(q)} \langle C \| U^{(J)} \| j_j^{-1} j_n;J \rangle \langle j_i^{-1} j_m;J \| U^{(J)} \| C \rangle$$

$$= \sum_{nj} |\langle j_j^{-1} j_n;J \| U^{(J)} \| C \rangle|^2 \tag{4.16}$$

where the closure property, eqn (4.15b), has been used to sum over q. This result (eqn (4.16)) is known as the non-energy-weighted sum rule, S_{NEW}. It states that the total transition probability for γ-ray absorption to a series of configuration mixed particle-hole states equals the sum of the transition probabilities to the unperturbed particle-hole basis states. Thus for any given particle-hole state, its transition probability can be expressed as a fraction of the sum, viz.

$$R_q = |\langle q \| U^{(J)} \| C \rangle|^2 / S_{NEW} \tag{4.17}$$

and if R_q for any particular state approaches unity then that state is said to be collective. Note that in TDA calculations, S_{NEW} is invariant with respect to changes in the model Hamiltonian.

A second sum rule is the energy-weighted sum rule defined as

$$S_{EW} = \sum_q E_q |\langle q \| U^{(J)} \| C \rangle|^2$$

$$= \frac{1}{2} \langle C | [U^{(J)}, [H, U^{(J)}]] | C \rangle, \tag{4.18}$$

which is in fact independent of the model space, depending only on the properties of the closed-shell state $|C\rangle$ and the chosen Hamiltonian H. However, unlike S_{NEW}, this sum rule is not invariant with respect to changes in the model Hamiltonian, and so a comparison with experiment might give additional insight into the appropiate model Hamiltonian to use in TDA theories.

4.4 DIGRESSION ON THE ELECTRIC DIPOLE SUM RULE

It is interesting to evaluate the energy-weighted sum rule S_{EW} for electric dipole (E1) radiation. The one-body operator in this case is

$$U^{(J)} = \sum_{i=1}^{A} e_i(\underset{\sim}{r}_i - \underset{\sim}{R})$$

where e_i is the charge on nucleon i, $\underset{\sim}{r}_i$ is the coordinate of nucleon i referred to some arbitrary origin, and $\underset{\sim}{R}$ is the centre of mass coordinate $\underset{\sim}{R} = A^{-1}\Sigma\,\underset{\sim}{r}_i$. We set e_i = e if i is a proton and e_i = 0 if i is a neutron, so that

$$U^{(J)} = e\sum_{i=1}^{Z}\underset{\sim}{r}_i - Ze\,\underset{\sim}{R}$$

$$= e\sum_{i=1}^{Z}\left(1-\frac{Z}{A}\right)\underset{\sim}{r}_i - \frac{Ze}{A}\sum_{i=Z+1}^{A}\underset{\sim}{r}_i$$

$$= \frac{e}{A}\left[N\sum_{i=1}^{Z}\underset{\sim}{r}_i - Z\sum_{i=Z+1}^{Z}\underset{\sim}{r}_i\right].$$

Next the double commutator is evaluated for the kinetic-energy part of the Hamiltonian, $T = \sum_{i=1}^{A} p_i^{\,2}/2M$:

$$\frac{1}{2}\langle C|[U^{(J)},[T,U^{(J)}]]|C\rangle = \frac{e^2\hbar^2}{2M}\cdot\frac{NZ}{A}.$$

This has been obtained by repeated use of the commutator $[p_i,r_j] = -i\hbar\,\delta_{ij}$. The potential-energy part of the Hamiltonian is harder to evaluate, so we put this to one side for the moment and write

$$S_{EW} = \frac{e^2\hbar^2}{2M}\,\frac{NZ}{A}\,(1+\kappa) \tag{4.19}$$

where κ is known as the enhancement factor in the electric

dipole sum rule

$$\kappa = \frac{A}{NZ}\,\frac{M}{e^2\hbar^2}\,\langle C|[U^{(J)},[V,U^{(J)}]]|C\rangle.$$

The result in eqn (4.19) is known as the Thomas-Reiche-Kuhn sum rule for dipole photo-absorption cross-sections.

If the nucleon-nucleon potential V were local and had no exchange or velocity dependent terms, then it would commute with $U^{(J)}$ and κ would be zero. This is the situation in atomic systems for which eqn (4.19) yields a rigorous model-independent sum rule. The value of S_{EW} with $\kappa = 0$ is known as the classical TRK limit.

Experimental information on the sum rule comes from the integrated photonuclear absorption cross-sections (Levinger 1960)

$$\int_0^\infty \sigma_\gamma(E)dE = (2\pi)^2 S_{EW}/\hbar c = 60\,\frac{NZ}{A}\,(1+\kappa)\text{MeV mb}.$$

These data when integrated up to photon energies of 30 MeV (i.e. through and beyond the giant dipole resonance region) are found to exceed the classical TRK limit by typically 40 to 50 per cent, i.e. κ is in the range of 0.4 to 0.5. Levinger and Bethe (1950) used this result to estimate the strength of exchange force mixtures in the nucleon-nucleon potential; their calculations were based on simple modelistic forms of V.

Recently this subject has been revived, largely due to the experiments of the Mainz group (Ahrens *et al.* 1972) who obtained the integrated photonuclear absorption cross-sections up to photon energies of 140 MeV (the meson threshold). They find that the enhancement factor κ in the sum rule is around .0 for ^{16}O, which is about twice as large as the value given y earlier experiments and theories. An explanation has been given by Weng, Kuo, and Brown (1973), and although this is ather a digression at this point we record the essential eatures of their calculation here.

Weng, Kuo, and Brown noted that the entire double commutator appearing in the definition of κ can be treated as a simple two-body operator, K_{ij}. The matrix element $\langle C|K|C\rangle$ can then be evaluated using the linked diagram expansion, in exactly the same way as the ground-state energy of a closed-shell system was calculated.

The zeroth-order and first order diagrams are illustrated in Fig. 4.2, where the coiled interaction line represents

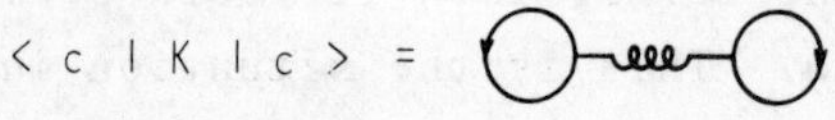

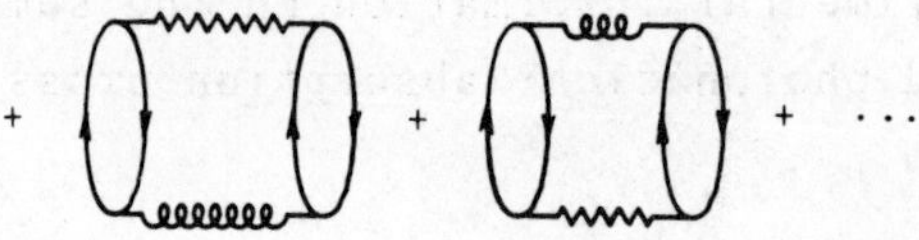

FIG. 4.2. The zeroth- and first-order corrections to the expectation value of a two-body operator K in the closed-shell state, $\langle C|K|C\rangle$. The K-interaction vertex is represented by a coiled line, and the perturbation, taken as the G-matrix, is represented by the jagged line.

the operator K and the jagged line represents the reaction G-matrix. The expansion is a perturbation series in G such that the nth-order terms comprise all closed linked diagrams containing one K-interaction line and n reaction matrix lines. (Ladder diagrams should be excluded, see section (3.2).) Weng, Kuo, and Brown calculated the zeroth-, first- and selected second-order diagrams using the Reid hard-core potential (Reid 1968) for V and constructing the appropriate G-matrix by a matrix inversion method. Their results for κ showed excellent order by order convergence. However, the calculation of the matrix elements of K was complicated by the presence of the hard-core potential V and required the use of correlated two-body wavefunctions, which they obtained using the reference spectrum method (see section (3.6)). Their results for ^{16}O and ^{40}Ca were that κ was close to unity. This increase in κ over the value of 0.5 obtained in previous calculations was due primarily to the two-body correlations, especially those induced by the tensor force.

However, more recent calculations (Fink, Gari, and Hebach 1974, Fink, Gari, Hebach and Zabolitzky 1974) using Breuckner-Hartree-Fock wavefunctions obtained with the correct treatment of the Pauli projection operator (rather than the reference spectrum method of Weng, Kuo, and Brown), and including approximately the three-body Bethe-Faddeev amplitudes, fail to give such a large enhancement. Their results for ^{16}O ranged from κ = 0.58 to 0.74 depending on the nucleon-nucleon potential used.

4.5 TDA CALCULATIONS

The first application of the Tamm-Dancoff approximation to nuclear spectra was the study by Elliott and Flowers (1957) of the negative parity states of ^{16}O. The residual interaction V was taken as a simple Yukawa potential of range 1.4 fm with a Rosenfeld exchange force mixture. No Hartree-Fock minimization was performed; instead the basis functions were taken as harmonic oscillator functions and the single-particle energies were taken in the usual shell model way from the energy levels of ^{17}O and ^{15}O.

Elliott and Flowers studied in detail the 1^-T=1 (electric dipole) states. Diagonalizing the TDA matrix produced two states considerably pushed up in the excitation spectrum from their unperturbed locations. These states were highly collective, taking 97 per cent of the model value for the non-energy-weighted sum rule, S_{NEW}. This agreed well with experimental total γ-ray absorption cross-sections.

Another feature of the calculation was its prediction of a highly collective 3^-T=0 state in ^{16}O at about 6 MeV. The nearest unperturbed 3^- particle-hole state is at 11 MeV so the residual interaction brings this state down by 5 MeV. Although this state was found to be collective, it was not collective enough and this to some extent was one failure of the TDA, which was subsequently remedied by the random phase approximation.

Other TDA calculations have been performed by Brown,

Castillejo, and Evans (1961). These authors used a zero-range residual interaction and showed that with reasonable exchange force mixtures, the particle-hole matrix elements were mainly repulsive in T=1 states and attractive in T=0 states. This corresponds to the collective T=1 states being pushed up and the collective T=0 states being brought down in the energy spectrum. To see how this result comes about, we will next consider a schematic model approach, which enables the TDA matrix diagonalization to be performed algebraically.

4.6 SCHEMATIC MODELS

Schematic models have proved themselves to be very useful for a qualitative understanding of particle-hole vibrations in closed-shell nuclei. They were first introduced by Brown and Bolsterli (1959) to illustrate the collective properties of the electric dipole resonance. The essence of their procedure is to find an approximate solution to the eigenvalue problem (eqns (4.7) and (4.12)).

$$\sum_{mi} \langle j_j^{-1} j_n; J|V|j_i^{-1} j_m; J\rangle \, x_{mi}^{(q)} = [E_q - \varepsilon_m + \varepsilon_j] x_{nj}^{(q)} \qquad (4.20)$$

by writing the particle-hole matrix element in a factorized form:

$$\langle j_j^{-1} j_n; J|V|j_i^{-1} j_m; J\rangle = \lambda \, a(j_j j_n J) a(j_i j_m J)$$

$$\equiv \lambda \, a_{nj} \, a_{mi} \qquad (4.21)$$

The superscript q in eqn (4.20) labels the different eigenvalue solutions. Such a factorization enables an algebraic solution to the eigenvalue problem to be found. However, before finding this solution, it is instructive to have some idea what the functions a_{nj} represent. For negative parity particle-hole states, a multipole-multipole force gives just this form of factorization. This multipole-multipole force is defined as

$$V_{12} = [\sum_k U^{(k)}(1) \cdot U^{(k)}(2)][a_0 + a_\tau \, \underset{\sim}{\tau}_1 \cdot \underset{\sim}{\tau}_2]$$

where $U^{(k)}(1)$ is a one-body operator of tensorial rank k in the coordinates of particle 1, and a_0 and a_τ are parameters characterizing the isospin dependence of the force. Then it is shown in the appendix that the particle-hole matrix element of V is

$$\langle j_j^{-1} j_n;J|V|j_i^{-1} j_m;J\rangle = (-)^p[a_0+a_\tau+(-)^T(a_0-a_\tau)]\delta(k,J) \times \\ \times \langle j_j^{-1} j_n;J\|U^{(J)}\|C\rangle\langle j_i^{-1} j_m;J\|U^{(J)}\|C\rangle \tag{4.22}$$

where T is the isospin quantum number for the particle-hole state. On comparing eqns (4.18) and (4.17) we identify

$$a_{nj} = \langle j_j^{-1} j_n;J\|U^{(J)}\|C\rangle$$

$$\lambda = (-)^p[a_0+a_\tau+(-)^T(a_0-a_\tau)] \tag{4.23}$$

If $U^{(J)}$ is the one-body operator describing γ-ray absorption, then the sign $(-)^p$ is always -1. [The sign $(-)^p$ originates in the definition of the Hermitian conjugate for the one-body operator U, viz. $U_m^{(J)\dagger} = (-)^{p-m}U_{-m}^{(J)}$. For electric operators, p is J and for magnetic operators p is J+1. However, when considering negative parity particle-hole states produced by γ-ray absorption on the closed-shell state $|C\rangle$, the selection rules require J to be odd for electric and even for magnetic absorption. Hence $(-)^p$ is -1 in both cases.] Thus we identify a_{nj} as the matrix element for γ-ray absorption by the closed-shell state to form the particle-hole basis state $|j_j^{-1} j_n;J\rangle$. Brown (1967) gives a diagrammatic interpretation in which the particle-hole matrix element, shown on the left in Fig. 4.3, is divided into a product of two one-body matrix elements.

The coefficient λ represents the strength of the residual interaction. For T=0 particle-hole states, λ, depending only on a_0, turns out to be negative for most two-body potentials, whereas for T=1 particle-hole states λ is positive. As we shall see, this sign change reflects the fact that the T=0

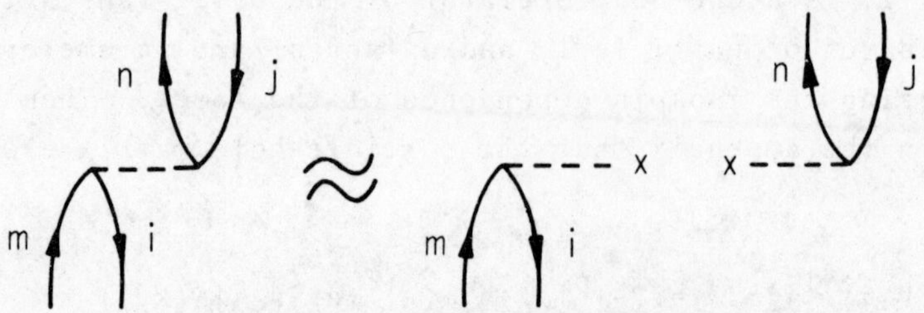

FIG. 4.3. The schematic model factorization in which a two-body particle-hole matrix element is replaced by a product of two one-body matrix elements.

collective particle-hole states are low-lying whereas the collective T=1 states are high-lying in the spectrum of closed-shell nuclei.

We return now to solving the secular problem eqn (4.20) using the factorization eqn (4.21) with the quantities λ, a_{nj} and $(\varepsilon_n - \varepsilon_j)$ all being known *a priori*. Inserting eqn (4.21) into (4.20),

$$[E_q - (\varepsilon_n - \varepsilon_j)]x_{nj}^{(q)} = \lambda\, a_{nj} \sum_{mi} x_{mi}^{(q)}\, a_{mi}$$

and defining α_q as being the sum $\sum_{mi} x_{mi}^{(q)}\, a_{mi}$, an expression for the wavefunction amplitude is obtained:

$$x_{nj}^{(q)} = \frac{\lambda\, \alpha_q\, a_{nj}}{E_q - (\varepsilon_n - \varepsilon_j)} \tag{4.24}$$

The α_q can be treated as a normalization constant which is determined by the requirement $\Sigma_{nj}\, |x_{nj}^{(q)}|^2 = 1$. Multiplying both sides of eqn (4.24) by a_{nj}, summing, and dividing throughout by α_q, a dispersion formula for the eigenenergies E_q is obtained:

$$1 = \lambda \sum_{nj} \frac{|a_{nj}|^2}{E_q - (\varepsilon_n - \varepsilon_j)} \tag{4.25}$$

The procedure then is to solve eqn (4.25) for these energies E_q and substitute in eqn (4.24) to find the wavefunction amplitudes.

It is informative to find a graphical solution for the eigen-energies by plotting the right-hand side of eqn (4.25) as a function of E. First mark off the positions of the unperturbed particle-hole energies $\varepsilon_n-\varepsilon_j$ and draw vertical lines at these positions (see Fig. 4.4). Next consider the asymptotic behaviour. As E tends to $+\infty$, the right-hand side of

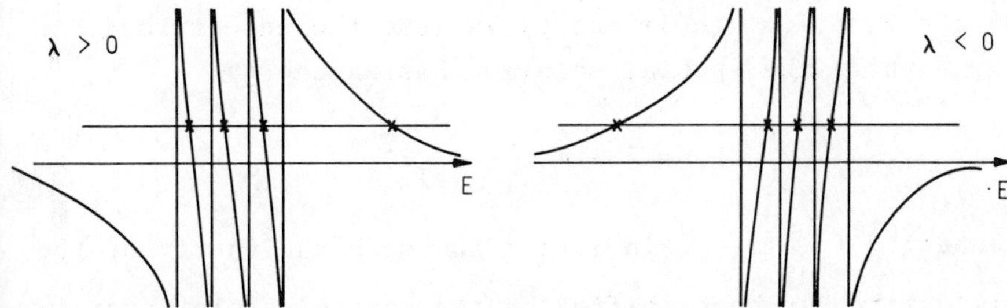

FIG. 4.4. A graphical solution to eqn (4.25) for the eigenenergies, E_q. Plotted is the right-hand side of eqn (4.25) as a function of E for two values of λ. The vertical lines correspond to values of $E = (\varepsilon_n-\varepsilon_j)$, the unperturbed particle-hole energies. The solutions are given by the intercept with the horizontal line drawn at the value unity and are marked with a cross.

eqn (4.25) tends to zero positively for $\lambda > 0$ and negatively for $\lambda < 0$. At the other extreme, E tending to $-\infty$, the converse situation holds. In Fig. 4.4, separate diagrams are drawn for the cases $\lambda > 0$ and $\lambda < 0$. Consider $\lambda < 0$: as E increases from $-\infty$, the right-hand side of eqn (4.25) increases in value until the first $\varepsilon_n-\varepsilon_j$ is reached, at which point the right-hand side becomes infinite. As E is increased a little further, the right-hand side starts negative, but very quickly becomes positive and goes to $+\infty$ again at the next $\varepsilon_n-\varepsilon_j$. This pattern continues until all the $\varepsilon_n-\varepsilon_j$ have been passed, whence the curve remains negative, tending to zero as E tends to $+\infty$. The eigenvalues are now given by the intercepts this curve has with a horizontal line drawn at the value unity, the left-hand side of eqn (4.25).

This graphical solution shows that all but one of the

eigenenergies are sandwiched between the unperturbed particle-hole energies $\varepsilon_n-\varepsilon_j$. One solution, which we shall label the q = 1 solution, is shifted away from the unperturbed position, upwards in the spectrum for $\lambda > 0$ and downwards for $\lambda < 0$. This special solution is the collective state containing most of the transition strength. To see this, consider the limit in which all the unperturbed particle-hole energies are degenerate, i.e. $(\varepsilon_n-\varepsilon_j) = \bar{\varepsilon}$ say. Then for all the trapped solutions q = 2, 3,..., their energy is just the unperturbed energy $\bar{\varepsilon}$ while the special solution has an energy

$$E_1 = \bar{\varepsilon} + \lambda \sum_{nj} |a_{nj}|^2.$$

The quantity $\lambda \sum_{nj} |a_{nj}|^2$ indicates how much the energy of the special state has been shifted by the particle-hole interaction. Furthermore the wavefunction amplitudes (eqn (4.24)) for this state, when appropriately normalized, are given by

$$x_{nj}^{(1)} = a_{nj}/[\sum_{mi} |a_{mi}|^2]^{1/2}.$$

Thus evaluating the fraction of the sum rule, eqn (4.17), for this special state gives

$$R_1 = |\sum_{nj} x_{nj}^{(1)} a_{nj}|^2/S_{NEW}$$

$$= \sum_{nj} |a_{nj}|^2/S_{NEW}$$

$$= 1,$$

indicating that this state is highly collective, exhausting 100 per cent of the sum rule in the limit of degenerate particle-hole energies. Hence in this limit, the trapped solutions, q = 2,3..., have zero transition strength. Thus the schematic model explains how the particle-hole interaction mixes the unperturbed states in such a way that one state is shifted appreciably from its unperturbed location and takes with it the bulk of the γ-ray absorption strength, leaving the other solutions considerably reduced in strength. This collective solution occurs at low energies for T=0 states, i.e. $\lambda < 0$, and at high energies for T=1 states.

The energy-weighted sum rule is also easily evaluated in this limit of the schematic model:

$$S_{EW} = \sum_q E_q |\langle q \| U^{(J)} \| C \rangle|^2$$

$$= E_1 |\sum_{mi} x_{mi}^{(1)} \; a_{mi}|^2$$

$$= E_1 \sum_{mi} |a_{mi}|^2. \qquad (4.26)$$

Notice that this result does not equal $\bar{\varepsilon}\ \Sigma_{mi}|a_{mi}|^2$ which is the value of the energy-weighted sum for the unperturbed particle-hole basis states. Rather S_{EW}, eqn (4.26), exceeds the unperturbed sum for $\lambda > 0$, i.e. for T=1 states, and falls short when $\lambda < 0$. Thus we see that the TDA diagonalization preserves the non-energy-weighted sum S_{NEW}, but does not preserve the energy-weighted sum. The opposite is true for the random phase approximation.

The first such collective state to be identified experimentally was the E1 (T=1) giant dipole resonance, which has been observed in γ-ray absorption studies. The resonance appears throughout the periodic table. In light nuclei it occurs in the energy spectrum at about 20 MeV but for medium and heavy nuclei it falls to around 12 MeV. The integrated absorption cross-section is roughly equal to the energy-weighted sum rule S_{EW}, thereby confirming that the resonance is collective. However, the resonance is unstable to particle emission and has a characteristic width ranging from 3 to 10 MeV.

In even-even, N=Z, nuclei there is a characteristic low-lying 3^- T=0 state in the spectrum which is strongly excited in inelastic (p,p') and (α,α') reactions. This is the octopole vibration, and is the collective T=0 solution predicted in the schematic model. The experimental transition strength is underestimated by the model.

There is little experimental evidence for magnetic collective states; however, recent inelastic electron scattering experiments (see survey by Satchler (1974)) show a strong peak in the cross-section data at excitation energies of around 7 to 9 MeV in medium and heavy nuclei. This peak shows up predominantly at backward-angle scattering and has the characteristics of an M1 giant resonance.

It is also possible to find indirect evidence for magnetic giant resonances, not by observing the resonance itself but by noting a depletion of transition strength in a low-lying state. One example concerns the β-decay of ^{40}K. The ground state of ^{40}K has quantum numbers $J^{\pi} = 4^{-}$, T=1 and can be interpreted as a $d_{3/2}$-hole, $f_{7/2}$-particle state in the shell model. This state beta decays to the ground state of ^{40}Ca, the transition operator being $[Y_3 \times \underset{\sim}{\sigma}]^{(4)}\underset{\sim}{\tau}$, which is the spin part of an isovector M4 gamma decay operator. The decay is strongly retarded, that is the experimental lifetime is five times longer than the shell model prediction using a $d_{3/2}{}^{-1}f_{7/2}$ particle-hole state decaying to core description. In the schematic model, the ground state of ^{40}K is treated as one of the trapped solutions and its transition probability is expected to be retarded. The fact that experimental observation shows this retardation provides some indirect evidence for the existence of a magnetic M4 T=1 resonance lying high in the spectrum. This result has been discussed by Towner (1970).

Finally we should like to mention one extension to the schematic model. If the particle-particle interaction V is taken to be a zero-range interaction, then the particle-hole matrix elements are factorizable in the form

$$\langle j_j{}^{-1} j_n; J | V | j_i{}^{-1} j_m; J \rangle = \lambda\, a_{nj}\, a_{mi} + \mu\, b_{nj}\, b_{mi} \qquad (4.27)$$

where the precise definitions of λ, μ, a_{nj} and b_{nj} can be found in the appendix. However, the interpretations are essentially as before: the λ and μ relate to the strength of the residual interaction, and a_{nj} and b_{nj} are reduced matrix elements of one-body operators connecting the particle-hole basis state $|j_j^{-1} j_n{:}J\rangle$ with the core. As before, an algebraic solution to the eigenvalue problem, eqn (4.20), can be found (Goswami and Pal 1962, Lane 1964), as we now show.

Substituting eqn (4.27) back into eqn (4.20) and defining

$$\alpha_q = \sum_{mi} x_{mi}^{(q)} a_{mi},$$

$$\beta_q = \sum_{mi} x_{mi}^{(q)} b_{mi},$$

an expression for the wavefunction amplitudes is obtained:

$$x_{nj}^{(q)} = \frac{\lambda \alpha_q a_{nj} + \mu \beta_q b_{nj}}{E_q - (\varepsilon_n - \varepsilon_j)} \tag{4.28}$$

Substituting this value of $x_{nj}^{(q)}$ back into the definitions of α_q and β_q, the following pair of equations is derived:

$$\begin{aligned} \alpha_q(1 - \lambda \sigma_{aa}^{(q)}) &= \mu \sigma_{ab}^{(q)} \beta_q \\ \beta_q(1 - \mu \sigma_{bb}^{(q)}) &= \lambda \sigma_{ab}^{(q)} \alpha_q \end{aligned} \tag{4.29}$$

where

$$\sigma_{aa}^{(q)} = \sum_{mi} \frac{|a_{mi}|^2}{E_q - (\varepsilon_m - \varepsilon_i)},$$

$$\sigma_{ab}^{(q)} = \sum_{mi} \frac{a_{mi} b_{mi}}{E_q - (\varepsilon_m - \varepsilon_i)},$$

and

$$\sigma_{bb}^{(q)} = \sum_{mi} \frac{|b_{mi}|^2}{E_q - (\varepsilon_m - \varepsilon_i)}.$$

The condition that the pair of equations (4.29) should be consistent, is that the determinant of the coefficients of α_q and β_q should vanish. This leads to an algebraic equation which can be solved for the eigenvalues E_q:

$$1 = [\lambda\, \sigma_{aa}^{(q)} + \mu\, \sigma_{bb}^{(q)}] + \lambda\mu[|\sigma_{ab}^{(q)}|^2 - \sigma_{aa}^{(q)}\sigma_{bb}^{(q)}]. \tag{4.30}$$

For each solution E_q of these equations, the ratio

$$\frac{\alpha_q}{\beta_q} = \frac{\mu\, \sigma_{ab}^{(q)}}{1 - \lambda\, \sigma_{aa}^{(q)}} = \frac{1-\mu\, \sigma_{bb}^{(q)}}{\lambda\, \sigma_{ab}^{(q)}} \tag{4.31}$$

is evaluated, and the wavefunction amplitudes, $x_{nj}^{(q)}$, are found by substituting eqn (4.31) into eqn (4.28) and normalizing $\Sigma_{nj}\, |x_{nj}^{(q)}|^2 = 1$. A graphical solution to eqn (4.30) can be mapped out in much the same way as Fig. 4.4 was constructed. However, it is now possible to have two special solutions which shift away from the positions of the unperturbed particle-hole energies. If both λ and μ have the same sign, then both special solutions move in the same direction (to higher energies if $\lambda > 0$, and to lower energies if $\lambda < 0$), while if λ and μ are of opposite sign, one special solution moves up and one moves down in the spectrum. Some typical examples are given in Fig. 4.5 in which the right-hand side of eqn (4.30) has been plotted as a function of E. The energy eigenvalues are given by the intercepts with the horizontal line drawn at the value unity.

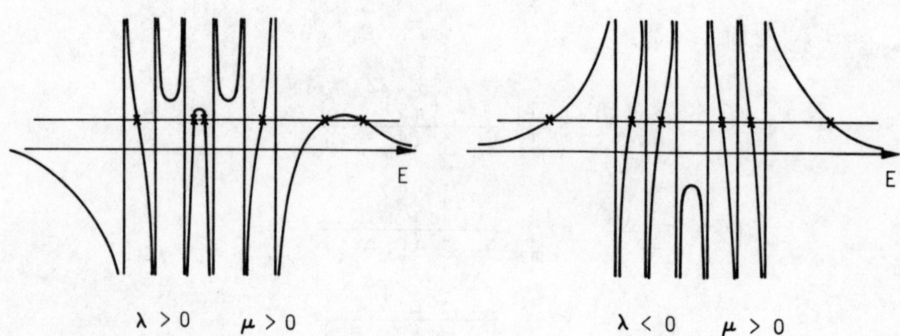

FIG. 4.5. A graphical solution to eqn (4.30) charcteristic of the extended schematic model. Plotted is the right-hand side of eqn (4.30) as a function of E for two typical values of λ and μ. The eigenenergies are given by the intercept with the horizontal line drawn at the value unity.

However, both special solutions do not necessarily become collective. For example, Towner (1970) has shown that in the limit of the unperturbed particle-hole energies all being degenerate, one of the special solutions takes 100 per cent of the transition strength and the other none. This result is proven using the zero range interaction for which $\lambda = \mu$ for magnetic states and $\lambda \neq \mu$ for electric. Also required in the proof is that the division between particle and hole states should occur at a closed shell in L-S coupling. Then both members of a spin-orbit pair $j = \ell \pm \frac{1}{2}$ are included in the sums over the particle states and in the sums over the hole states. In more realistic circumstances it is probable that the transition strength is shared between the two special solutions, but with one solution taking considerably more strength than the other.

4.7 THE RANDOM PHASE APPROXIMATION

In the Tamm-Dancoff approximation, the ground-state wavefunction is unaffected by the presence of 1p-1h excitations and remains as the Hartree-Fock solution. At closed major shells this solution is spherical. Consequently when discussing transition strengths between the ground state and excited 1p-1h states, all collective phenomena are attributed to coherence properties in the wavefunction of the excited state. This is a somewhat asymmetric description and might be considered a weakness in the TDA theory; a more reasonable theory would have the 'collectiveness' shared between the ground state and the excited state. Another weakness in the TDA is the neglect of 2p-2h excitations even though they too can exhibit a collective phenomenon capable of bringing them down in the spectrum to excitation energies comparable with the 1p-1h energies.

Apart from the formal objections just stated, a number of deficiencies show up when the results of TDA calculations are compared with experiment. For example, the energy-weighted sum S_{EW} for electric dipole absorption, when evaluated as in eqn (4.26) by explicitly summing the appropriate TDA eigenfunctions, is found to exceed the model-independent value.

eqn (4.19), and the experimental integrated photonuclear absorption cross-sections integrated over the giant resonance region (i.e. up to photon energies of 30 MeV). Since the sum rule S_{EW} depends primarily on the properties of the ground-state wavefunction, we conclude that the closed-shell assumption used in TDA is not sufficiently accurate. The TDA model also fails to push the collective T=0 solutions to sufficiently low energies and fails to endow them with sufficient transition strength. All these deficiencies are removed in the Random Phase Approximation (RPA) which improves upon the ground-state wavefunction by building in 2p-2h correlations.

There are a number of different ways in which the formulae of the RPA can be derived; we will describe just one method which has been called the 'linearization of the equations of motion' method (Brown 1967, Rowe 1970).

Let $|\Psi\rangle$ be the exact ground-state wavefunction for the Hamiltonian, H. It will be more general than the Hartree-Fock ground state wavefunction $|C\rangle$, in that it will have a structure

$$|\Psi\rangle = |C\rangle + |2p\text{-}2h\rangle + \ldots.. \tag{4.32}$$

with $|2p\text{-}2h\rangle$ correlations being explicitly built into it. Note that in labelling particle and hole states, we are still using the Hartree-Fock basis, hence no 1p-1h terms appear in Ψ. Now suppose there exists an operator $Q_q^\dagger$ such that the commutator of $Q_q^\dagger$ with H satisfies

$$[H, Q_q^\dagger]_- = E_q\, Q_q^\dagger. \tag{4.33}$$

Then because $H|\Psi\rangle = E_0|\Psi\rangle$, we have that

$$\begin{aligned} H\, Q_q^\dagger|\Psi\rangle &= [H, Q_q^\dagger]_-|\Psi\rangle + Q_q^\dagger\, H|\Psi\rangle \\ &= (E_q + E_0)\, Q_q^\dagger|\Psi\rangle. \end{aligned} \tag{4.34}$$

Thus if a set of operators satisfying eqn (4.33) can be found then a procedure, eqn (4.34), for generating the excited states of the system has been constructed.

Since we are interested in particle-hole excitations, we choose the operator $Q_q{}^\dagger$ to have the form

$$Q_q{}^\dagger = \sum_{mi} (x_{mi}{}^{(q)}\ a_m{}^\dagger b_i{}^\dagger - y_{mi}{}^{(q)}\ b_i a_m) \tag{4.35}$$

which is the most general form linear in the particle-hole operators. Here $b_i{}^\dagger$ creates a hole in the Hartree-Fock state $|C\rangle$ and $a_m{}^\dagger$ inserts a particle into an unoccupied orbit and the coefficients $x_{mi}{}^{(q)}$ and $y_{mi}{}^{(q)}$ have yet to be determined. This form shows that when $Q_q{}^\dagger$ acts on $|\Psi\rangle$, eqn (4.32), the first term, the x-term, acting on $|C\rangle$ produces a 1p-1h state, as does the y-term acting on $|2p\text{-}2h\rangle$. However, in addition the x-term acts on $|2p\text{-}2h\rangle$ to produce a 3p-3h state, and so forth; hence a constraint has to be introduced, such that these 3p-3h and higher np-nh components, which are generated when $Q_q{}^\dagger$ acts of $|\Psi\rangle$, will all be eliminated. That is, we must force eqn (4.35) to satisfy eqn (4.33) by making some linear approximation. There are a number of ways in which this can be done. The 'linearization' method assumes that when evaluating the commutators of the particle-hole operators with the Hamiltonian H, only terms linear in particle-hole operators are kept. Specifically this requires that

$$[H, a_m{}^\dagger b_i{}^\dagger]_- = \sum_{nj} (A_{nj,mi}\ a_n{}^\dagger b_j{}^\dagger + B^*_{nj,mi}\ b_j a_n) \tag{4.36}$$

where $A_{nj,mi}$ and $B_{nj,mi}$ are matrix elements whose values are known once H has been specified. The Hermitian conjugate of eqn (4.36) is

$$[H, b_i a_m]_- = -\sum_{nj} (A^*_{nj,mi}\ b_j a_n + B_{nj,mi}\ a_n{}^\dagger b_j{}^\dagger). \tag{4.37}$$

Using these relations, the commutator $[H, Q_q{}^\dagger]$ is now evaluated:

$$[H, Q_q{}^\dagger]_- = \sum_{mi,nj} [(A_{nj,mi}\ x_{mi}{}^{(q)} + B_{nj,mi}\ y_{mi}{}^{(q)}) a_n{}^\dagger b_j{}^\dagger + + (B^*_{nj,mi}\ x_{mi}{}^{(q)} + A^*_{nj,mi}\ y_{mi}{}^{(q)}) b_j a_n],$$

and this expression is required to equal the right-hand side of eqn (4.33), namely

$$E_q \sum_{nj} (x_{nj}^{(q)} a_n^{\dagger} b_j^{\dagger} - y_{nj}^{(q)} b_j a_n).$$

Thus equating the coefficients of $a_n^{\dagger} b_j^{\dagger}$ and $b_j a_n$ leads to a pair of coupled equations

$$\sum_{mi} (A_{nj,mi} x_{mi}^{(q)} + B_{nj,mi} y_{mi}^{(q)}) = E_q x_{nj}^{(q)}$$

$$\sum_{mi} (B^*_{nj,mi} x_{mi}^{(q)} + A^*_{nj,mi} y_{mi}^{(q)}) = -E_q y_{nj}^{(q)},$$

which can be summarized in matrix form as

$$\begin{pmatrix} A & B \\ -B^* & -A^* \end{pmatrix} \begin{pmatrix} X_q \\ Y_q \end{pmatrix} = E_q \begin{pmatrix} X_q \\ Y_q \end{pmatrix} \tag{4.38}$$

This is known as the RPA matrix equation. It takes the form of an eigenvalue problem for a non-Hermitian matrix, so that the eigenvalues may not be real and the eigenvectors may not be normalizable.

However, before discussing this problem, we must first evaluate the matrix elements $A_{nj,mi}$ and $B_{nj,mi}$. These are defined in eqns (4.36) and (4.37). First we require the commutator results

$$[a_m^{\dagger} b_i^{\dagger}, a_n^{\dagger} b_j^{\dagger}]_- = [b_i a_m, b_j a_n]_- = 0$$

$$[b_i a_m, a_n^{\dagger} b_j^{\dagger}]_- = \delta_{ij}\delta_{mn} - a_n^{\dagger} a_m \delta_{ij} - b_j^{\dagger} b_i \delta_{mn},$$

and remembering that $a_m|C\rangle = 0$ and $b_i|C\rangle = 0$, the latter can be cast in the form

$$[b_i a_m, a_n^{\dagger} b_j^{\dagger}]_- |C\rangle = \delta_{ij}\delta_{mn} |C\rangle.$$

These results are very reminiscent of the Bose-Einstein commutation relations. In fact another way of stating the linearization approximation introduced above is to require the particle-hole operators $a_m^\dagger b_i^\dagger$ to behave as boson operators. Next we construct the double commutator, $[b_j a_n, [H, a_m^\dagger b_i^\dagger]]$ and operate on the Hartree-Fock ground state

$$[b_j a_n, [H, a_m^\dagger b_i^\dagger]]|C\rangle$$

$$= \sum_{pk} \{A_{pk,mi}[b_j a_n, a_p^\dagger b_k^\dagger] + B^*_{pk,mi}[b_j a_n, b_k a_p]\}|C\rangle$$

$$= A_{nj,mi}\ |C\rangle$$

to obtain the result that

$$\begin{aligned} A_{nj,mi} &= \langle C|[b_j a_n, [H, a_m^\dagger b_i^\dagger]]|C\rangle \\ &= \langle C|b_j a_n\ H\ a_m^\dagger b_i^\dagger|C\rangle - \langle C|b_j a_n\ a_m^\dagger b_i^\dagger\ H|C\rangle. \end{aligned} \tag{4.39}$$

Finally the expansions (1.87) and (1.90) for the one-body and two-body parts of the Hamiltonian, H = T+V, are inserted in eqn (4.39) to give

$$\begin{aligned} A_{nj,mi} &= \langle n|T|m\rangle\delta_{ij} + \sum_k \langle kn|V|km\rangle_A\ \delta_{ij} - \\ &\quad - \left\{\langle\tilde{i}|T|\tilde{j}\rangle\delta_{mn} + \sum_k \langle k\tilde{i}|V|k\tilde{j}\rangle_A\ \delta_{mn}\right\} - \\ &\quad - \langle\tilde{i}n|V|\tilde{j}m\rangle_A \\ &= (\varepsilon_m - \varepsilon_i)\delta_{mn}\delta_{ij} - \langle\tilde{i}n|V|\tilde{j}m\rangle_A \end{aligned} \tag{4.40}$$

where the ε are the Hartree-Fock single-particle energies. Notice that this result is identical (to within a constant core on the diagonal elements) to the TDA matrix elements, eqn (4.9), so that the random phase approximation reduces to the Tamm-Dancoff approximation if all the y-terms vanish, and only the A-terms are operable in the RPA matrix.

Using similar methods, the matrix element $B_{nj,mi}$ is expressed in terms of a double commutator as

$$\begin{aligned} B_{nj,mi} &= \langle C|[[H,b_i a_m],b_j a_n]|C\rangle \\ &= \langle C|b_j a_n\ b_i a_m\ H|C\rangle. \end{aligned} \tag{4.41}$$

Notice that the matrix elements of B connect the closed-shell Hartree-Fock state with 2p-2h states, viz. $B \sim \langle 2p\text{-}2h|H|C\rangle$. Therefore only the two body part of H, in fact only the term (1.90d) in the expansion of V, contributes in the evaluation of eqn (4.41), and the final expression becomes

$$B_{nj,mi} = \langle mn|V|\tilde{i}\tilde{j}\rangle_A\ . \tag{4.42}$$

Thus the generalization of the TDA to include 2p-2h correlations in the ground state has been the introduction of the matrix B into the eigenvalue problem.

4.8 STABILITY OF THE HARTREE-FOCK GROUND STATE

We digress for a moment to discuss the problem of the stability of the Hartree-Fock solution since, as we now show, this is intimately connected with the RPA equations. In Hartree-Fock theory the ground-state wavefunction is a single determinant $|C\rangle$ satisfying the minimization condition that

$$E_0 = \frac{\langle C|H|C\rangle}{\langle C|C\rangle} = \text{minimum}.$$

Thouless (1960) proves that any other Slater determinant $|\Phi\rangle$ which is not orthogonal to $|C\rangle$ can be written in the form

$$\begin{aligned} |\Phi\rangle &= \exp\Big(\sum_{mi} C_{mi}\ a_m^\dagger b_i^\dagger\Big)|C\rangle \\ &= |C\rangle + \sum_{mi} C_{mi}\ a_m^\dagger b_i^\dagger|C\rangle + \frac{1}{2}\sum_{\substack{mi\\nj}} C_{mi}C_{nj}\ a_m^\dagger b_i^\dagger a_n^\dagger b_j^\dagger|C\rangle + \dots \end{aligned}$$

The procedure is to evaluate the energy expectation value E for $|\Phi\rangle$ up to second order in the coefficients C_{mi}, and then to find the conditions for which E exceeds E_0. Thus we must evaluate the normalization $\langle\Phi|\Phi\rangle$ and the Hamiltonian expectation value $\langle\Phi|H|\Phi\rangle$, bearing in mind that no terms of first order in C_{mi} are retained, since these are removed by the stationary conditions of the Hartree-Fock theory, eqn (2.13). Thus we have that

$$\langle\Phi|\Phi\rangle = \langle C|C\rangle \{1 + \sum_{mi} |C_{mi}|^2\}$$

$$\begin{aligned}\langle\Phi|H|\Phi\rangle = \langle C|H|C\rangle &+ \sum_{mi,nj} C_{mi}C_{nj}^* \langle C|b_j a_n\ H\ a_m^\dagger b_i^\dagger|C\rangle + \\ &+ \frac{1}{2}\sum_{mi,nj} C_{mi}C_{nj} \langle C|H\ a_m^\dagger b_i^\dagger a_n^\dagger b_j^\dagger|C\rangle + \\ &+ \frac{1}{2}\sum_{mi,nj} C_{mi}^*C_{nj}^* \langle C|b_j a_n b_i a_m\ H|C\rangle \\ = \langle C|H|C\rangle &+ \sum_{mi,nj} C_{mi}C_{nj}^* A_{nj,mi} + \sum_{mi} |C_{mi}|^2 \langle C|H|C\rangle + \\ &+ \frac{1}{2}\sum_{mi,nj} C_{mi}C_{nj} B_{nj,mi}^* + \frac{1}{2}\sum_{mi,nj} C_{mi}^*C_{nj}^* B_{nj,mi},\end{aligned}$$

where eqns (4.39) and (4.41) have been used for the matrix elements of A and B. The energy E is now written as

$$E = \frac{\langle\Phi|H|\Phi\rangle}{\langle\Phi|\Phi\rangle} = E_0 + \delta_2$$

where

$$\begin{aligned}\delta_2 &= \sum_{mi,nj} C_{mi}C_{nj}^* A_{nj,mi} + \frac{1}{2}\sum_{mi,nj} C_{mi}C_{nj} B_{nj,mi}^* + \frac{1}{2}\sum_{mi,nj} C_{mi}^*C_{nj}^* B_{nj,mi} \\ &= \frac{1}{2}(C^*\ \ C)\begin{pmatrix} A & B \\ B^* & A^* \end{pmatrix}\begin{pmatrix} C \\ C^* \end{pmatrix}\end{aligned}$$

in an obvious matrix notation. The condition for the Hartree-Fock solution to be stable is that δ_2 should be positive, which implies that the matrix must have non-negative definite eigenvalues, that is

$$\begin{pmatrix} A & B \\ B^* & A^* \end{pmatrix} \begin{pmatrix} C \\ C^* \end{pmatrix} = \lambda \begin{pmatrix} C \\ C^* \end{pmatrix}, \ \lambda \geqslant 0. \tag{4.43}$$

Returning for the moment to the definitions of the matrix elements of A and B, eqns (4.39) and (4.41), we have

$$A^{\dagger}_{nj,mi} = \langle C|b_i a_m \, H^{\dagger} \, a_n{}^{\dagger} b_j{}^{\dagger}|C\rangle - \langle C|H^{\dagger} \, b_i a_m \, a_n{}^{\dagger} b_j{}^{\dagger}|C\rangle$$

$$= A_{mi,nj},$$

$$B_{nj,mi} = \langle C|b_j a_n b_i a_m H|C\rangle = \langle C|b_i a_m b_j a_n H|C\rangle$$

$$= B_{mi,nj},$$

so that the matrix A is Hermitian (assuming H is Hermitian), and the matrix B is symmetric. This in turn implies that the matrix $\begin{pmatrix} A & B \\ B^* & A^* \end{pmatrix}$ is Hermitian, and the eigenvalues λ are therefore real.

Notice the close similarity between this stability condition, eqn (4.43), and the RPA matrix equation (4.38). If the stability condition admits a zero eigenvalue, i.e. $\lambda = 0$, then the RPA equation also has a zero eigenvalue. If the stability equation has all positive eigenvalues, i.e. the Hartree-Fock solution is stable, then all eigenvalues of the RPA equation are real (Thouless, 1961). However, if the Hartree-Fock solution is not a stable minimum, i.e. eqn (4.43) has a negative eigenvalue, then the RPA equation may have an imaginary eigenvalue. The proof follows in the next section.

4.9 PROPERTIES OF THE RPA MATRIX EQUATION

(i) If the stability condition eqn (4.43) is satisfied, then the eigenvalues of the RPA matrix equation, eqn (4.38), are real.

Proof: The first part of the proof requires the theorem:

If A is a Hermitian matrix with no non-negative eigenvalues, and x is an arbitrary vector, then $x^{\dagger}Ax$ is real and positive.

The proof is as follows. Let S be the unitary matrix ($S^{\dagger} = S^{-1}$) that reduces A to diagonal form, i.e.

$$S^{\dagger}AS = \text{diag}\,(\lambda_i) \qquad \lambda_i,\ \text{real}$$

where λ_i are the non-negative eigenvalues, then

$$\begin{aligned} x^{\dagger}Ax &= x^{\dagger}S^{-1}S\,A\,S^{-1}Sx \\ &= (Sx)^{\dagger}\,\text{diag}(\lambda_i)\,Sx \\ &= \sum_i \lambda_i\,|(Sx)_i|^2 = \text{real} \\ &\geqslant \lambda_{min}\sum_i |(Sx)_i|^2 \\ &\geqslant 0 \qquad \text{if } \lambda_{min} \geqslant 0. \end{aligned} \tag{4.44}$$

The inequality in eqn (4.44) is just a statement of the Rayleigh quotient theorem, with λ_{min} being the lowest eigenvalue of the matrix A. The equality only applies when x is the eigenvector of A corresponding to the eigenvalue λ_{min}. The second part of the proof applies this theorem to the matrix $\begin{pmatrix} A & B \\ B^* & A^* \end{pmatrix}$ which is Hermitian and satisfies the stability condition. Let $\binom{X}{Y}$ be an arbitrary vector, then

$$(X^{\dagger}\ Y^{\dagger})\begin{pmatrix} A & B \\ B^* & A^* \end{pmatrix}\begin{pmatrix} X \\ Y \end{pmatrix} \geqslant \lambda_{min}(X^{\dagger}\ Y^{\dagger})\binom{X}{Y}.$$

If $\binom{X}{Y}$ is an eigenvector of the RPA matrix equation corresponding to an eigenvalue E, then the inequality reads

$$E(X^{\dagger}X - Y^{\dagger}Y) \geqslant \lambda_{min}(X^{\dagger}X + Y^{\dagger}Y),\ \text{real}. \tag{4.45}$$

Since $(X^{\dagger}X - Y^{\dagger}Y)$ is real, it follows that E is real. The only possibility for which E can still be complex occurs when $(X^{\dagger}X - Y^{\dagger}Y) = 0$. If this is so, the left-hand side of eqn

(4.45) takes its minimum value, implying that $\binom{X}{Y}$ is now an eigenvector of $\begin{pmatrix} A & B \\ B^* & A^* \end{pmatrix}$ corresponding to the minimum eigenvalue λ_{min}, that is

$$\begin{pmatrix} A & B \\ B^* & A^* \end{pmatrix} \begin{pmatrix} X \\ Y \end{pmatrix} = \lambda_{min} \begin{pmatrix} X \\ Y \end{pmatrix} = E \begin{pmatrix} X \\ -Y \end{pmatrix} = 0,$$

so that $\lambda_{min} = 0$ and hence $E = 0$. Thus we have shown that E cannot be complex.

If λ_{min} is negative (i.e. the stability condition is *not* satisfied) then $(X^{\dagger}X - Y^{\dagger}Y)$ can be zero without $\binom{X}{Y}$ having to be an eigenvector of $\begin{pmatrix} A & B \\ B^* & A^* \end{pmatrix}$, implying that E can be complex in this case.

(ii) If the stability condition is satisfied, then for each eigenvalue E associated with the eigenvector $\binom{X}{Y}$, there is also an eigenvalue -E corresponding to the eigenvector $\binom{Y^*}{X^*}$.

Proof: Taking the complex conjugate of the pair of equations (4.38) gives a second set of equations

$$A^*X^* + B^*Y^* = E^*X^*$$
$$-BX^* - AY^* = E^*Y^*$$

which can be reordered into the form

$$\begin{pmatrix} A & B \\ -B^* & -A^* \end{pmatrix} \begin{pmatrix} Y^* \\ X^* \end{pmatrix} = -E^* \begin{pmatrix} Y^* \\ X^* \end{pmatrix},$$

showing that $\binom{Y^*}{X^*}$ also satisfies the RPA equation with eigenvalue $-E^*$. If the stability condition is satisfied, then E is real, thus proving the statement (ii).

This -E solution, however, is usually discarded in nuclear physics applications. The reason is as follows. If $Q_q^{\dagger}|\Psi\rangle$ represents a physical state of the system, then the operator $Q_q^{\dagger}$ must satisfy eqn (4.33). The Hermitian conjugate of this equation is

$$[H,\ Q_q]_- = -E_q\ Q_q,$$

so that

$$H\ Q_q|\Psi\rangle = [H, Q_q]_-|\Psi\rangle + Q_q\ H|\Psi\rangle$$

$$= (E_0 - E_q) Q_q|\Psi\rangle,$$

implying that $Q_q|\Psi\rangle$ is an eigenvector of the Hamiltonian with an eigenvalue less than the ground state energy, E_0. This is considered unphysical, so a subsidiary condition is imposed on the RPA solutions, namely that

$$Q_q|\Psi\rangle = 0.$$

(iii) The solutions, $E_q \neq 0$, of the RPA equations can be orthogonalized and normalized.

Proof: Consider the RPA equation

$$\begin{pmatrix} A & B \\ B^* & A^* \end{pmatrix} \begin{pmatrix} X_q \\ Y_q \end{pmatrix} = E_q \begin{pmatrix} X_q \\ -Y_q \end{pmatrix}$$

and its Hermitian conjugate

$$(X_r^{\dagger}\ Y_r^{\dagger}) \begin{pmatrix} A & B \\ B^* & A^* \end{pmatrix} = E_r (X_r^{\dagger}\ -Y_r^{\dagger}).$$

Multiply the former by $(X_r^{\dagger}\ Y_r^{\dagger})$ and the latter by $\binom{X_q}{Y_q}$ and subtract:

$$0 = (E_q - E_r)(X_r^{\dagger}\ X_q - Y_r^{\dagger}\ Y_q).$$

Thus if E_q and E_r are different, we have

$$(X_r^\dagger X_q - Y_r^\dagger Y_q) = 0$$

as the orthogonality condition on two eigenvectors with different eigenvalues. If E_q equals E_r, then from the stability condition, eqn (4.45) with $\lambda_{min} > 0$, we have that

$$E_q(X_q^\dagger X_q - Y_q^\dagger Y_q) > 0,$$

so that the sign of the normalization is positive if the eigenvalue is positive and negative if the eigenvalue is negative. If there is an eigenvalue $E_q = 0$, this indicates the RPA equations have two solutions that are degenerate with zero energy. Two corresponding eigenvectors, which are orthogonal, can be constructed but they cannot be normalized. All these results can be summed up by the one relation:

$$(X_r^\dagger X_q - Y_r^\dagger Y_q) = \mathrm{sgn}(E_q)\delta_{qr}, \tag{4.47}$$

where

$$\mathrm{sgn}(E_q) = \begin{cases} +1 & \text{if} \quad E_q > 0 \\ 0 & \text{if} \quad E_q = 0 \\ -1 & \text{if} \quad E_q < 0 \end{cases}$$

We now wish to show that the normalization of the physical solutions $Q_q^\dagger|\Psi\rangle$ is compatible with eqn (4.47). Recalling the definition, eqn (4.35), and the subsidiary condition $Q_q|\Psi\rangle = 0$, eqn (4.46), we have that

$$\begin{aligned}
\langle q|q\rangle &= \langle\Psi|Q_q Q_q^\dagger|\Psi\rangle \\
&= \langle\Psi|[Q_q, Q_q^\dagger]_-|\Psi\rangle \\
&= \sum_{mi,nj} \{x_{mi}^{(q)*} x_{nj}^{(q)} \langle\Psi|[b_i a_m, a_n^\dagger b_j^\dagger]_-|\Psi\rangle + \\
&\quad + y_{mi}^{(q)*} y_{nj}^{(q)} \langle\Psi|[a_m^\dagger b_i^\dagger, b_j a_n]_-|\Psi\rangle\}.
\end{aligned}$$

To continue, a further condition has to be introduced, namely,

$$\langle\Psi|[b_i a_m, a_n^\dagger b_j^\dagger]_-|\Psi\rangle = \delta_{ij}\delta_{mn}; \tag{4.48}$$

that is, the particle-hole operators are required to satisfy the Bose-Einstein commutation relations not only when operating on the Hartree-Fock ground state $|C\rangle$ as discussed earlier, but also when operating on the RPA ground state $|\Psi\rangle$. With this assumption, the normalization becomes

$$\langle q|q\rangle = \sum_{mi,nj} (x_{mi}^{(q)*} x_{nj}^{(q)} - y_{mi}^{(q)*} y_{nj}^{(q)})\delta_{ij}\delta_{mn}$$

$$= X_q^\dagger X_q - Y_q^\dagger Y_q$$

which is now compatible with eqn (4.47).

(iv) If there are no zero energy solutions to the RPA equations, then the eigenvectors form a complete set (Thouless 1961):

$$\sum_q \begin{pmatrix} X_q \\ Y_q \end{pmatrix} (X_q^\dagger - Y_q^\dagger) = \begin{pmatrix} I & 0 \\ 0 & I \end{pmatrix} \tag{4.49a}$$

where I is the unit matrix. The sum is over both positive and the negative energy solutions.

By using the symmetry of the RPA equations, the closure expression can be rewritten as

$$\sum_{q>0} \left\{ \begin{pmatrix} X_q \\ Y_q \end{pmatrix} (X_q^\dagger - Y_q^\dagger) - \begin{pmatrix} Y_q^* \\ X_q^* \end{pmatrix} (Y_q^{*\dagger} - X_q^{*\dagger}) \right\} = \begin{pmatrix} I & 0 \\ 0 & I \end{pmatrix}, \tag{4.49b}$$

involving now only a sum over the positive energy solutions. However, for each zero-energy eigenvalue, the set of solutions will be one short of completeness, and eqn (4.49) cannot be used as it now stands. Instead an additional vector, not necessarily an eigenvector of the RPA equations, has to be found which is orthogonal and linearly independent of the RPA eigenvectors. If found, this additional vector can be used to complete the set. Rowe (1970) gives some discussion to this.

4.10 ANGULAR MOMENTUM COUPLED PARTICLE-HOLE REPRESENTATION

In the previous sections on RPA theory, the formulae were presented in the simple uncoupled particle-hole representation. However, in practice it is necessary to use a representation in which the angular momenta of the particle-hole states have been coupled to good total angular momentum, J, and total isospin, T. This reduces the size of the matrices which have to be handled. Thus the operator $Q_q^\dagger$, eqn (4.35), which when operating on the RPA ground state produces an excited state of the system, is now defined in terms of angular momentum coupled particle-hole operators, namely

$$Q_q^\dagger(JM) = \sum_{mi} \{x_{mi}^{(q)} A_{mi}^\dagger(JM) - y_{mi}^{(q)} A_{mi}(J\tilde{M})\} \qquad (4.50)$$

where

$$A_{mi}^\dagger(JM) = \sum_{m_m m_i} \langle j_i m_i j_m m_m | JM \rangle a_m^\dagger b_i^\dagger$$

$$A_{mi}(J\tilde{M}) = (-)^{J+M} A_{mi}(J{-}M)$$

$$= \sum_{m_m m_i} (-)^{J+M} \langle j_i m_i j_m j_m | J{-}M \rangle b_i a_m .$$

We are again using an isospin representation, but for economy of notation we have not written down explicitly the isospin labels or the isospin Clebsch-Gordan coefficients in eqn (4.51), although their presence should be understood. These operators obey the Bose-Einstein commutation relations when operating on $|C\rangle$, viz.

$$\langle C|[A_{mi}(JM), A_{nj}^\dagger(J'M')]_-|C\rangle = \delta_{mn}\delta_{ij}\delta_{JJ'}\delta_{MM'}$$
$$\langle C|[A_{mi}(JM), A_{nj}(J'M')]_-|C\rangle = 0, \qquad (4.52)$$

and it is assumed that the same commutation relations hold when evaluated between the RPA ground state $|\Psi\rangle$, see eqn (4.48). The RPA eigenvalue problem has the identical structure to eqn (4.38) and the eigenvectors have all the properties discussed in the last section; however, there is now a separate eigenvalue problem for each angular momentum J and isospin T. The elements of the RPA matrix, in analogy to eqns (4.39) and (4.41), are found to be

$$\begin{aligned} A_{nj,mi} &= \langle C|[\mathcal{A}_{nj}(JM),[H,\mathcal{A}_{mi}{}^{\dagger}(JM)]]|C\rangle \\ B_{nj,mi} &= \langle C|[[H,\mathcal{A}_{mi}(J\tilde{M})],\mathcal{A}_{nj}(JM)]|C\rangle, \end{aligned} \tag{4.53}$$

and written out in terms of coupled antisymmetric two-body matrix elements, they are

$$\begin{aligned} A_{nj,mi} &= (\varepsilon_m-\varepsilon_i)\delta_{mn}\delta_{ij} - \sum_{\bar{J}} \frac{\hat{\bar{J}}}{\hat{J}} U(j_j j_n j_m j_i; J\bar{J})\langle j_n j_i;\bar{J}|V|j_m j_j;\bar{J}\rangle_A \\ B_{nj,mi} &= [(1+\delta_{mn})(1+\delta_{ij})]^{\frac{1}{2}} \sum_{\bar{J}} \frac{\hat{\bar{J}}\,\hat{\bar{J}}}{\hat{j}_i \hat{j}_n} U(j_m \bar{J} J j_j; j_n j_i)\langle j_m j_n;\bar{J}|V|j_i j_j;\bar{J}\rangle_A \end{aligned} \tag{4.54}$$

where $\hat{J} = (2J+1)^{\frac{1}{2}}$. The matrix element $A_{nj,mi}$ is of course identical to the Tamm-Dancoff approximation, eqn (4.13).

We will not discuss the methods available for solving the RPA eigenvalue problem. However, it is not necessary to resort to a direct diagonalization of a $2N\times 2N$ non-Hermitian matrix, where N is the dimension of the matrices A and B. Both Chi (1970) and Ullah and Rowe (1971) have given algorithms which exploit the symmetry of the RPA matrix such that the problem can be reduced to the diagonalization of a real symmetric $N\times N$ matrix, whose eigenvalues are the squares of the energies sought. In Chi's method, it is assumed that all the eigenvalues E are real, that is a Hartree-Fock basis is in use, with the Hartree-Fock ground state satisfying the stability condition. In practice one often works with harmonic oscillator or Saxon-Woods basis functions which satisfy neither the consistency nor the stability criteria of Hartree-Fock functions; then the RPA equations can admit complex eigenvalues. Ullah and Rowe's method allows the RPA matrix to have not more than one pair of non-real eigenvalues; then if the matrices A and B are real with A positive-definite they show that these non-real eigenvalues are pure imaginary, so that the squares of all eigenvalues are again real, and the problem again reduces to the diagonalization of a real symmetric N×N matrix.

4.11 TRANSITION RATES AND SUM RULES IN RPA

We now wish to calculate the transition matrix element of a one-body operator, U, between the RPA ground state and an excited state, $|q\rangle = Q_q^{\dagger}|\Psi\rangle$. The calculation follows very closely the normalization procedure outlined earlier. We have, using the condition (4.46), that

$$\langle q|U|\Psi\rangle = \langle\Psi|Q_q U|\Psi\rangle$$

$$= \langle\Psi|[Q_q,U]_-|\Psi\rangle.$$

Writing Q_q as in eqn (4.35), and expanding the one-body operator U as in eqn (1.87), keeping only terms (1.87b) and (1.87c), we find

$$\langle q|U|\Psi\rangle = \sum_{mi,nj} x_{mi}^{*(q)}\langle n|U|\tilde{j}\rangle\langle\Psi|[b_i a_m, a_n^{\dagger} b_j^{\dagger}]_-|\Psi\rangle -$$

$$- y_{mi}^{*(q)}\langle\tilde{j}|U|n\rangle\langle\Psi|[a_m^{\dagger} b_i^{\dagger}, b_j a_n]_-|\Psi\rangle\}.$$

The approximate commutation relations, eqn (4.48), are again used to evaluate this expression with the result:

$$\langle q|U|\Psi\rangle = \sum_{mi}\{x_{mi}^{*(q)}\langle m|U|\tilde{i}\rangle + y_{mi}^{*(q)}\langle\tilde{i}|U|m\rangle\}$$

$$= (X_q^{\dagger}\ Y_q^{\dagger})\begin{pmatrix} U_x \\ U_y \end{pmatrix}, \tag{4.55}$$

where we have introduced a column vector of one-body matrix elements, viz.

$$\begin{pmatrix} U_x \\ U_y \end{pmatrix} = \begin{pmatrix} \langle m|U|\tilde{i}\rangle \\ \langle n|U|\tilde{j}\rangle \\ \vdots \\ \langle \tilde{i}|U|m\rangle \\ \langle \tilde{j}|U|n\rangle \\ \vdots \end{pmatrix}$$

Notice that if the y-coefficients are all zero, the result here just reduces to the Tamm-Dancoff expression. If the one-body operator is Hermitian ($U^\dagger = U$), or anti-Hermitian ($U^\dagger = -U$), then

$$\langle \tilde{i}|U|m\rangle = \langle m|U^\dagger|\tilde{i}\rangle^* = \pm\langle m|U|i\rangle^*$$

that is $U_y = \pm U_x^*$. The sign conventions can usually be chosen so that these one-body matrix elements are real, in which case $U_y = \pm U_x$.

In an angular momentum coupled representation, the operator Q_q as defined in eqn (4.50) is used, and following an analogous procedure the final result is

$$\langle q\|U^{(J)}\|\Psi\rangle = \sum_{mi} \{x_{mi}^{*(q)} (-)^{j_m+j_i-J} \frac{\hat{j}_m}{\hat{J}} \langle j_m\|U^{(J)}\|j_i\rangle + + y_{mi}^{*(q)} (-)^{2j_i} \frac{\hat{j}_i}{\hat{J}} \langle j_i\|U^{(J)}\|j_m\rangle\}. \tag{4.56}$$

Recalling that in the TDA discussion, we introduced the matrix element a_{mi}, eqn (4.23), defined as

$$a_{mi} = \langle j_i^{-1} j_m; J\|U^{(J)}\|C\rangle = (-)^{j_m+j_i-J} \frac{\hat{j}_m}{\hat{j}} \langle j_m\|U^{(J)}\|j_i\rangle,$$

then eqn (4.56) can be rewritten as

$$\langle q\|U^{(J)}\|\Psi\rangle = \sum_{mi} \{x_{mi}^{*(q)} + S\, y_{mi}^{*(q)}\} a_{mi} \tag{4.57}$$

where S is a sign factor $S = (-)^{J+p}$ with the $(-)^p$ originating in the definition of the Hermitian conjugate for the one-body operator U, viz. $U_M^{(J)\dagger} = (-)^{p+M} U_{-M}^{(J)}$.

We next evaluate the non-energy-weighted S_{NEW} and the energy weighted S_{EW} sum rules. First, consider

$$S_{NEW} = \sum_q |\langle q|U|\Psi\rangle|^2$$

where the sum is over both the positive and negative energy solutions of the RPA equations. We assume there are no zero-energy eigenvalues so that the completeness relation eqn (4.49) holds; then

$$S_{NEW} = \sum_q (U_x^\dagger \;\; \pm U_x^\dagger) \begin{pmatrix} X_q \\ Y_q \end{pmatrix} (X_q^\dagger - Y_q^\dagger) \begin{pmatrix} U_x \\ \mp U_x \end{pmatrix}$$

$$= (U_x^\dagger \pm U_x^\dagger) \begin{pmatrix} I & 0 \\ 0 & I \end{pmatrix} \begin{pmatrix} U_x \\ \mp U_x \end{pmatrix}$$

$$= \text{zero},$$

that is the non-energy-weighted sum over *all* the eigenvalue solutions sums to zero in the RPA. This implies that the sum over all the positive energy solutions is exactly equal but opposite in sign to the sum over the negative energy solutions. In practice, however, the negative energy solutions are discarded as being non-physical, and one is only interested in the sum rules defined as the sums over the positive energy solutions. In this case, S_{NEW} is defined as

$$S_{NEW} = \sum_{q>0} |(X_q^\dagger \;\; Y_q^\dagger) \begin{pmatrix} U_x \\ \pm U_x \end{pmatrix}|^2 \tag{4.58}$$

which cannot be simplified. Furthermore this sum now varies as the eigenvectors X_q and Y_q are varied as a result of changes in the Hamiltonian. This is in contrast to TDA where the summed strength was invariant to such changes and was equal to the sum of the unperturbed particle-hole excitations

On the other hand, the energy-weighted sum rule

$$S_{EW} = \sum_{q>0} |\langle q|U|\Psi\rangle|^2 E_q$$

can be summed over the positive energy solutions. This is because E_q changes sign as one goes from the positive energy to the negative energy solutions, such that the sum over the positive energy solutions exactly equals the sum over the negative energy solutions. Thus we have

$$\begin{aligned} S_{EW} &= \sum_{q>0} (U_x^{\dagger} \mp U_x^{\dagger}) \begin{pmatrix} X_q \\ -Y_q \end{pmatrix} E_q \, (X_q^{\dagger} - Y_q^{\dagger}) \begin{pmatrix} U_x \\ \mp U_x \end{pmatrix} \\ &= \frac{1}{2} \sum_q (U_x^{\dagger} \mp U_x^{\dagger}) \begin{pmatrix} A & B \\ B^* & A^* \end{pmatrix} \begin{pmatrix} X_q \\ Y_q \end{pmatrix} (X_q^{\dagger} - Y_q^{\dagger}) \begin{pmatrix} U_x \\ \mp U_x \end{pmatrix} \\ &= \frac{1}{2} (U_x^{\dagger} \mp U_x^{\dagger}) \begin{pmatrix} A & B \\ B^* & A^* \end{pmatrix} \begin{pmatrix} U_x \\ \mp U_x \end{pmatrix} \end{aligned} \tag{4.59}$$

where again closure is used, eqn (4.49), on the assumption that there are no zero energy solutions.

This sum rule can also be written in terms of the double commutator

$$S_{EW} = \frac{1}{2} \langle C | [U,[H,U]] | C \rangle$$

calculated between ground-state wavefunctions. This is easily shown using the commutation relations of the particle-hole creation and annihilation operators with the Hamiltonian, eqns (4.36) and (4.37). Thus by writing

$$U = \sum_{mi} \{ \langle m|U|\tilde{i}\rangle \, a_m^{\dagger} b_i^{\dagger} + \langle \tilde{i}|U|m\rangle \, b_i a_m \}$$

and explicitly evaluating

$$\begin{aligned} [H,U]_- |C\rangle = \sum_{mi,nj} \{ &\langle m|U|\tilde{i}\rangle A_{nj,mi} \, a_n^{\dagger} b_j^{\dagger} + \langle m|U|\tilde{i}\rangle B^*_{nj,mi} \, b_j a_n - \\ &- \langle \tilde{i}|U|m\rangle A^*_{nj,mi} \, b_j a_n - \langle \tilde{i}|U|m\rangle B_{nj,mi} \, a_n^{\dagger} b_j^{\dagger} \} |C\rangle, \end{aligned}$$

one obtains

$$\frac{1}{2}\langle C|[U,[H,U]]|C\rangle = \frac{1}{2}\sum_{mi,nj} \{\langle\tilde{j}|U|n\rangle A_{nj,mi}\langle m|U|\tilde{i}\rangle - \langle n|U|\tilde{j}\rangle B^*_{nj,mi}\langle m|U|\tilde{i}\rangle -$$

$$- \langle\tilde{j}|U|n\rangle B_{nj,mi}\langle\tilde{i}|U|m\rangle + \langle n|U|\tilde{j}\rangle A^*_{nj,mi}\langle\tilde{i}|U|m\rangle\}.$$

This proves a very important result, first derived by Thouless (1961), that

$$2\sum_{q>0} |\langle q|U|\Psi\rangle|^2 E_q = \langle C|[U,[H,U]]|C\rangle$$

$$= (U_x^{\dagger} \mp U_x^{\dagger}) \begin{pmatrix} A & B \\ B^* & A^* \end{pmatrix} \begin{pmatrix} U_x \\ \mp U_x \end{pmatrix} \tag{4.60}$$

where the upper sign is used if U is Hermitian and the lower sign if U is anti-Hermitian. The result is important since it forms the basis of the following theorem:

To the extent that the RPA is executed exactly (i.e. Hartree-Fock basis functions are used), all spurious states are separated out exactly, and occur as zero-energy solutions in the RPA eigenvalue problem.

4.12 SPURIOUS STATES

In the chapter on Hartree-Fock theory (chapter 2), it was noted that some symmetries, the 'consistent' symmetries, were conserved in the iteration procedure, and some symmetries wer[e] not conserved. An example of the latter concerns the total linear momentum operator $\underset{\sim}{P}$. The original Hamiltonian is translationally invariant, i.e.

$$[H, \underset{\sim}{P}] = 0;$$

however, the Hartree-Fock single-particle Hamiltonian, h, or rather the Hartree-Fock potential is *not* translationally invariant. Thus the Hartree-Fock ground state $|C\rangle$ is not an eigenstate of $\underset{\sim}{P}$, but rather the state $\underset{\sim}{P}|C\rangle$ is a 1p-1h excitation with quantum numbers $J^{\pi} = 1^-$ and T=0. This excitation i[s] however, spurious since it corresponds merely to the translation of the centre of mass of the nucleus and as such is not considered as a true excitation of the nucleus. Nevertheless, in the Tamm-Dancoff approximation the state $\underset{\sim}{P}|C\rangle$ can get mixed up with all the other non-spurious states of the same quantum numbers, $J^{\pi} = 1^{-}$ and T=0, and this leads to a difficulty in the TDA theory. However, in RPA theory these spurious states are separated out exactly.

To see this, consider the result (4.60) where the one-body operator U is now the Hermitian operator $\underset{\sim}{P}$. Since $[H,\underset{\sim}{P}] = 0$, the double commutator vanishes so that

$$\begin{pmatrix} A & B \\ B^{*} & A^{*} \end{pmatrix} \begin{pmatrix} P_x \\ -P_x \end{pmatrix} = 0 \qquad (4.62)$$

where P_x is a column vector of matrix elements $\langle m|\underset{\sim}{P}|i\rangle$. This shows that $\begin{pmatrix} P_x \\ -P_x \end{pmatrix}$ is an eigenvector of the RPA matrix equation (and also of the Hartree-Fock stability equation) with eigenvalue zero. Since the state $\underset{\sim}{P}|\Psi\rangle$ is a spurious state, this allows us to eliminate spurious states by simply rejecting all solutions to the RPA equations having zero energy eigenvalues. This exact separation is one advantage the RPA theory has over the TDA theory; however, the exact separation only occurs when using true Hartree-Fock basis states, and furthermore its presence destroys the completeness relation, eqn (4.49), thereby affecting the sum rules calculated in the last section.

.13 SCHEMATIC MODEL IN RPA

n order to obtain some insight into the properties of the RPA quations, the schematic model introduced in section (4.6) can e generalized so that the RPA eigenvalue problem can be olved algebraically. The required eigenvalues, E_q, and igenvectors, $x_{mi}^{(q)}$ and $y_{mi}^{(q)}$, are solutions to the equaions

$$\sum_{mi} (A_{nj,mi}\, x_{mi}^{(q)} + B_{nj,mi}\, y_{mi}^{(q)}) = E_q\, x_{nj}^{(q)}$$
$$\sum_{mi} (B_{nj,mi}\, x_{mi}^{(q)} + A_{nj,mi}\, y_{mi}^{(q)}) = -E_q\, y_{nj}^{(q)}, \qquad (4.63)$$

where the matrix elements are now written in a factorized form

$$A_{nj,mi} = (\varepsilon_n - \varepsilon_j)\delta_{mn}\delta_{ij} + \lambda a_{nj} a_{mi}$$
$$B_{nj,mi} = \lambda S\, a_{nj} a_{mi} \qquad (4.64)$$

with S just being a sign factor $S = (-)^{J+p}$, introduced in eqn (4.57). This factorization was used in eqn (4.21) for the TDA matrix element $A_{nj,mi}$, and a diagrammatic interpretation due to Brown (1967) was illustrated in Fig. 4.3. Similarly the matrix element $B_{nj,mi}$, which connects a 2p-2h state with the closed shell, is illustrated in Fig. 4.6. To the extent that the factorization of the matrix element $A_{nj,mi}$ can be justified diagramatically in Fig. 4.3, so too can the matrix element $B_{nj,mi}$ in Fig. 4.6 be represented by the same product of two one-body matrix elements. This is the schematic model assumption leading to eqn (4.64).

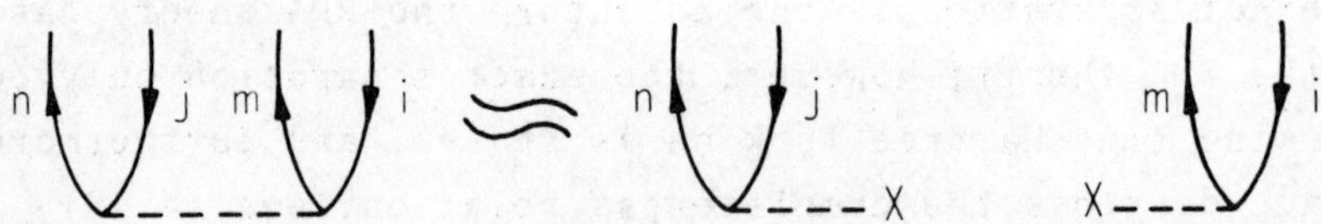

FIG. 4.6. The schematic model factorization in which the two-body matrix element $B_{nj,mi}$ is replaced by a product of two one-body matrix elements.

The coefficient λ represents the strength of the residua interaction (negative for T=0 states and positive for T=1 states) and a_{nj} represents the matrix element of a one-body operator $U^{(J)}$ connecting the closed-shell state with the particle-hole basis state $|j_j^{-1} j_n; J\rangle$. These definitions are the same as those introduced in eqn (4.23), which followed from using a multipole-multipole force to represent the residual interaction. Similarly this multipole-multipole force can be used to evaluate the matrix element $B_{nj,mi}$. By substituting eqn (A.90) into eqn (4.54) and maintaining only the direct part of the multipole-multipole two-body matrix element, the factorization of eqn (4.64) is obtained with the sign factor $S = (-)^{J+p}$ being introduced in connection with the Hermitian property of the one-body matrix elements a_{mi}.

To solve, substitute eqn (4.64) back into eqn (4.63) and define

$$\alpha_q = \sum_{mi} x_{mi}^{(q)} a_{mi} \quad ; \quad \beta_q = \sum_{mi} y_{mi}^{(\bar{q})} a_{mi} \tag{4.65}$$

to obtain the following expressions for the wavefunctions:

$$\begin{aligned} x_{nj}^{(q)} &= \lambda a_{nj}(\alpha_q + S\beta_q)/(E_q - \varepsilon_n + \varepsilon_j) \\ y_{nj}^{(q)} &= -\lambda S a_{nj}(\alpha_q + S\beta_q)/(E_q + \varepsilon_n - \varepsilon_j). \end{aligned} \tag{4.66}$$

Here $(\alpha_q + S\beta_q)$ can be treated as a constant, which is determined from the normalization condition

$$\sum_{nj} \{|x_{nj}^{(q)}|^2 - |y_{nj}^{(q)}|^2\} = \mathrm{sgn}(E_q). \tag{4.67}$$

Next multiply both equations (4.66) by a_{nj}, sum over the particle-hole states nj, and then add the two equations together. On dividing both sides by $(\alpha_q + S\beta_q)$ one obtains a dispersion equation for the energy eigenvalues:

$$1 = 2\lambda \sum_{nj} |a_{nj}|^2 \frac{(\varepsilon_n - \varepsilon_j)}{E_q{}^2 - (\varepsilon_n - \varepsilon_j)^2} . \tag{4.68}$$

A graphical solution to eqn (4.68) is shown in Fig. 4.7; however, in contrast to the TDA solution illustrated in Fig. 4.4, the right-hand side of eqn (4.68) is plotted as a function of E over both positive and negative values of E. If one concentrates on just the positive energy solutions, then the right-hand half of each diagram in Fig. 4.7 looks qualitatively very similar to the TDA diagrams in Fig. 4.4. Thus all but one of the solutions are trapped between the unperturbed

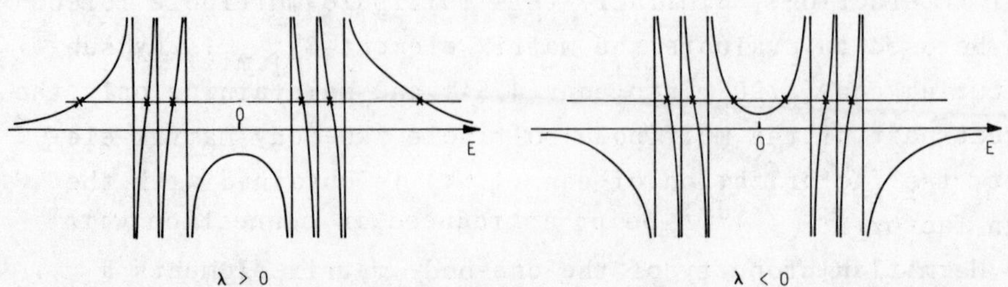

FIG. 4.7. Graphical solutions to the RPA equations for two typical values of λ. Plotted is the right-hand side of eqn (4.68) as a function of the energy E. The eigenenergies are given by the intercept with the horizontal line drawn at the value unity.

particle-hole energies, and the one untrapped solution, the collective state, is pushed to higher energies when $\lambda > 0$ and to lower energies when $\lambda < 0$. The main difference between the RPA and TDA graphical solutions is found in the behaviour of the collective solution corresponding to an attractive force, $\lambda < 0$. As $-\lambda$ increases, this solution goes much faster to zero energy than did the corresponding TDA solution and there is a critical value of the interaction strength λ beyond which the lowest eigenvalue becomes purely imaginary.

To show the collective behaviour of the untrapped solution, we again consider the limit in which all the unperturbed particle-hole energies are degenerate, i.e. $(\varepsilon_n - \varepsilon_j) = \bar{\varepsilon}$. Then the energy of the trapped solutions, $q = 2,3,4...$, is just $\bar{\varepsilon}$, while the untrapped solution, $q = 1$, has an energy

$$E_1 = (\bar{\varepsilon}^2 + 2\lambda\bar{\varepsilon} \sum_{nj} |a_{nj}|^2)^{1/2}.$$

This is to be contrasted with the TDA value $E_1 = \bar{\varepsilon} + \lambda\Sigma|a_{nj}|^2$ which shows that the RPA energy for the collective state is always lower than the TDA energy, both for an attractive and a repulsive force.

The wavefunction for the collective state is

$$x_{nj}^{(1)} = \lambda\, a_{nj}(\alpha_1 + S\beta_1)/(E_1 - \bar{\varepsilon})$$

$$y_{nj}^{(1)} = -\lambda S a_{nj}(\alpha_1+S\beta_1)/(E_1+\bar{\varepsilon}),$$

using eqn (4.66), with the normalization being

$$(\alpha_1+S\beta_1)^2 = \frac{\bar{\varepsilon}}{E_1}\sum_{nj} |a_{nj}|^2.$$

The transition rate for exciting the collective state by γ-ray absorption is, from eqn (4.57),

$$T_1 = |\langle 1|U|\Psi\rangle|^2 = (\alpha_1+S\beta_1)^2 \tag{4.69}$$

where the definitions of α_1 and β_1 are in eqn (4.65). The final expression for the transition rate therefore becomes

$$T_1 = \frac{\bar{\varepsilon}}{E}\sum_{nj} |a_{nj}|^2 \tag{4.70}$$

which is to be compared with the TDA value $T_1 = \sum_{nj} |a_{nj}|^2$. We see that for a state shifted down, $\lambda < 0$ with $E > \bar{\varepsilon}$, the RPA increases the transition strength over the TDA value, while for states shifted up, $\lambda > 0$ and $E > \bar{\varepsilon}$, the converse is true. The enhancement of the transition rate of the low-lying states such as the 3^- T=0 state may be quite appreciable since E can be several times smaller than $\bar{\varepsilon}$.

Another interesting result is that, whereas the TDA satisfies the non-energy-weighted sum rule S_{NEW} but violates the S_{EW}, the RPA satisfies the energy-weighted sum rule but augments or diminishes the S_{NEW} according as $\bar{\varepsilon}/E$ is greater or less than unity. The RPA energy-weighted sum is given in eqn (4.59), which written out in component form becomes

$$S_{EW} = \sum_{mi,nj} (a_{nj} A_{nj,mi} a_{mi} - S a_{nj} B_{nj,mi} a_{mi}).$$

Inserting the schematic model factorizations for the matrix elements $A_{nj,mi}$ and $B_{nj,mi}$, eqn (4.64), the sum rule becomes

$$S_{EW} = \bar{\varepsilon}\sum_{nj} |a_{nj}|^2, \tag{4.71}$$

which just equals the value of the energy-weighted sum for the unperturbed particle-hole basis states, thereby indicating that S_{EW} is conserved in RPA. This value, eqn (4.71), also equals E_1T_1 from eqn (4.70), indicating that the untrapped solution is highly collective, exhausting 100 per cent of the energy-weighted sum rule in the limit of degenerate particle-hole energies. Thus in this limit, the trapped solutions have zero transition strength.

This schematic model shows how the deficiencies in the TDA theory mentioned at the start of section (4.7) are remedied in the RPA. The first deficiency was that the energy-weighted sum S_{EW} for electric dipole absorption exceeded experimental measurements. RPA reduces this sum. The second deficiency was that the TDA failed to push the collective T=0 solutions to sufficiently low energies and failed to endow them with sufficient transition strength. Again RPA corrects this fault by enhancing the transition strength by $\bar{\varepsilon}/E$ (see eqn (4.70), where this factor is typically in the range 2 to 4).

4.14 RPA CALCULATIONS

Among the first RPA calculations to be performed were those by Brown, Evans and Thouless (1961) who studied the 3^- T=0 octopole vibrations in ^{16}O and ^{40}Ca using the TDA and RPA methods. A zero-range residual interaction was chosen, and the particle-hole matrix elements exhibit the schematic model factorization. The effect of replacing the TDA by the RPA was to lower the energy of the 3^- T=0 state by 1 MeV and to increase its collectivity by a factor of two.

Some of the most extensive RPA calculations using simple effective residual interactions have been those of Gillet and co-workers (Gillet 1964, Gillet and Vinh Mau 1964, Gillet and Sanderson, 1964, Gillet and Melkanoff 1964, Gillet, Green and Sanderson 1966). The closed-shell nuclei studied were ^{12}C, ^{16}O, ^{40}Ca, and ^{208}Pb. The interaction used had a Gaussian radial dependence; the strength of the interaction and the parameters of the exchange force mixtures were obtained by a least squares fit to selected states in ^{12}C and ^{16}O. Both the TDA and RPA approximations were used; however, the parameters obtained showed a smaller variation from ^{12}C to ^{16}O in RPA than in TDA. Both approximations gave an equally good fit to the experimental energy levels; however, the transition rates for the low-lying collective states were better fitted by RPA. For the non-collective states the RPA and TDA gave essentially the same results.

More recently particle-hole calculations have been performed using realistic interactions derived from free two-nucleon scattering data. Mavromatis, Markiewicz, and Green (1967) calculated the spectra for the odd-parity states in ^{16}O using the Kallio-Kolltveit (1964) and Kuo-Brown (Kuo and Brown 1966) interactions. Considering that there were no adjustable parameters, the fit to the energy levels was reasonable. However, one problem with using a realistic reaction matrix to calculate the properties of finite nuclei is that the truncation in the size of the shell model space which necessarily has to be introduced requires the reaction matrix to be renormalized. This problem is discussed in the next chapter. Nevertheless Kuo (1968) has calculated the particle-hole states in ^{208}Bi using the reaction matrix elements deduced from the Hamada-Johnston potential and finds that the energy levels calculated both with and without these model space renormalizations are in very close agreement with each other and with 36 experimentally identified low-lying energy levels. However, in a subsequent calculation (Kuo, Blomqvist, and Brown 1970) for ^{208}Pb, the position and concentration of strength in the giant dipole resonance were poorly reproduced. This was true also of the earlier results of Gillet, Green and Sanderson (1966).

All the above-mentioned calculations used harmonic oscillator single-particle wavefunctions rather than Hartree-Fock functions. One calculation by Perez (1969) has employed Saxon-Woods functions. In this case an effective interaction of Yukawa radial dependence was chosen, with the strength and exchange force parameters being determined by least squares fits to energy levels in ^{16}O and ^{40}Ca, along the lines adopted by Gillet (1964). However, when this interaction was used in ^{208}Pb, Perez (1970) was able to get the giant dipole resonance at the correct position, providing the neutron radius parameter in the Saxon-Woods well was smaller than the corresponding proton radius parameter. Perez argued that using oscillator functions generated from a common oscillator potential implied a neutron r.m.s. radius some 10 per cent larger than the proton distribution, whereas his prescription yields similar neutron and proton r.m.s. radii, a result which has some experimental support.

Other calculations with Saxon-Woods functions have been performed by Ring and Speth (1973). Their technique is to diagonalize the Saxon-Woods potential in a finite space of harmonic oscillator functions. This procedure does allow for the possibility that particle states may be unbound and lie in the continuum. Up to now we have only discussed particle-hole calculations in which the particle states are bound with discrete energies. Ring and Speth's technique amounts to a discretization of the continuum and so, in common with previous calculations, does not allow one to calculate the width of the resonances.

Raynal, Melkanoff, and Sawada (1967) and Buck and Hill (1967) have given a generalization of the standard spectroscopy calculation to include continuum states. Their generalizations lead to coupled integral equations, which can be solved easily, and can predict the width of a giant resonance. In practice, however, many channels have to be coupled and this ultimately limits the practicality of the method.

5
EFFECTIVE INTERACTIONS

5.1 INTRODUCTION

So far we have only discussed closed-shell nuclei, their ground state properties and simple particle-hole excitations. Now let us introduce one or more additional particles, realizing that the lowest energy configurations inevitably will comprise some partially-occupied shells, and let us attempt to calculate the properties of these so-called open-shell nuclei.

Again, we could proceed as with closed-shell nuclei, and assume that the sought eigenfunction of the Hamiltonian, H, can be approximated by a single Slater determinant, and then use the Hartree-Fock procedure to find the optimum determinant that minimizes the energy. There is, however, a difficulty for open-shell nuclei in that the determinant does not in general have definite total angular momentum. Since the Hamiltonian, H, is rotationally invariant, its eigenfunctions are eigenfunctions of the angular momentum operator. The Hartree-Fock solution, on the other hand, frequently is deformed, so a projection procedure has to be invoked in order to recover states of specific angular momentum. Much work has been channelled into this direction, but we shall not discuss it further here.

An alternative procedure is the shell model approach. First a simple one-body Hamiltonian H_0 is selected, whose many-body eigenfunctions are, as before, single Slater determinants. Next, a set of basis states with definite angular momentum are constructed by taking the appropriate linear combinations of determinants. Finally, the true eigenstates of the system are expanded in terms of these basis states, and degenerate perturbation theory in the residual interaction, $H-H_0$, used to find the mixing amplitudes. No variation principle is involved and no energy minimization is performed.

The difficulty with the shell model approach is that the

complete set of basis states span an infinite-dimensioned Hilbert space and some form of truncation is inevitable. Exactly how the truncation is chosen is a matter of expedience. For example, in a nucleus of A particles, it might be assumed that N of these particles arrange themselves into closed shells and form an inert core. The remaining n nucleons are allowed to populate a few selected orbitals and all interactions between the n nucleons are actively incorporated into the calculation. The truncation has been two-fold. First the number of nucleons active in the problem has effectively been reduced from A to n, and secondly the number of orbitals being populated has been reduced from an infinite number to a small finite number. Now one has a tractable calculation and the hope is that the calculated eigenfunctions closely resemble the true eigenfunctions, or rather that the calculated matrix elements for observable quantities should correspond with experiment.

It is quite obvious that a Hamiltonian acting in an infinite sized Hilbert space will produce different eigenfunctions from the same Hamiltonian acting in a finite model space. So inevitably there will be differences; but the hope is that for any calculated observable involving the lowest few eigenfunctions, these differences will be small. Unfortunately this cannot always be guaranteed.

Consider an extreme example. The nucleus ^{17}O might be considered in a shell model calculation as just one neutron outside a closed-shell ^{16}O core. Since this neutron carries no charge, one has the model prediction that the electric quadrupole moment of ^{17}O should be zero. In practice, this is not so. The nucleus has a quadrupole moment comparable to that of a single proton.

There are two ways to resolve this dilemma. One is, obviously, to enlarge the model space and break the ^{16}O core, but this very quickly becomes unmanageable. The second way is to try and salvage the small model space calculation by asserting that in a model calculation, one should be using model operators rather than true operators.

This now poses an interesting existence problem. Is it

possible to replace the true Hamiltonian H acting in an infinite-dimensioned Hilbert space by a modified or effective Hamiltonian, H_{eff}, acting only in a finite-sized model space, such that the eigenvalues of H_{eff} inside the model space are the same as some of the eigenvalues of the true H in the entire space? In addition, can an effective transition operator F_{eff} be found such that the matrix elements of F_{eff} between eigenstates of H_{eff} are the same as the matrix elements of the true operator F between the corresponding eigenstates of the true H?

Bloch and Horowitz (1958) examined just this question. They demonstrated that degenerate perturbation theory could be used for open-shell nuclei to produce a secular equation involving an effective Hamiltonian in a truncated model space whose solutions were the true energies of the system. The effective interaction, however, depended on the energy eigenvalue being sought. The results of Bloch and Horowitz were generalized by Brandow (1967), who showed how the energy dependence of the effective interaction could be eliminated by introducing a special type of diagram, the so-called 'folded diagram'. The resulting energy-independent effective interaction is non-Hermitian but has the same eigenenergies and eigenvectors as the Bloch-Horowitz effective interaction.

These theories give a justification for the standard shell model procedure in which effective interactions are used in truncated model spaces. They also trace via perturbation theory how the effective interaction is derived from the 'real' nucleon-nucleon interaction. There are, however, many problems and uncertainties remaining in the theories and these are the subject of current research. In this chapter we cannot do justice to all this endeavour; instead we give a very schematic treatment based on the review articles of Macfarlane (1967) and of Barrett and Kirson (1973). For a more rigorous approach, the reader is referred to Brandow (1967, 1969). These reviews all use a time-independent perturbation theory. An alternative time-dependent derivation has been given by Johnson and Baranger (1971), by Kuo, Lee and Ratcliff (1971), and Kuo (1974). However, we will not discuss these time-dependent theories at all, but will follow the Bloch-Horowitz-Brandow procedure.

5.2 THE BLOCH-HOROWITZ EQUATIONS

Let us denote the true Hamiltonian for the system by $H = T+V$ where T is the total kinetic energy and V the bare nucleon-nucleon interaction. An alternative division of the Hamiltonian is

$$H = H_0 + H_1, \tag{5.1}$$

where H_0 is a simple Hamiltonian whose energies and eigenvectors are known, i.e.

$$H_0 \Phi_i = E_i \Phi_i, \tag{5.2}$$

and whose lowest eigensolution E_0 and Φ_0 represents the unperturbed ground state of the system. It is convenient to take H_0 as a purely one-body Hamiltonian $H_0 = T+U$ with U being some single-particle potential; then the residual interaction in eqn (5.1) is

$$H_1 = V - U.$$

The Schrödinger equation for the exact state of the system is

$$H\Psi = E\Psi \tag{5.3}$$

and the aim is to use perturbation theory to relate the exact solution to the known unperturbed solutions of H_0.

The characteristic method of solving eqn (5.3) is to expand Ψ in terms of the complete set of solutions to the simpler Hamiltonian H_0, namely

$$\Psi = \sum_i^{\infty} a_i \Phi_i. \tag{5.4}$$

Inserting eqn (5.4) back into eqn (5.3) and left-multiplying by Φ_i, one obtains

$$(E-E_i)a_i = \langle \Phi_i | H_1 | \Psi \rangle$$

$$= \sum_j^\infty \langle \Phi_i | H_1 | \Phi_j \rangle \, a_j. \tag{5.5}$$

This is an infinite set of linear equations for the coefficients a_i and some form of truncation has to be introduced. Accordingly a certain number d of the Φ_i are selected. This reduced subset spans a much smaller (and now finite) model space, which will be denoted by D. This division of the complete Hilbert space into a model space and an excluded space can be accomplished by means of projection operators:

$$P = \sum_{i \in D} |\Phi_i\rangle\langle\Phi_i|$$

$$Q = \sum_{i \notin D} |\Phi_i\rangle\langle\Phi_i|. \tag{5.6}$$

The operator P projects on to the model space D and the complementary operator Q projects out of D such that $P+Q = I$ (infinite unit operator), $PQ = QP = 0$, $P^2 = P$ and $Q^2 = Q$. Since P and Q are defined in terms of the eigenstates of H_0, they commute with H_0, so that $PH_0Q = QH_0P = 0$, and PH_0P and QH_0Q are diagonal. That is, there are no matrix elements of H_0 connecting the model space with the excluded space.

Next we introduce a 'Green's function' defined as

$$\mathscr{G} = \sum_{i \notin D} \frac{|\Phi_i\rangle\langle\Phi_i|}{E-E_i} = \frac{Q}{E-H_0} \tag{5.7}$$

and note the identity

$$\mathscr{G} H_1 \Psi = \sum_{i \notin D} \frac{|\Phi_i\rangle\langle\Phi_i | H_1 | \Psi\rangle}{E-E_i}$$

$$= \sum_{i \notin D} a_i |\Phi_i\rangle$$

$$= Q\Psi \tag{5.8}$$

where eqn (5.5) has been used. Then the total wavefunction can be written in the form

$$\Psi = (P+Q)\Psi$$

$$= \Psi_D + \mathcal{G}H_1 \Psi \tag{5.9}$$

where

$$\Psi_D = P\Psi = \sum_{i \in D} a_i \Phi_i \tag{5.10}$$

is the 'model wavefunction', that is the projection of the true wavefunction on to the model space. Finally we introduce a wave operator Ω, such that

$$\Psi = \Omega \Psi_D. \tag{5.11}$$

and an effective interaction

$$\mathcal{V} = H_1 \Omega. \tag{5.12}$$

Substituting eqn (5.11) in eqn (5.9) leads to an integral equation for Ω,

$$\Omega = 1 + \mathcal{G}H_1 \Omega,$$

which can be iterated:

$$\Omega = 1 + \mathcal{G}H_1 + \mathcal{G}H_1 \mathcal{G}H_1 + \ldots$$

$$= \sum_{r=0}^{\infty} (\mathcal{G}H_1)^r. \tag{5.13}$$

Similarly an integral equation for $\mathcal{V}$ can be found

$$\mathcal{V} = H_1 + H_1\mathcal{G}\mathcal{V} \tag{5.14}$$

which also can be iterated

$$\mathscr{V} = H_1 + H_1 \frac{Q}{E-H_0} H_1 + H_1 \frac{Q}{E-H_0} H_1 \frac{Q}{E-H_0} H_1 + \cdots$$

$$= \sum_{r=0}^{\infty} H_1(\mathscr{G}H_1)^r. \tag{5.15}$$

Returning to eqn (5.5) one sees that

$$\begin{aligned}(E-E_i)a_i &= \langle \Phi_i | H_1 | \Psi \rangle \\ &= \langle \Phi_i | H_1 \Omega | \Psi_D \rangle \\ &= \langle \Phi_i | \mathscr{V} | \Psi_D \rangle \\ &= \sum_{j \in D} \langle \Phi_i | \mathscr{V} | \Phi_j \rangle a_j \end{aligned} \tag{5.16}$$

The equation holds for all a_i, in particular it holds for the d model amplitudes a_i, $i \in D$. These d equations may be combined to give the d×d matrix equation

$$[H_0 + \mathscr{V}(E) - EP]\underset{\sim}{A} = 0 \tag{5.17}$$

where $\underset{\sim}{A}$ is the column vector of the model amplitudes a_i, and P just represents the unit operator within the model space D. Remember that $P\Psi_D = P^2\Psi = P\Psi = \Psi_D$. The energy dependence of the effective interaction $\mathscr{V}(E)$ has been explicitly displayed. It is useful to rewrite eqn (5.14) in terms of the basis spanned by the eigenstates of H_0. Then the matrix elements of the effective interaction required in eqn (5.16) satisfy

$$\langle \Phi_i | \mathscr{V}(E) | \Phi_j \rangle = \langle \Phi_i | H_1 | \Phi_j \rangle + \sum_{k \notin D} \frac{\langle \Phi_i | H_1 | \Phi_k \rangle \langle \Phi_k | \mathscr{V}(E) | \Phi_j \rangle}{E-E_k}. \tag{5.18}$$

Note the similarity between this equation for the effective interaction and the Bethe-Goldstone equation for the G-matrix.

Equations (5.14) and (5.17) are the Bloch-Horowitz equations. They are the desired result; namely they represent a shell model secular equation in a truncated model space whose eigensolutions give the true energies of the system. The equations are exact, all the problems being hidden in the construction of the effective interaction $\mathscr{V}$, the most troublesome being that $\mathscr{V}$ depends on the energy E. Bloch and Horowitz (1958) have studied the analytic properties of the matrix $[H_0 + \mathscr{V}(z) - zP]$ for complex z. Their most important result is that the eigenvalues are all real. The eigenvectors $\Psi_D = P\Psi$, however, are *not* orthogonal, because they are eigenvectors of different Hamiltonians, i.e. $\mathscr{V}(E)$ differs for each eigenvalue E.

There have been very few quantitative studies of the energy-dependence of the effective interaction. The most common procedure has been to replace the eigenvalues E in the energy denominator of eqn (5.15) by a suitable average; the same effective interaction is then obtained for all eigenstates in the model space and the eigenvectors become orthogonal. This averaging process is reasonable only if the excitation energies $E-E_k$ of all important states Φ_k in eqn (5.18) are large compared with the spread in the desired eigenvalues E. The approximation breaks down (Macfarlane 1967) whenever in the midst of energy levels to be treated within a certain model space there lies a highly collective state arising predominantly from higher configurations. The same conclusion is reached by Schucan and Weidenmuller (1972) who make use of an algebraic model to assert further that whenever the eigenvalues of the excluded space QHQ overlap the spectrum of eigenvalues for the effective Hamiltonian, then the iterative solution to eqn (5.14) for the effective interaction $\mathscr{V}$ may well diverge. Unfortunately this condition occurs in almost all physical cases of interest. The only remedy is to expand the model space, but this may not be practical.

5.3 GOLDSTONE EXPANSION

The form of degenerate perturbation theory presented in the last section reduces to the Brillouin-Wigner (BW) perturbation theory when the model space D contains only one state, say Φ_0. This is the limit in which degenerate perturbation theory reduces to the non-degenerate theory. It is useful to consider this limit using the example of the ground state energy of a closed-shell system. This example was extensively discussed in section 3.1, where the Rayleigh-Schrödinger (RS) form of perturbation theory was used. The purpose of this section is to illustrate the relationship between the two perturbation theories.

From eqn (5.16), restricted now to only one state Φ_0 in the model space, the BW expression for the energy shift becomes

$$\Delta E = E - E_0 = \langle \Phi_0 | \mathscr{V} | \Phi_0 \rangle$$

$$= \langle \Phi_0 | H_1 + H_1 \frac{Q}{E-H_0} H_1 + H_1 \frac{Q}{E-H_0} H_1 \frac{Q}{E-H_0} H_1 \ldots | \Phi_0 \rangle. \quad (5.19)$$

Contrast this with eqn (3.3) for the RS expression. The various terms here may be conveniently represented by diagrams of the Goldstone form, as shown in Fig. 5.1. Hartree-Fock self-

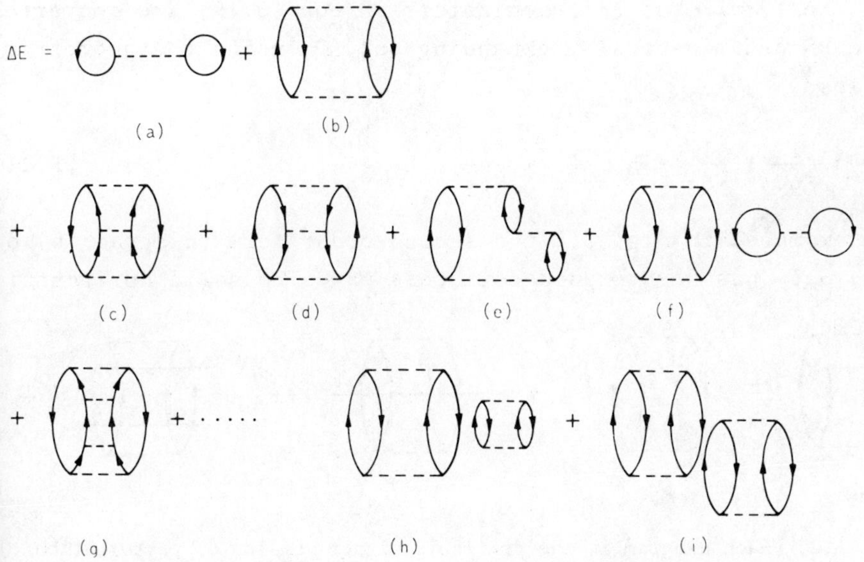

IG. 5.1. Diagram expansion for the shift in the energy of a closed-shell tate evaluated using the Brillouin-Wigner version of perturbation theory. ote the presence of unlinked diagrams, e.g. diagrams (f), (h), and (i). he energy denominators associated with the intermediate states are xpressed in terms of the true energy rather than the unperturbed energy.

consistency has been assumed, so no bubble insertion diagrams appear. Note the occurrence of unlinked diagrams in third and higher orders. The rules for interpreting these diagrams are as discussed in appendix (B), but one exception should be noted. Here we are using BW perturbation theory, consequently BW energy denominators occur. True Goldstone diagrams use RS energy denominators. They are related as follows:

$$e_{BW} \equiv E-H_0 = E_0+\Delta E-H_0 = e_{RS}+\Delta E \tag{5.20}$$

where $e_{RS} = E_0-H_0$ represents the RS denominators, which according to the rules for closed-shell diagrams have the form

$$e_{RS} = \sum \varepsilon_h - \sum \varepsilon_p,$$

with ε_h being the unperturbed single-hole energies and ε_p the unperturbed single-particle energies.

Next, all the BW denominators in eqn (5.19) are converted into RS denominators by expanding out ΔE, using the geometric series

$$\frac{1}{e_{BW}} = \frac{1}{e_{RS}} + \frac{1}{e_{RS}}(-\Delta E)\frac{1}{e_{RS}} + \frac{1}{e_{RS}}(-\Delta E)\frac{1}{e_{RS}}(-\Delta E)\frac{1}{e_{RS}} + \dots \tag{5.21}$$

For example, in Fig. 5.2 the second-order term (diagram (b) of Fig. 5.1) has been expanded in this way. The solid horizontal

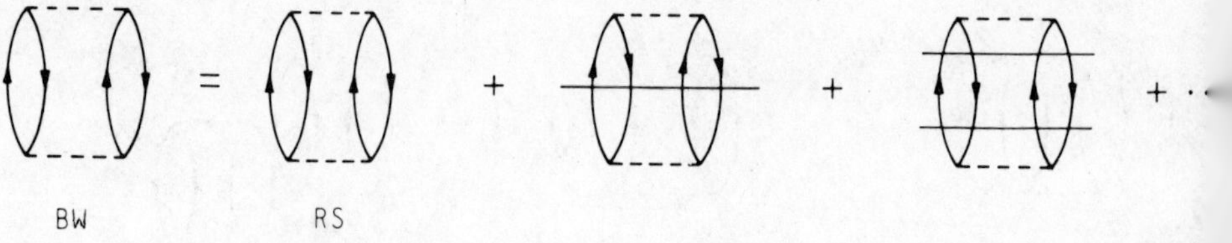

BW = RS + ... + ... + ...

FIG. 5.2. Each diagram in the Brillouin-Wigner version of perturbation theory can be represented as a sum of conventional Rayleigh-Schrödinger diagrams by expressing the BW energy denominators $(e_{RS} + \Delta E)^{-1}$ in terms of the RS energy denominators, e_{RS}^{-1}. The solid horizontal bars represent 'insertions' of $-\Delta E$.

bars here represent 'insertions' of $(-\Delta E)$. When this conversion is carried out systematically for every denominator in the original BW expansion, one finds that diagrams with $(-\Delta E)$ insertions cancel the unlinked diagrams. The only diagrams which survive are those consisting of a single linked part with no $(-\Delta E)$ insertions. This is just the famous Goldstone result.

As an example, consider the second diagram on the right-hand side in Fig. 5.2. The $(-\Delta E)$ insertion is replaced by the expansion in Fig. 5.1 and the result is shown schematically in Fig. 5.3. Diagram (a) here exactly cancels diagram (f) of Fig. 5.1 and diagram (b) exactly cancels diagrams (h) and (i) of Fig. 5.1. Note that the BBP factorization theorem (see section (3.3)) is used with diagrams (h) and (i) to effect the required cancellation.

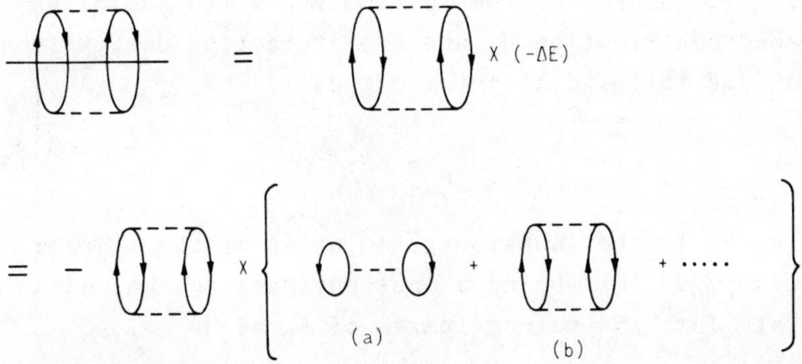

FIG. 5.3. An illustrative example of the Goldstone theorem, in which a diagram with a $-\Delta E$ insertion exactly cancels an unlinked diagram. For example, diagram (a) here cancels diagram (f) of Fig. 5.1.

5.4 EXTENSION TO OPEN-SHELL NUCLEI

We now apply the expansion (5.18) to a degenerate many-body system such as an open-shell nucleus. The model subspace is chosen in the standard shell model way. Consider a nucleus with A = N+n particles. The unperturbed wavefunctions Φ_i are defined to be all those A-body Slater determinants in which the N core particles all occupy a corresponding set of N core orbitals, while the remaining n valence particles are all located somewhere within a specified band of valence orbitals. Our goal is the calculation of the matrix elements $\langle \Phi_i | \mathscr{V}(E) | \Phi_j \rangle$.

The approach adopted by Brandow (1967, 1969) is to analyse the perturbation expansion obtained by an iterative solution of eqn (5.18) in terms of diagrams. In drawing these diagrams, our chosen closed-shell core represents the vacuum state, so that only the history of the n valence nucleons, represented initially and finally as incoming and outgoing external lines, need be drawn. That is, one only indicates the deviations from the filled-core situation. As in the last section these diagrams differ from true Goldstone diagrams in that they contain E rather than E_0 in their energy denominators.

The total energy of the system, E, is the sum of an unperturbed contribution E_0 and an interaction energy (energy shift) ΔE due to residual interactions:

$$E = E_0 + \Delta E. \tag{5.22}$$

The energy E_0 is the lowest eigensolution of the simpler Hamiltonian H_0. This being a one-body Hamiltonian, it is easy to separate core and valence parts of E_0 as

$$E_0 = E_0{}^c + E_0{}^v \tag{5.23}$$

where $E_0{}^c$ is the unperturbed energy of the N-body core and $E_0{}^v$ is a sum of the n unperturbed single-particle energies of the valence nucleons. Similarly the energy shift is separated into core and valence terms

$$E = \Delta E^c + \Delta E^v. \tag{5.24}$$

We specify the core interaction energy ΔE^c below, and eqn (5.24) is used to determine uniquely the valence interaction energy ΔE^v. The elimination of the energy dependence in the effective interaction now proceeds in two stages:

(i) The core parts are entirely separated from valence

parts and a reduced shell model secular equation is obtained in which the effective interaction depends only on the energies of the valence nucleons.

(ii) This lingering energy dependence is then eliminated with the introduction of 'folded diagrams'.

These steps are achieved by manipulating the diagrams representing the perturbation expansion of eqn (5.18). Four types of diagrams can occur and these are illustrated in Fig. 5.4 for the case of three valence particles. Each diagram is

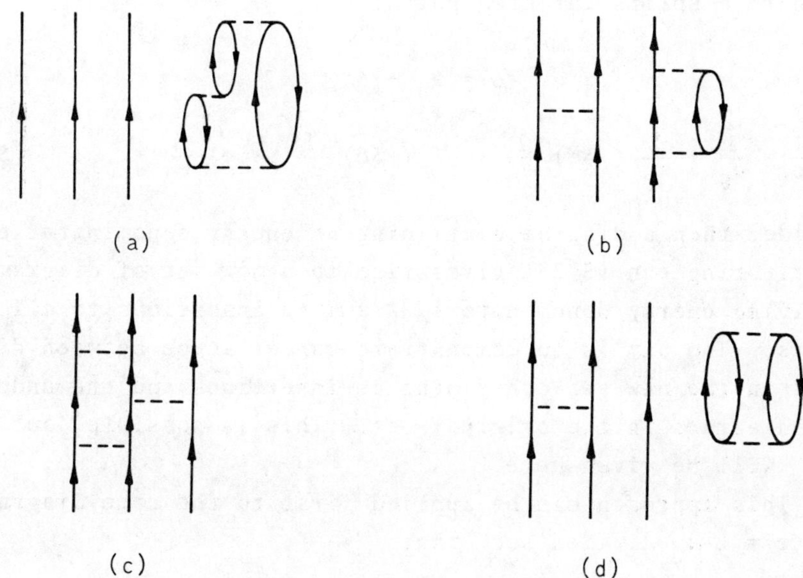

FIG. 5.4. A variety of Goldstone diagrams for open-shell nuclei. Diagram (a) is a 'core diagram', (b) a 'disconnected valence diagram', (c) a 'linked valence diagram', and (d) a 'mixed diagram'.

third order in the perturbation $H_1 = V-U$. Note that the operator Q in eqn (5.15) excludes any intermediate state which is an eigenstate of H_0 and which is contained in the model space, D.

In Fig. 5.4, diagram (a) is a 'core diagram' in which none of the V-interactions is attached to an external line; the valence particles are simply spectators, while all the interactions occur in the core. Diagram (b) is a 'disconnected valence diagram' in which all the V-interactions are attached either directly or indirectly to external lines, but the diagram divides into two distinct pieces. Diagram (c) is a 'linked valence diagram' and diagram (d) is a 'mixed diagram' in which some of the V-interactions are attached to external lines and some are not.

The first manipulation of these diagrams leads to the elimination of mixed diagrams. The basic mathematical tool is once again the algebraic identity for an energy denominator e, where e splits into two parts:

$$e = e_0 + \delta e$$

$$\frac{1}{e} = \frac{1}{e_0} + \frac{1}{e_0}(-\delta e)\frac{1}{e_0} + \frac{1}{e_0}(-\delta e)\frac{1}{e_0}(-\delta e)\frac{1}{e_0} + \ldots \qquad (5.25)$$

Consider then a diagram containing an energy denominator e. Substituting eqn (5.25) gives rise to a new set of diagrams involving energy denominators e_0 and δe-insertions to all orders. The aim is to demonstrate cancellation between diagrams in the new set containing δe-insertions and the undesirable diagrams in the original set. This is possible, but no proof will be given here.

This approach can be applied first to the core diagrams with $e = E-H_0$ divided such that

$$\delta e = \Delta E^V \qquad e_0 = E_0 + \Delta E^c - H_0,$$

and second to the valence diagrams with

$$\delta e = \Delta E^c \qquad e_0 = E_0 + \Delta E^V - H_0.$$

Then core diagrams with ΔE^V-insertions and valence diagrams with ΔE^c-insertions exactly cancel the mixed diagrams of the original expansion. One is now left with an expansion for $\mathcal{V}$ containing only core diagrams with energy denominators $(E_0 + \Delta E^c - H_0)$ and valence diagrams with energy denominators $(E_0 + \Delta E^V - H_0)$.

Consider first all the surviving core diagrams. They are characterized by having n external valence lines propagating undisturbed through the diagrams (as in diagram (a), Fig. 5.4). The core interaction energy ΔE^c is now specified. It is defined as the ground-state energy shift of the core nucleus in the absence of the n valence nucleons - that is, ΔE^c is given by the Goldstone linked cluster expansion for the N-body core with standard Rayleigh-Schrödinger energy denominators. With this definition, the surviving core diagrams are trivially summed, resulting in an effective interaction $\mathcal{V}(E)$ given by

$$\mathcal{V}(E) = \mathcal{V}(E_v) + \Delta E^c P. \tag{5.26}$$

The matrix P comes from the non-interacting valence lines; these amount to the unit operator in the model space, D (eqn (5.6)). Substituting this result into eqn (5.17), one obtains a 'reduced' secular equation

$$[H_0{}^v + \mathcal{V}(E_v) - E_v P]\underset{\sim}{A} = 0 \tag{5.27}$$

with $H_0{}^v = H_0 - E_0{}^c$ and $E_v = E_0{}^v + \Delta E^v$.

The effective interaction $\mathcal{V}(E)$ has been replaced by a reduced effective interaction $\mathcal{V}(E_v)$ from which the contributions from the core have been eliminated. Whereas $\mathcal{V}(E)$ was a solution of an integral equation of the Lippmann-Schwinger type (eqn (5.14)), no such integral equation exists for $\mathcal{V}(E_v)$. Instead $\mathcal{V}(E_v)$, depending on ΔE^v, which in turn depends on the Goldstone expansion for ΔE^c via eqn (5.24), can only be defined in terms of a perturbation expansion:

$$\mathcal{V}(E_v) = \sum_{n=0}^{\infty} H_1\left(\frac{Q}{E_v - H_0{}^v} H_1\right)^n. \tag{5.28}$$

That is, $\mathcal{V}(E_v)$ is determined as the sum of all valence diagrams, both linked and unlinked, with energy denominators given by $(E_v - H_0{}^v)$. The reduced effective interaction, therefore, is still energy-dependent.

This completes the first stage of the analysis, the

desired separation of core and valence contributions to the effective interaction having been achieved. The advantage of this separation is that if one is only interested in the nuclear excitation energies (i.e., energies relative to the closed-shell core), then it is never necessary to calculate ΔE^c. This quantity does not appear in the reduced secular equation (5.27) for the total valence eigenvalue $E_v = E_0^v + \Delta E^v$.

The determination of the reduced secular equation may not seem to be much of an advance (except for the elimination of core and mixed diagrams), since the two main difficulties associated with the full interaction $\mathscr{V}(E)$ are still present in the reduced effective interaction $\mathscr{V}(E_v)$. These difficulties are (i) that both linked and unlinked valence diagrams appear in the diagram representation, and (ii) that the perturbation expansion for $\mathscr{V}(E_v)$, eqn (5.28), is of the Brillouin-Wigner type with energy denominators $(E_v - H_0^v)$. It would be nice if these two problems could be made to cancel each other by using the standard trick of expanding ΔE_v out of the denominators. This is what happens for non-degenerate systems where diagrams with ΔE_v-insertions exactly cancel the unlinked diagrams. However, for degenerate systems this cancellation is not complete. There are some remaining terms which can be represented by linked diagrams of a special type — the so-called folded diagrams. The essence of the difficulties encountered with degenerate systems is that there is no simple diagram expansion for a ΔE_v-insertion. Instead ΔE_v is given by $\underset{\sim}{A}^\dagger\, \mathscr{V}(E_v)\underset{\sim}{A}$ indicating that the problems of calculating $\mathscr{V}(E_v)$ are intermixed with those of finding the eigenvectors $\underset{\sim}{A}$.

Brandow's solution to the problem proceeds as follows. Set E_v equal to one of the desired eigenvalues, say $E_{v\alpha}$, and solve the Hermitian secular problem

$$[H_0^v + \mathscr{V}(E_{v\alpha}) - E_\beta^{(\alpha)}P]\underset{\sim}{A}_\beta^{(\alpha)} = 0 \qquad (5.29)$$

to obtain a set of d eigenvalues $E_\beta^{(\alpha)}$, $\beta = 1 \ldots d$, and a set of d eigenvectors $\underset{\sim}{A}_\beta^{(\alpha)}$ which are now strictly orthogonal. Remember that $\underset{\sim}{A}$ is the column vector of amplitudes a_i in the expansion of the required wavefunction Ψ_D in terms of the model basis states, i.e. $\Psi_D = \sum_i a_i \Phi_i$. One of the eigenvalues, say $E_\alpha^{(\alpha)}$, corresponds to the original choice $E_{v\alpha}$, all the other solutions having no physical significance. The eigenvectors $\underset{\sim}{A}_\beta^{(\alpha)}$ can be normalized and form a complete set, that is

$$\begin{aligned} \underset{\sim}{A}_\beta^{(\alpha)\dagger} \underset{\sim}{A}_\gamma^{(\alpha)} &= \delta_{\beta\gamma} \\ \sum_\beta \underset{\sim}{A}_\beta^{(\alpha)} \underset{\sim}{A}_\beta^{(\alpha)\dagger} &= P, \end{aligned} \tag{5.30}$$

where P is the unit operator in the model space, D.

At this point it is assumed that all the model basis states Φ_i, eigenfunctions of H_0^v, are degenerate. (This, in practice, does not constitute a restriction in the theory, since a state-dependent term can always be added to H_0^v such that its eigenfunctions are degenerate, and subsequently this state-dependent term is subtracted from the residual interaction H_1. Most calculations to date have confined the valence space D to one major oscillator shell, and taking H_0^v as the one-body oscillator Hamiltonian, all the Φ_i are automatically degenerate). With this assumption H_0^v becomes a constant diagonal matrix

$$H_0^v \underset{\sim}{A}_\beta^{(\alpha)} = E_0^v \underset{\sim}{A}_\beta^{(\alpha)} \qquad \text{(any } \alpha,\beta)$$

and as a result of solving eqn (5.29), the matrix $\mathscr{V}(E_{v\alpha})$ is diagonal in the $\underset{\sim}{A}_\beta^{(\alpha)}$ basis, viz.

$$\underset{\sim}{A}_\beta^{(\alpha)\dagger} \mathscr{V}(E_{v\alpha}) \underset{\sim}{A}_\alpha^{(\alpha)} = 0, \qquad \beta \neq \alpha. \tag{5.31}$$

Thus an interaction energy can be defined by

$$\Delta E_\alpha = E_\alpha^{(\alpha)} - E_0^v = \underset{\sim}{A}_\alpha^{(\alpha)\dagger} \mathscr{V}(E_{v\alpha}) \underset{\sim}{A}_\alpha^{(\alpha)}. \tag{5.32}$$

Now expand ΔE_α out of the energy denominators in the perturbation series for $\mathscr{V}(E_{v\alpha})$ and collect together all terms containing exactly r insertions of $(-\Delta E_\alpha)$. We denote this sum by

(r) and write

$$\mathscr{V}(E_{v\alpha}) = \sum_{r=0}^{\infty} \mathscr{V}^{(r)} [-\Delta E_\alpha]^r. \quad (5.33)$$

This is just the Taylor series expansion for the function $\mathscr{V}(E_{v\alpha})$ about the point $E_{v\alpha} = E_0{}^v$, and $\mathscr{V}^{(r)}$ is expressible in terms of the rth derivative of $\mathscr{V}$:

$$\mathscr{V}(r) = \frac{(-)^r}{r!} \left.\frac{\partial^r \mathscr{V}(E_{v\alpha})}{\partial E_{v\alpha}^r}\right|_{E_{v\alpha}=E_0{}^v}. \quad (5.34)$$

The energy denominators in $\mathscr{V}^{(r)}$ are of Rayleigh-Schrödinger form, $(E_0{}^v - H_0)$. For example, the r=0 term is just the original perturbation expansion for $\mathscr{V}(E_{v\alpha})$ with RS denominators replacing the BW ones, viz.

$$\mathscr{V}^{(0)} = H_1 + H_1 \frac{Q}{E_0{}^v - H_0{}^v} H_1 + H_1 \frac{Q}{E_0{}^v - H_0{}^v} H_1 \frac{Q}{E_0{}^v - H_0{}^v} H_1 + \ldots. \quad (5.35)$$

The first derivative term has a series expansion

$$\mathscr{V}^{(1)} = H_1 \frac{Q}{(E_0{}^v - H_0{}^v)^2} H_1 + H_1 \frac{Q}{(E_0{}^v - H_0{}^v)^2} H_1 \frac{Q}{(E_0{}^v - H_0{}^v)} H_1 + $$

$$+ H_1 \frac{Q}{(E_0{}^v - H_0{}^v)} H_1 \frac{Q}{(E_0{}^v - H_0{}^v)^2} H_1 + \ldots \quad (5.36)$$

and so on.

Returning now to eqn (5.33) and inserting eqn (5.32) for ΔE_α, we have

$$\mathscr{V}(E_{v\alpha}) = \sum_{r=0}^{\infty} \mathscr{V}^{(r)} [-\underset{\sim}{A}_\alpha{}^{(\alpha)\dagger} \mathscr{V}(E_{v\alpha}) \underset{\sim}{A}_\alpha{}^{(\alpha)}]^r$$

$$= \sum_{r=0}^{\infty} \mathscr{V}^{(r)} [(-)^r \underset{\sim}{A}_\alpha{}^{(\alpha)\dagger} \mathscr{V}(E_{v\alpha}) \underset{\sim}{A}_\alpha{}^{(\alpha)} A_\alpha{}^{(\alpha)\dagger} \mathscr{V}(E_{v\alpha}) \underset{\sim}{A}_\alpha{}^{(\alpha)} \underset{\sim}{A}_\alpha{}^{(\alpha)\dagger} \ldots \mathscr{V}(E_{v\alpha}) \underset{\sim}{A}_\alpha{}^{(\alpha)}$$

$$= \sum_{r=0}^{\infty} \mathscr{V}^{(r)} [(-)^r \underset{\sim}{A}_\alpha{}^{(\alpha)\dagger} \mathscr{V}(E_{v\alpha}) \sum_\beta \underset{\sim}{A}_\beta{}^{(\alpha)} \underset{\sim}{A}_\beta{}^{(\alpha)\dagger} \ldots \mathscr{V}(E_{v\alpha}) \underset{\sim}{A}_\alpha{}^{(\alpha)}]$$

$$= \sum_{r=0}^{\infty} \mathscr{V}^{(r)} [(-)^r \underset{\sim}{A}_\alpha{}^{(\alpha)\dagger} \mathscr{V}(E_{v\alpha}) P \mathscr{V}(E_{v\alpha}) P \ldots \mathscr{V}(E_{v\alpha}) \underset{\sim}{A}_\alpha{}^{(\alpha)}],$$

where the diagonal property of $\mathscr{V}(E_{v\alpha})$, eqn (5.31), and the closure property of $\underset{\sim}{A}_{\beta}^{(\alpha)}$, eqn (5.30), have been exploited. Thus post-multiplying by $\underset{\sim}{A}_{\alpha}^{(\alpha)}$ and observing that the quantity in square brackets is a scalar $(-\Delta E_{\alpha})$, we get

$$\mathscr{V}(E_{v\alpha})\underset{\sim}{A}_{\alpha}^{(\alpha)} = \sum_{r=0}^{\infty} \mathscr{V}^{(r)}\underset{\sim}{A}_{\alpha}^{(\alpha)}[(-)^r \underset{\sim}{A}_{\alpha}^{(\alpha)\dagger}\mathscr{V}(E_{v\alpha})P\ldots(E_{v\alpha})\underset{\sim}{A}_{\alpha}^{(\alpha)}]$$

$$= \sum_{r=0}^{\infty} \mathscr{V}^{(r)}[-P\mathscr{V}(E_{v\alpha})]^r\,\underset{\sim}{A}_{\alpha}^{(\alpha)}. \tag{5.37}$$

Finally introducing a new matrix $\mathscr{W}_1(\alpha)$ as the matrix product

$$\mathscr{W}_1(\alpha) = \sum_{r=0}^{\infty} \mathscr{V}^{(r)}[-P\mathscr{V}(E_{v\alpha})]^r \tag{5.38}$$

We have that

$$\mathscr{V}(E_{v\alpha})A_{\alpha}^{(\alpha)} = \mathscr{W}_1(\alpha)A_{\alpha}^{(\alpha)} = (\Delta E_{\alpha})A_{\alpha}^{(\alpha)}. \tag{5.39}$$

That is, $\mathscr{V}(E_{v\alpha})$ may be replaced by $\mathscr{W}_1(\alpha)$ in the original secular equation (5.29) without altering the desired eigensolution $\underset{\sim}{A}_{\alpha}^{(\alpha)}$ and $E_{\alpha}^{(\alpha)}$. All the other eigensolutions $\underset{\sim}{A}_{\beta}^{(\alpha)}$ are affected by this replacement, but this is of no consequence to us. Thus the matrix elements of $\mathscr{V}(E_{v\alpha})$ can be replaced by

$$\langle\Phi_i|\mathscr{V}(E_{v\alpha})|\Phi_j\rangle = \langle\Phi_i|\mathscr{W}_1(\alpha)|\Phi_j\rangle$$

$$= \sum_{r=0}^{\infty}\sum_{\substack{k_1k_2\ldots k_r\\ \in D}} (-)^r\langle\Phi_i|\mathscr{V}^{(r)}|\Phi_{k_1}\rangle\langle\Phi_{k_1}|\mathscr{V}(E_{v\alpha})|\Phi_{k_2}\rangle\ldots$$

$$\ldots\langle\Phi_{k_r}|\mathscr{V}(E_{v\alpha})|\Phi_j\rangle.$$

Note that in contrast to usual perturbation expansions, the intermediate states here are entirely within the model space.

Let us see what has been accomplished. In the original expansion for $\mathscr{V}(E_v)$ (eqn (5.28)), the appearance of E_v in the energy denominator first occurs in the term which is second order in H_1. In contrast, its explicit appearance in $\mathcal{W}_1(\alpha)$ is in a term of order $H_1^{\,4}$ arising from $\mathcal{V}^{(1)}[-P\mathcal{V}(E_{v\alpha})]$, namely

$$-H_1 \frac{Q}{(E_0^{\,v}-H_0^{\,v})^2} H_1 \, P \, H_1 \frac{Q}{(E_v - H_0)} H_1 .$$

Thus, by an appropriate reordering of terms, the energy dependence of the effective interaction has been pushed to higher orders in the perturbation series. Repeating this process, i.e. putting $\mathcal{W}_1(\alpha)$ in the secular equation (5.29), and resolving for a particular $E_{v\alpha}$, a new matrix $\mathcal{W}_2(\alpha)$ can be constructed:

$$\mathcal{W}_2(\alpha) = \sum_{r=0}^{\infty} \mathcal{V}^{(r)}[-P\mathcal{W}_1(\alpha)]^r \tag{5.40}$$

in which the first explicit appearance of E_v is now in a term of order $H_1^{\,6}$. This new matrix, when inserted in the original secular equation, still reproduces the desired eigensolution $\underset{\sim}{A}_\alpha^{(\alpha)}$ and $E_\alpha^{(\alpha)}$. Clearly these arguments can be repeated *ad infinitum*, using

$$\mathcal{W}_n(\alpha) = \sum_{r=0}^{\infty} \mathcal{V}^{(r)}[-P\mathcal{W}_{n-1}(\alpha)]^r$$

and in the limit $\mathcal{W}_\infty = \mathcal{W}$, the interaction no longer depends on the label α. Furthermore the ΔE_α-dependence in $\mathcal{W}$ has been completely removed, resulting in a degenerate version of the Rayleigh-Schrödinger perturbation expansion.

To see how this effective interaction is built up, we write out the lowest few terms, starting at the second iteration (eqn (5.40)) and retaining just the r = 0, 1 and 2 insertions. Thus

$$\mathcal{W} \approx \mathcal{V}^{(0)} - \mathcal{V}^{(1)}P\mathcal{W}_1 + \mathcal{V}^{(2)}P\mathcal{W}_1 P\mathcal{W}_1 - \ldots$$

$$\approx \mathcal{V}^{(0)} - \mathcal{V}^{(1)}P\mathcal{V}^{(0)} + \mathcal{V}^{(1)}P\mathcal{V}^{(1)}P\mathcal{V}^{(0)} + \mathcal{V}^{(2)}P\mathcal{V}^{(0)}P\mathcal{V}^{(0)} + \ldots$$

where $\mathcal{W}_1 \approx \mathcal{V}^{(0)} - \mathcal{V}^{(1)}P\mathcal{V}^{(0)}$ has been used. The first term is just the original sum of valence diagrams (both linked and unlinked) but with RS denominators replacing the BW ones. The second term in eqn (5.41) is called the 'once-folded' term and represents a matrix multiplication of eqns (5.35) and (5.36). The third and fourth terms in eqn (5.41) are called 'twice-folded' terms etc.

There is considerable cancellation in the expansion (5.41). For example, those unlinked diagrams in $\mathscr{V}^{(0)}$ having only two separated pieces are cancelled by some of the 'once-folded' diagrams. We illustrate this with a specific example. In Fig. 5.5, diagram (a) corresponds to an unlinked valence diagram whose value is

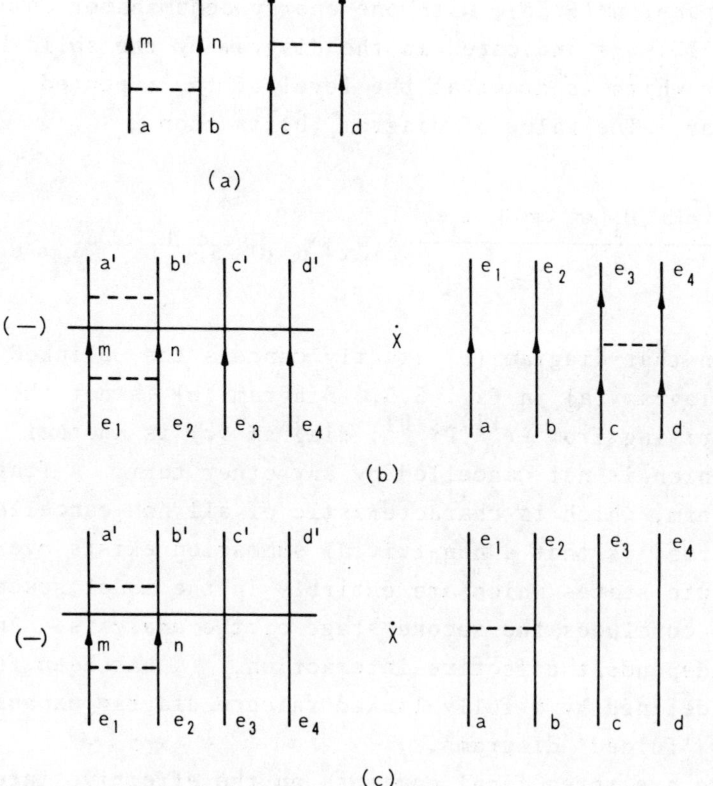

IG. 5.5. An example of an unlinked valence diagram being cancelled by a folded' diagram. Diagram (a) is exactly cancelled by diagram (b). iagram (c) is an example of a folded diagram *not* cancelled by any nlinked valence diagram.

$$\frac{\langle a'b'|H_1|mn\rangle\langle c'd'|H_1|cd\rangle\langle mn|H_1|ab\rangle}{(2\varepsilon-\varepsilon_m-\varepsilon_n)(2\varepsilon-\varepsilon_m-\varepsilon_n)}.$$

We use the notation that a,b,c,d,e all represent single-particle states within the model space D and by the assumptions of the theory, these are all degenerate with $\varepsilon_a = \varepsilon_b = \ldots\ldots = \varepsilon$. Ths single particle states m and n are outside the model space so that $\varepsilon_m \neq \varepsilon$ and $\varepsilon_n \neq \varepsilon$. Diagram (b) of Fig. 5.5 represents the term $-\mathscr{V}^{(1)}\,P\mathscr{V}^{(0)}$ in eqn (5.41), with the left-hand part of the diagram representing $\mathscr{V}^{(1)}$, the right-hand part $\mathscr{V}^{(0)}$, and the symbol $\dot{x}$ denoting the matrix multiplication. Remember that $\mathscr{V}^{(1)}$ is evaluated as indicated by the expansion (5.36), with one energy denominator coming in squared. This is indicated in the diagram by the solid horizontal bar which is drawn at the level of the repeated denominator. The value of diagram (b) is then

$$-\sum_{e_1e_2e_3e_4}\frac{\langle a'b'|H_1|mn\rangle\langle mn|H_1|e_1e_2\rangle}{(2\varepsilon-\varepsilon_m-\varepsilon_n)^2}\delta_{e_3,c'}\delta_{e_4,d'}\langle e_3e_4|H_1|cd\rangle\delta_{e_1,a}\delta_{e_2,b}.$$

It is seen that diagram (b) exactly cancels the unlinked valence diagram (a) in Fig. 5.5. Diagram (b) is not the only diagram arising from $-\mathscr{V}^{(1)}P\mathscr{V}^{(0)}$; diagram (c) is another example which is not cancelled by any other term. A feature of this term, which is characteristic of all non-cancelled folded terms, is that a non-trivial summation exists over intermediate states which are entirely in the model space, D.

This concludes the second stage of the analysis. An energy-independent effective interaction, $\mathscr{W}$, has been found which is defined by a fully-linked valence diagram expansion, including 'folded' diagrams.

There are three final comments on the effective interaction $\mathscr{W}$ that we should like to make.

(1) The eigenvectors $\underset{\sim}{A}_\alpha^{(\alpha)}$ corresponding to energy $E_{v\alpha}$ and the eigenvectors $\underset{\sim}{A}_\beta^{(\beta)}$ corresponding to other desired energies $E_{v\beta}$ are generally *not* orthogonal. This implies tha the effective interaction $\mathscr{W}$ is not Hermitian. This arises,

the diagram notation, from an asymmetry in the folded diagrams; interchanging the labels at the top with those at the bottom does not produce the same result.

(2) The effective interaction has been presented as a perturbation series in the residual interaction $H_1 = V-U$, where V is the nucleon-nucleon interaction. The singular nature of V can be compensated for by first solving the Bethe-Goldstone equation to obtain a reaction matrix G, and then regrouping the series expansion for $\mathscr{W}$ in terms of G such that no two-particle ladder diagrams are retained. All the practical difficulties associated with obtaining G, such as the treatment of the Pauli operator and the choice of intermediate-state spectrum (see chapter 3) are present here. Thus calculations of the effective interaction $\mathscr{W}$ are considerably affected by the methods used for computing G, as well as by the choice of the bare nucleon-nucleon interaction, V.

(3) There does not exist a satisfactory criterion for choosing the single-particle potential U. Most popular has been the Hartree-Fock definition in which every U-interaction in a diagram is cancelled by an appropriate V-interaction in a similar diagram. Thus in the perturbation series, no diagram containing a bubble insertion is retained in the diagram expansion and this results in a great saving in computational labour. When V is replaced by G, the diagram cancellation argument still holds but only for particle-hole and hole-hole matrix elements of U. One difficulty with Hartree-Fock definitions is that every diagram depends on U through the single-particle wavefunctions used in the matrix elements and through the single-particle energies used in the energy denominators. Another difficulty for the effective interaction $\mathscr{W}$ is that the unperturbed states in the model space D have to be degenerate. Hartree-Fock definitions may not satisfy this criterion. For these reasons, it is almost universal to use a harmonic oscillator potential for U, but now there is no justification for omitting the bubble-insertion diagrams other than the hope that oscillator functions resemble Hartree-Fock functions in such a way that the contribution from bubble-insertion diagrams is small.

5.5 RULES FOR FOLDED DIAGRAMS

The lowest few terms in the perturbation expansion for the effective interaction $\mathcal{W}$ are given by:

$$\mathcal{W} = \mathcal{V}^{(0)} - \mathcal{V}^{(1)} P \mathcal{V}^{(0)} + \mathcal{V}^{(1)} P \mathcal{V}^{(1)} P \mathcal{V}^{(0)} + \mathcal{V}^{(2)} P \mathcal{V}^{(0)} P \mathcal{V}^{(0)} \ldots \tag{5.41}$$

Here $\mathcal{V}^{(0)}$ is a sum of all valence diagrams with Rayleigh-Schrödinger energy denominators. Only fully-linked diagrams are retained. The second term represents a matrix multiplication of $\mathcal{V}^{(1)}$ and $\mathcal{V}^{(0)}$ and is called the 'once-folded' term. The third and fourth terms in eqn (5.41) are called 'twice-folded' terms, and in general there will be 'f-folded' terms. Some of these folded diagrams have been used to cancel the unlinked valence diagrams. A prescription for enumerating which folded diagrams remain has been given by Brandow (1965, 1967).

Before discussing the general case, we will first attempt to explain the ideas behind the diagram representation of $\mathcal{W}$ by considering the simple closed-shell-plus-one system. This is a simple case because there is only one valence line and consequently all valence diagrams are necessarily linked; we can therefore forget the problem of cancelling the unlinked diagrams.

The first term in eqn (5.41) for $\mathcal{W}$ is $\mathcal{V}^{(0)}$. This term is illustrated in Fig. 5.6 as the sum of all valence diagrams with Rayleigh-Schrödinger energy denominators. The sum will be denoted collectively by a single box with one incoming and one outgoing external line. These boxes are called valence or interaction blocks by Brandow. Similarly the series $\mathcal{V}^{(1)}$ is denoted by a box with a horizontal bar, the bar denoting the level in the diagram at which the energy denominator come in squared.

Next the term $-\mathcal{V}^{(1)} P \mathcal{V}^{(0)}$ is represented in two ways. First the blocks $\mathcal{V}^{(1)}$ and $\mathcal{V}^{(0)}$ are placed in a vertical column and the valence lines between them are joined. A loop is drawn round this joining line as shown in Fig. 5.6. This is known as the unfolded version of the diagram. The second representation is obtained by bending the joining line to

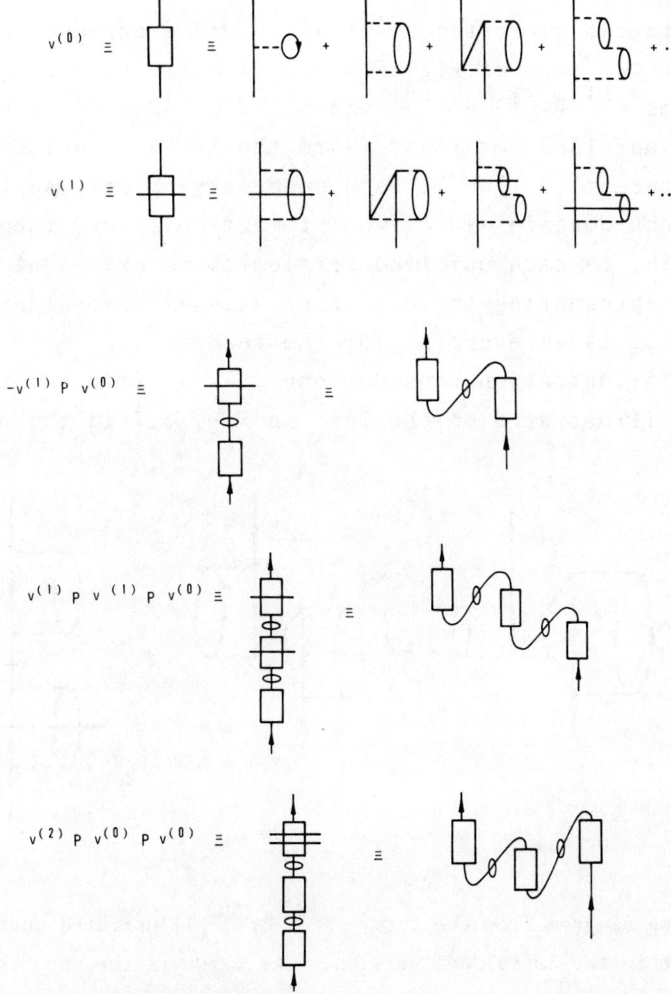

IG. 5.6. A schematic diagrammatic representation of the lowest few terms n the fully-linked expansion for the energy-independent effective interction $\mathscr{W}$, eqn (5.41).

ring the *top* interaction line in the block $\mathscr{V}^{(0)}$ to the same evel as the horizontal bar in the block $\mathscr{V}^{(1)}$. The bar may hen be dropped from the diagram. This is known as the folded ersion of the diagram. All possible time-orderings of the nteraction lines in the block $\mathscr{V}^{(0)}$ relative to the interacion lines in the block $\mathscr{V}^{(1)}$ are allowed, the only restriction eing that the top interaction line of $\mathscr{V}^{(0)}$ occurs at the same level as the repeated energy denominator in $\mathcal{V}^{(1)}$. Thus one unfolded diagram generates a set of folded diagrams. Finally in the last two rows of Fig. 5.6 are illustrated the twice-folded terms $\mathcal{V}^{(1)}P\,\mathcal{V}^{(1)}P\,\mathcal{V}^{(0)}$ and $\mathcal{V}^{(2)}P\,\mathcal{V}^{(0)}P\,\mathcal{V}^{(0)}$ in both folded and unfolded versions. Note the level at which the topmost interaction line in each successive block is placed. If each block contains just two interaction lines, then corresponding to each unfolded version there are eight folded diagrams, representing the different allowed time-orderings. We leave this as an exercise for the reader.

As an illustration consider one diagram from the term $-\mathcal{V}^{(1)}P\mathcal{V}^{(0)}$ illustrated on the left in Fig. 5.7 in the unfolded

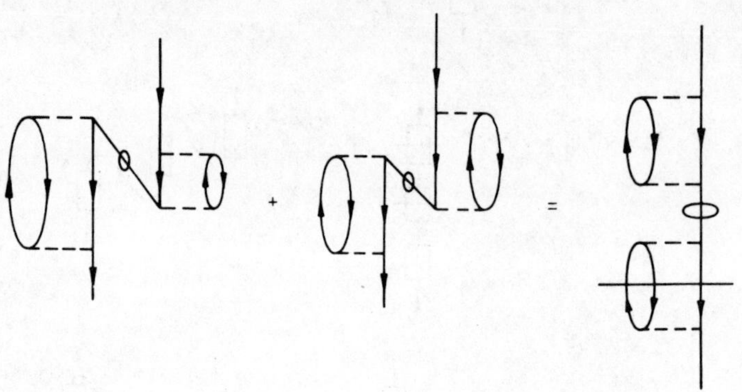

FIG. 5.7. One diagram from the term $-\mathcal{V}^{(1)}P\mathcal{V}^{(0)}$ illustrated on the left of the figure in the 'unfolded' version. Its value is the same as the sum of two 'folded' diagrams on the right of the figure.

version. The energy denominators occurring in this diagram we write as

$$\frac{1}{e_A^{\;2}} \cdot \frac{1}{e_B}$$

where e_A is the energy of the intermediate state in the block $\mathcal{V}^{(1)}$ and e_B that in the block $\mathcal{V}^{(0)}$. This unfolded diagram is equivalent to the two-folded diagrams illustrated on the righ in Fig. 5.7. The folded versions differ in value from each other only through their energy denominators, which are given by the standard Goldstone diagram rules, and are

$$\frac{1}{e_A} \cdot \frac{1}{e_A+e_B} \cdot \frac{1}{e_B} + \frac{1}{e_A} \cdot \frac{1}{e_A+e_B} \cdot \frac{1}{e_A} .$$

The sum of these two terms gives back the value of the unfolded diagram. This is just the factorization theorem (see section 3.3)), used in reverse.

We now quote the diagram rules for more general systems as enounced by Brandow (1965, 1967). Consider a general f-folded diagram. It comprises (f+I) valence or interaction blocks, which may be linked or unlinked, but with no unlinked core parts and no intermediate states lying within the degenerate model space, D. Arrange these blocks in a vertical column, and connect up the valence lines between successive blocks to form an *unfolded* diagram. For example, valence blocks (a) and (b) in Fig. 5.8 are so joined in diagram (c). Now discard all diagrams which are not completely connected, and erase any completely non-interacting valence lines. Draw a loop around each set of valence lines passing between successive blocks. The (f+1) blocks are thus connected by f *bundles* of valence lines. Finally *fold* each of these bundles to form diagrams like (d) and (e) in Fig. 5.8.

As a result of this folding, the interactions of each block can have any relative time order with respect to those in other blocks, subject to two restrictions:

(i) The top interaction of each block must occur above the bottom of the previous block. Neighbouring blocks must either overlap or completely overshoot each other.

(ii) The topmost interaction of the final folded diagram must be identical with the topmost interaction of the original unfolded diagram.

All diagrams satisfying these criteria must be considered.

In drawing the folded diagrams, those external valence lines which pass through one or more blocks without interacting can be removed from the bundles and allowed to leave the diagram by the most direct route.

This completes the prescription for enumerating the

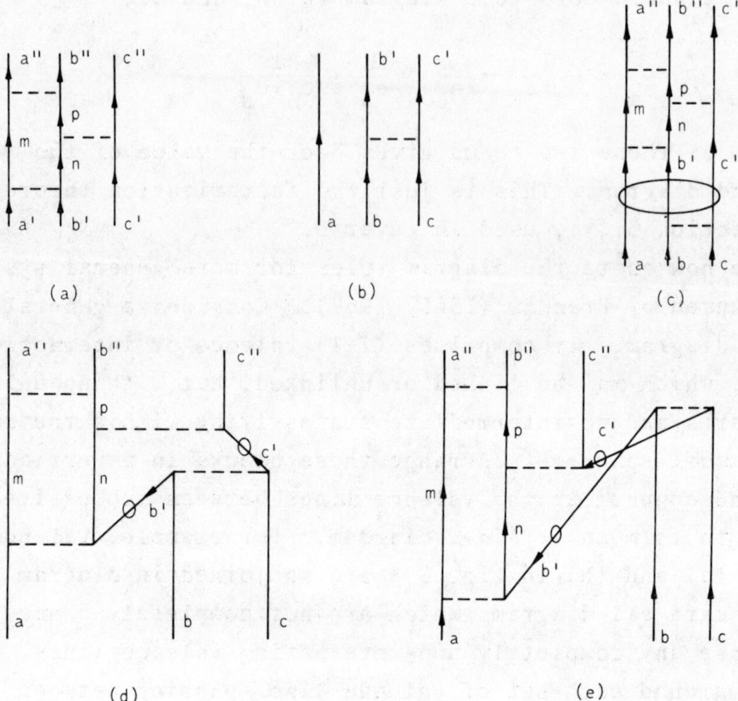

FIG. 5.8. An illustration of the procedure for constructing folded diagrams. Linked valence diagrams (a) and (b) are joined as in (c) to form the unfolded version, which is then subsequently folded to give diagrams such as (d) and (e).

folded diagrams. The set of rules for their evaluation has been relegated to appendix (B.3).

5.6 APPLICATION TO CLOSED-SHELL-PLUS-ONE AND CLOSED-SHELL-PLUS-TWO NUCLEI

The standard calculation for effective interactions takes the following form. For a given nucleus with n particles (usually n=1 or 2) outside some chosen closed-shell core, all the matrix elements of $\mathscr{W}$ between the unperturbed basis states Φ_i, $i \in D$ are evaluated by using the series expansion (5.41) to whatever order required. This resulting set of elements forms a d×d matrix which is diagonalized to find the energies and wavefunctions for the states in question. The success or failure of the calculation is then discussed in terms of comparison with experiment.

As a first application of Brandow's linked-cluster expansion, we consider the energy of a single-particle state in a closed-shell-plus-one nucleus. The relevant diagrams up to third order in the residual interaction, $H_1 = V-U$, are illustrated in Fig. 5.9. The only first order diagram, (a),

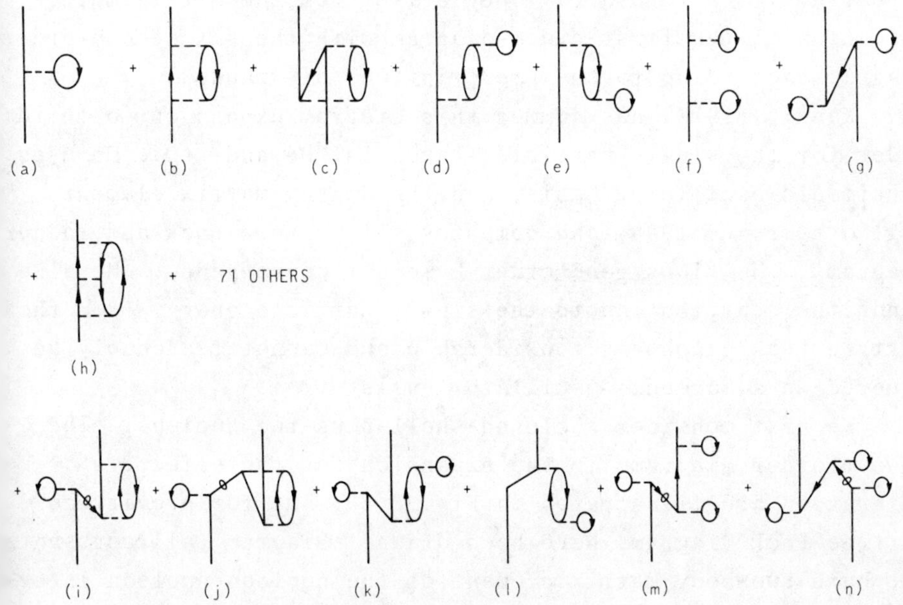

(a) + (b) + (c) + (d) + (e) + (f) + (g)

+ (h) + 71 OTHERS

+ (i) + (j) + (k) + (l) + (m) + (n)

FIG. 5.9. Diagram expansion for the effective interaction in closed-shell-plus-one nuclei. Diagram (a) is the only first-order (in the residual interaction H_1 = V-U) term. Diagrams (b) to (g) are the second-order terms, (h) is a typical third-order term, and (i) to (n) are all the folded diagrams that appear in third order.

represents the interaction of the single-particle with the closed-shell core. If one is using a Hartree-Fock basis, then U has been determined so that

$$\sum_i \langle ai|V|ai\rangle_A = \langle a|U|a\rangle$$

and all diagrams containing bubble-insertions, such as (a), would give zero contribution. We call these Hartree-Fock diagrams. There are six second-order diagrams (b) to (g), four of which are Hartree-Fock diagrams. Diagram (b) is a ladder diagram and does not arise if one is using a reaction matrix G instead of the nucleon-nucleon potential V. There are 72 linked valence third-order diagrams (Kassis 1972), 54 of which are eliminated in a Hartree-Fock basis, and a further 5 are ladder diagrams. There are 6 folded diagrams, (i) to (n), appearing in third order. They arise from the matrix multiplication of the first-order diagram with the six second-order diagrams according to the prescriptions of Brandow.

Kassis (1972) has summed this diagram expansion to third order for the single-particle states in ^{5}He and ^{17}O. He uses a harmonic oscillator basis, and the Sussex matrix elements (Elliott *et al.* 1968) and computes all Hartree-Fock and ladder diagrams. He finds good order-by-order convergence. He also finds the contribution to the single-particle energy from the Hartree-Fock graphs is considerable and cannot reasonably be ignored in a harmonic oscillator basis.

We next consider a closed-shell-plus-two nucleus. The lowest-order diagrams in the expansion for the effective interation are illustrated in Fig. 5.10; but for brevity no Hartree-Fock diagrams have been drawn. Diagram (a) represents the bare two-body matrix element of the nucleon-nucleon interaction, V, or of the reaction G-matrix if the Bethe-Goldstone equation has first been solved. Diagrams (b) to (f) are second order in V (or G). Diagram (b) involves two-particle intermediate states (outside the model space) and is part of the ladder sum used in the calculation of the G-matrix; it should therefore be omitted in any perturbation expansion involving G-matrices. However, there are other procedures. One possibility is to subdivide the space spanned by the two-particle intermediate states into two groups: that spanned by low-lying and that spanned by high-lying intermediate states. The G-matrix is first evaluated using only the high-lying states, making approximations which might now be plausible, such as regarding these intermediate states as plane waves an

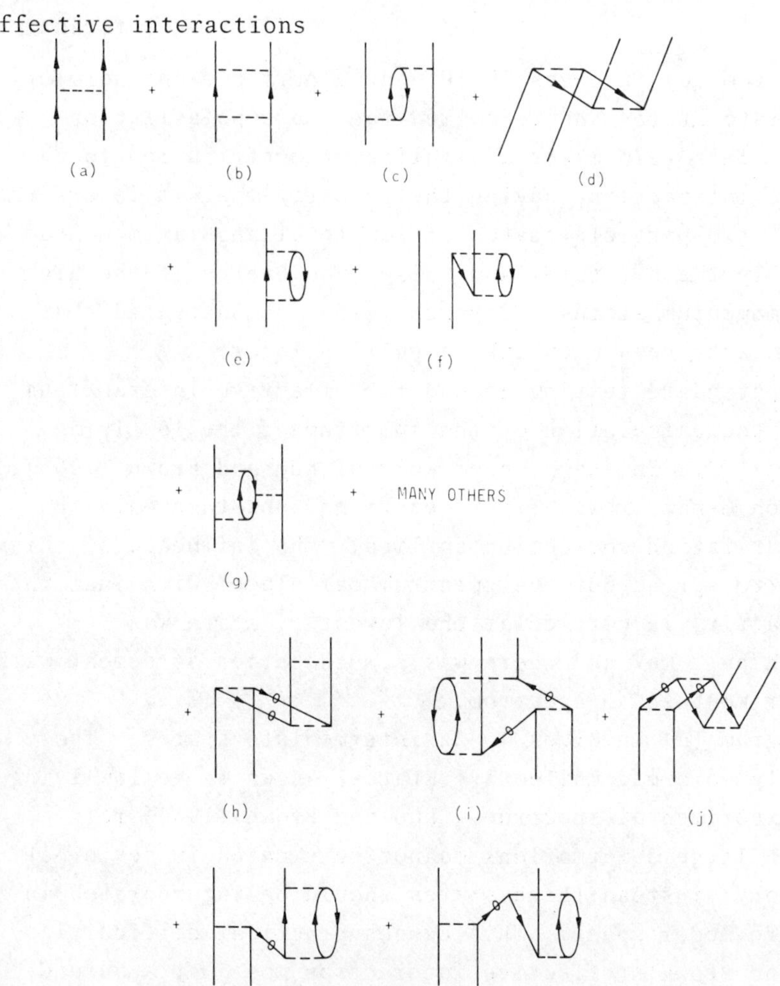

FIG. 5.10. Diagram expansion for the effective interaction in closed-shell-shell-plus-two nuclei. Hartree-Fock self consistency has been assumed. Diagram (a) is first order in V, diagrams (b) to (f) are second order, diagram (g) is typical of third order, and (h) to (ℓ) are the folded diagrams that appear in third order.

neglecting the Pauli operator. The remaining low-lying intermediate states are then specifically incorporated in the perturbation expansion. Arguments along these lines have been presented by Kuo (1967) to justify the inclusion of diagram (b) in his effective interaction. The so-called 'double-partitioned' G-matrix of Barrett (1974) follows this philosophy.

Diagram (c) involves 3p-1h (three particle-one hole) intermediate states and is called the 'core polarization' diagram. This term makes a significant contribution to the effective interaction, having the property that it lowers the energy of two-particle states of low total angular momentum (especially the 0^+ state) and raises the energy of the high angular momentum states. Bertsch (1965) demonstrated that this term acts very much like a pairing force.

The standard testing ground for effective interactions has been the calculation of the spectrum of the low-lying states in ^{18}O. In the pioneer work of Kuo and Brown (1966), a reaction G-matrix was first determined and then only the core-polarization correction applied. The influence of this term was to spread out the spectrum calculated with just the bare G-matrix, in particular the lowest 0^+ state was depressed by 2 MeV and there was a much better agreement with the experimental ^{18}O spectrum.

Diagram (d) involves 4p-2h intermediate states. These are highly deformed collective states, known to be low-lying in the experimental spectrum. Kuo and Brown (1966) felt that such large deformations cannot be treated in perturbation theory; instead these states should be incorporated in the active model space. This causes practical difficulties, so Kuo and Brown's effective interaction was just a bare G-matrix with core polarization corrections only, viz. $G + G_{3p,1h}$. Kuo (1967), in a subsequent calculation in which the bare G-matrix was calculated more carefully found that $G + G_{3p,1h}$ alone gave a very poor fit to the ^{18}O spectrum, but the interaction $G + G_{3p,1h} + G_{2p} + G_{4p,2h}$ restored the agreement. These two interactions, $G + G_{3p,1h}$ of Kuo and Brown and $G + G_{3p,1h} + G_{2p} + G_{4p,2h}$ of Kuo, have been used in a wide range of shell model calculations in the sd-shell nuclei. They each produce reasonably good agreement with the low-lying spectra, and spectroscopic factors for one-nucleon transfer reactions (Halbert *et al.* 1971).

Diagrams (e) and (f) also involve 3p-1h and 4p-2h intermediate states; however, in this case one valence line passes through the diagram undisturbed. Note that these diagrams are not classed as unlinked valence diagrams; a non-interacting valence line is a legitimate part of a linked diagram. However, diagrams (e) and (f) do not contribute to off-diagonal matrix elements of the effective interaction, and in diagonal matrix elements they act only as an effective one-body Hamiltonian. In fact, the sum of all diagrams with one non-interacting valence line merely represents the energy of a closed-shell-plus-one nucleus expressed relative to the closed-shell core. The standard practice in shell-model calculations, therefore, has been to take these so-called single-particle energies from experiment and add them to the calculated diagonal matrix elements of the effective interaction before diagonalization. Thus diagrams (e) and (f) are not considered part of the effective interaction itself. Note that these experimental single-particle energies are *not* used in the evaluation of energy denominators; here the eigenvalues of the unperturbed Hamiltonian H_0^V must still be used.

There are many linked valence diagrams in third order, one of which is given in diagram (g) of Fig. 5.10, and there are five linked folded diagrams, (h) to (ℓ), which arise from the matrix multiplication of the first-order term, diagram (a), with the five second-order terms, diagrams (b) to (f). Although there are reasons for omitting diagrams (b), (e), and (f) in calculating the effective interaction to second order, they must still be included in the matrix multiplication which produces the folded diagrams in third order. None of these folded diagrams contributes to the single-particle energies or to the ladder summations defining the G-matrix, which were the reasons for excluding (b), (e), and (f) in second order.

Barrett and Kirson (1970) have calculated all the third-order diagrams contributing to the effective interaction in ^{18}O. They used Kuo's bare G-matrices and omitted all ladder and Hartree-Fock diagrams. Only the shell-model states of excitation energy 2ħω relative to the shell-model configuration of two valence particles and a closed-shell core were considered. Their results indicated that the total third-order contribution (typically of order 500 keV) can be as large as or larger than the second-order contribution and is of opposite sign. There appears to be no apparent convergence of the perturbation expansion of the effective interaction in powers of G.

This is in contrast to the results obtained in the closed-shell-plus-one system by Kassis (1972); however, different nucleon-nu leon potentials, V, have been used in each case. Nevertheless, one can wonder about the omission of the Hartree-Fock diagrams. Ellis and Mavromatis (1971) have calculated these to second-order in G and obtain a non-negligible result. However, they also present a qualitative argument suggesting that the third-order Hartree-Fock diagrams will give a contribution to the effective interaction having the *same* sign as the second-order contribution. Thus they feel that the conclusion of Barrett and Kirson, namely that the perturbation series does not converge in terms of the number of interactions, would be unchanged when Hartree-Fock graphs are included.

One limitation of the Barrett-Kirson calculation was that they considered only J=0 T=1 matrix elements of the effective interaction. It is natural to ask whether third-order effects are also large for other J,T values. Goode and Koltun (1972) tackled this question by evaluating the weighted average of the matrix elements, weighted according to (2J+1)(2T+1). This gives a tremendous simplification to the calculation, since evaluating weighted averages is equivalent to evaluating a single vacuum or closed-shell diagram. Goode and Koltun find that these weighted averages in third order are smaller and the same sign as second order, and typically of magnitude 50 keV. The contrast with the Barrett-Kirson result can be traced to the core polarization diagrams (diagram (c) of Fig. 5.10, plus similar diagrams in third order), which are weakly repulsive on the average, but strongly attractive for J=0. Goode and Koltun (1972) therefore tentatively concluded that the series expansion for the effective interaction behaves well on the average through to third order in G. However, this result was shattered when the calculations were extended to fourth order. Goode and Koltun (1975) found that the fourth-order averages were as large as those in the second and third orders. There now appears to be little hope of convergence, at least order by order in G.

It should be remembered that Goode and Koltun only considered intermediate states of energy $2\hbar\omega$ above the model states. Vary, Sauer, and Wong (1973) have shown in a study of the second order (core polarization, diagram (c) of Fig. 5.10), that particle-hole excitations up to $12\hbar\omega$ make a significant contribution. This is due to the short range of the tensor force. Probably such excitations would also contribute significantly in third and fourth order.

The whole question of convergence of the effective interaction expansion is very much an open issue. The best hope for resolving the problem is to search for a class of diagrams that can be summed to all orders (the G-matrix itself is such an example, being a sum of all ladder diagrams), leaving a residual interaction that is sufficiently weak that an order-by-order perturbation expansion now has some meaning.

5.7 WAVEFUNCTION NORMALIZATION AND TRANSITION MATRIX ELEMENTS

The 'model' wavefunctions $\Psi_{D\alpha}$, eigenfunctions of the effective interaction $H_0+\mathscr{W}$, are projections from the 'true' wavefunctions Ψ_α, $\Psi_{D\alpha} = P\Psi_\alpha$, and it is clear that $\Psi_{D\alpha}$ and Ψ_α cannot both be normalized simultaneously. It is convenient in the formalism to choose $\langle\Psi_{D\alpha}|\Psi_{D\alpha}\rangle = 1$, and to set about the calculation of $N_\alpha = \langle\Psi_\alpha|\Psi_\alpha\rangle$ rather than vice versa. This calculation is not required in the determination of the energy eigenvalues, but becomes necessary when searching for an effective operator F_{eff} whose matrix elements between model states $\langle\Psi_{D\alpha}|F_{eff}|\Psi_{D\beta}\rangle$ are to be the same as matrix elements of the true operator between the normalized eigenstates of the true Hamiltonian. Our first task, then, is to find an energy-independent perturbation expansion for the normalization of Ψ_α similar to the valence linked-cluster expansion obtained for the effective interaction $\mathscr{W}$.

Consider the same system as discussed in section (5.4), namely that of an open-shell nucleus comprising N core and n valence nucleons. It was shown there that the influence of the core particles on the calculation of the effective interaction, $\mathscr{W}$, could be separated from that of the valence particles and a 'reduced' Bloch-Horowitz equation was derived. This had the advantage that all core diagrams could be dropped in the expansion for the effective interaction $\mathscr{W}$, providing one was only interested in calculating energies relative to the closed-shell core. A similar result holds in the calculation of the normalization and transition matrix elements, as we now demonstrate.

Since it is possible to separate the core and valence parts of the Hamiltonian, the true wavefunction can be factorized, and the normalization is written

$$N_\alpha = \langle \Psi_\alpha | \Psi_\alpha \rangle = \langle \Psi | \Psi \rangle_c \langle \Psi_\alpha | \Psi_\alpha \rangle_v \equiv N_c N_{v\alpha}$$

The subscripts c and v indicate an integration over N and n coordinates respectively. The label α is not necessary for the Ψ in the core factors, since these merely consist of closed-shell diagrams. The matrix elements of F are similarly factorized. Defining $|\alpha\rangle$ as the normalized state corresponding to Ψ_α:

$$\langle \alpha | F | \beta \rangle \equiv N_\alpha^{-\frac{1}{2}} N_\beta^{-\frac{1}{2}} \langle \Psi_\alpha | F | \Psi_\beta \rangle$$

then

$$\langle \alpha | F | \beta \rangle = N_c^{-1} N_{v\alpha}^{-\frac{1}{2}} N_{v\beta}^{-\frac{1}{2}} [\langle \Psi_\alpha | F | \Psi_\beta \rangle_v \langle \Psi | \Psi \rangle_c + \langle \Psi | F | \Psi \rangle_c \langle \Psi_\alpha | \Psi_\beta \rangle_v]$$

$$= N_{v\alpha}^{-\frac{1}{2}} N_{v\beta}^{-\frac{1}{2}} \langle \Psi_\alpha | F | \Psi_\beta \rangle_v + \langle F \rangle_c \, \delta_{\alpha\beta} \qquad (5.42)$$

where $\langle F \rangle_c$ is the expectation value of F between normalized core states, viz. $\langle F \rangle_c = \langle \Psi | F | \Psi \rangle_c N_c^{-1}$. In a diagram language, the diagrams characterizing the matrix element $\langle \alpha | F | \beta \rangle$ have been divided into two sets: those in which F interacts with a valence group and those in which it does not If one is not interested in the expectation value of F in the core, or as often happens $\langle F \rangle_c$ is zero (for example, if F is spherical tensor of rank other than zero), then we have an expression for $\langle\alpha|F|\beta\rangle$ which does not explicitly involve the core.

However, the unsymmetrical treatment of core and valence particles causes the valence normalization factors $N_{v\alpha}$, $N_{v\beta}$ to appear in eqn (5.42). These factors are non-trivial even in the simple closed-shell-plus-one case, because they include interactions between the valence and core particles. We now proceed to find a perturbation expansion for these valence normalization factors.

The procedure follows very closely that used to find the energy-independent effective interaction, $\mathscr{W}$ (eqn (5.33) et seq.). One starts by assuming that the model wavefunctions, eigenfunctions of the reduced secular equation

$$[H_0^{\mathrm{V}} + \mathscr{W} - E_{v\alpha}\, P]\underset{\sim}{A}_{\alpha} = 0,$$

are normalized, orthogonal, and form a complete set

$$\langle\Psi_{D\alpha}|\Psi_{D\beta}\rangle \equiv \sum_{\substack{i\in D\\ j\in D}} a^{*}_{j\alpha}\langle\Phi_j|\Phi_i\rangle a_{i\beta} \equiv \underset{\sim}{A}^{\dagger}_{\alpha}\,\underset{\sim}{A}_{\beta} = \delta_{\alpha\beta}$$

$$\sum_{\beta} \underset{\sim}{A}_{\beta}\,\underset{\sim}{A}^{\dagger}_{\beta} = P.$$

This assumption, however, is not strictly correct, since $\mathscr{W}$ is not Hermitian and the $\underset{\sim}{A}_{\alpha}$ are not orthogonal. We will assume that the error introduced by this assumption is not serious and continue regardless. The correct version has been worked out by Des Cloizeaux (1960) and is discussed in Brandow's articles.

The normalization of the true wavefunction is given by

$$\begin{aligned} N_{v\alpha} &= \langle\Psi_{\alpha}|\Psi_{\alpha}\rangle_{v} = \langle\Psi_{D\alpha}|\Omega^{\dagger}(E_{v\alpha})\Omega(E_{v\alpha})|\Psi_{D\alpha}\rangle \\ &= \underset{\sim}{A}^{\dagger}_{\alpha}\,\Omega^{\dagger}(E_{v\alpha})\Omega(E_{v\alpha})\underset{\sim}{A}_{\alpha} \end{aligned} \qquad (5.43)$$

where Ω is the wave operator, introduced in eqn (5.11), $\Psi_{\alpha} = \Omega\Psi_{D\alpha}$, and satisfies the series expansion (see eqn (5.13))

$$\Omega(E_{v\alpha}) = \frac{Q}{E_{v\alpha}-H_0{}^v} H_1 + \frac{Q}{E_{v\alpha}-H_0{}^v} H_1 \frac{Q}{E_{v\alpha}-H_0{}^v} H_1 + \ldots..$$

Next $\Delta E_{v\alpha} = E_{v\alpha} - E_0{}^v$ is expanded out of the energy denominators using Taylor's theorem:

$$\Omega(E_{v\alpha}) = \sum_{r=0}^{\infty} \Omega^{(r)} [-\Delta E_{v\alpha}]^r \tag{5.44}$$

where

$$\Omega^{(r)} = \frac{(-)^r}{r!} \frac{\partial^r}{\partial E_{v\alpha}{}^r} \Omega(E_{v\alpha}) \Bigg|_{E_{v\alpha} = E_0{}^v} . \tag{5.45}$$

The $\Omega^{(r)}$ are represented by series expansions with Rayleigh-Schrödinger denominators. For example,

$$\Omega^{(0)} = 1 + \frac{Q}{E_0{}^v-H_0{}^v} H_1 + \frac{Q}{E_0{}^v-H_0{}^v} H_1 \frac{Q}{E_0{}^v-H_0{}^v} H_1 + \ldots.$$

$$\Omega^{(1)} = \frac{Q}{(E_0{}^v-H_0{}^v)^2} H_1 + \frac{Q}{(E_0{}^v-H_0{}^v)^2} H_1 \frac{Q}{E_0{}^v-H_0{}^v} H_1 +$$

$$+ \frac{Q}{(E_0{}^v-H_0{}^v)} H_1 \frac{Q}{(E_0{}^v-H_0{}^v)^2} H_1 + \ldots. .$$

Inserting eqn (5.44) back into eqn (5.43) and using the fact that $\Delta E_{v\alpha} = \underset{\sim}{A}_\alpha^\dagger \mathscr{W} \underset{\sim}{A}_\alpha$, we obtain

$$N_{v\alpha} = \sum_{r,s=0}^{\infty} [-\underset{\sim}{A}_\alpha^\dagger \mathscr{W}^\dagger \underset{\sim}{A}_\alpha]^r \underset{\sim}{A}_\alpha^\dagger \Omega^{\dagger(r)} \Omega^{(s)} \underset{\sim}{A}_\alpha [-\underset{\sim}{A}_\alpha^\dagger \mathscr{W} \underset{\sim}{A}_\alpha]^s$$

$$= \underset{\sim}{A}_\alpha^\dagger \sum_{r,s=0}^{\infty} \{[-\mathscr{W}^\dagger P]^r \Omega^{\dagger(r)} \Omega^{(s)} [-P\mathscr{W}]^s\} \underset{\sim}{A}_\alpha, \tag{5.46}$$

using the (approximate) closure properties. The product $\Omega^{\dagger(r)}\Omega^{(s)}$ can be simply evaluated for small r,s by multiplyin out the appropriate series, and not retaining any term which has Q as the first or last factor. These latter terms are dropped since Q would either become adjacent with P and by definition QP = PQ = 0; or would operate on $\underset{\sim}{A}_{\alpha}$ which is also zero by definition. Thus we obtain

$$\Omega^{\dagger(0)}\Omega^{(0)} = 1 + \mathcal{V}^{(1)}$$

$$\Omega^{\dagger(1)}\Omega^{(0)} = \Omega^{\dagger(0)}\Omega^{(1)} = \mathcal{V}^{(2)}$$

and in general (but tricky to prove),

$$\Omega^{\dagger(r)}\Omega^{(s)} = \mathcal{V}^{(r+s+1)} \tag{5.47}$$

where $\mathcal{V}^{(r)}$ was defined in eqn (5.34). Introducing a matrix θ such that

$$N_{v\alpha} = \underset{\sim}{A}^{\dagger}_{\alpha}\,(1 + \theta)\underset{\sim}{A}_{\alpha}, \tag{5.48}$$

then a perturbation expansion for θ is

$$\theta = \sum_{r,s=0}^{\infty} [-\mathcal{W}^{\dagger}P]^{r}\,\mathcal{V}^{(r+s+1)}[-P\mathcal{W}]^{s}$$

$$= \mathcal{V}^{(1)} - \mathcal{W}^{\dagger}P\mathcal{V}^{(2)} - \mathcal{V}^{(2)}P\mathcal{W} + \mathcal{W}^{\dagger}P\mathcal{W}^{\dagger}P\mathscr{V}^{(3)} + \mathcal{W}^{\dagger}P\mathscr{V}^{(3)}P\mathcal{W} + \mathcal{V}^{(3)}P\mathcal{W}P\mathcal{W} + \ldots$$

$$= \mathscr{V}^{(1)} - \mathscr{V}^{(0)}P\mathscr{V}^{(2)} - \mathscr{V}^{(2)}P\mathscr{V}^{(0)} + \mathscr{V}^{(0)}P\mathscr{V}^{(1)}P\mathscr{V}^{(2)} + \mathscr{V}^{(2)}P\mathscr{V}^{(1)}P\mathscr{V}^{(0)} +$$

$$+ \mathscr{V}^{(0)}P\mathscr{V}^{(0)}P\mathscr{V}^{(3)} + \mathscr{V}^{(0)}P\mathscr{V}^{(3)}P\mathscr{V}^{(0)} + \mathscr{V}^{(3)}P\mathscr{V}^{(0)}P\mathscr{V}^{(0)} + \ldots\,. \tag{5.49}$$

Note that the matrix θ is Hermitian, even though $\mathcal{W}$ is not. The diagram expansion for θ contains both linked and unlinked terms, and as was the case for $\mathcal{W}$, the unlinked graphs cancel with some of the 'folded terms' leaving a fully linked expansion.

The diagram representation for θ is very similar in construction to that for the effective interaction $\mathcal{W}$. To illustrate the procedure, we again consider just the closed-shell-plus-one case, where the problem of cancelling the unlinked graphs does not occur, and again use the valence blocks notation introduced in Fig. 5.6. Then the construction of the unfolded versions of the diagrams is as before, namely the valence blocks are arranged in a vertical column, and the valence lines between them are joined and circled with a loop. The folding procedure, however, is slightly modified.

Consider the nth block in the vertical column. It may be folded in either of the following two ways: either the nth block is raised until the top-most interaction line is level with the first horizontal bar encountered in a previous block, or the nth block is lowered until the bottom-most interaction line is level with the first horizontal bar encountered in a subsequent block. After the folding, that particular horizontal bar is removed from the diagram. Each diagram is folded block by block in this way. At the end of the procedure, one horizontal bar remains in the diagram, and this bar is then extended horizontally through the entire diagram. No folded lines are allowed to intersect this bar; the folds should be either entirely above or entirely below this level. All time-orderings of the interaction lines of one block relative to all the others are allowed, subject only to the constraint placed on either the top-most or bottom-most interaction line at each fold. In particular, the top-most interaction line of the final folded diagram does not necessarily have to be the top-most interaction line of the unfolded diagram as was the case in the $\mathscr{W}$-expansion. This is the essential difference that makes the θ-expansion Hermitian and the $\mathscr{W}$-expansion not. In Fig. 5.11, the θ-expansion corresponding to eqn (5.49) for the closed-shell-plus-one system has been illustrated in the valence block notation.

The rules for evaluating the diagrams in the θ-expansion are the same as listed in paragraphs (1") to (5") in appendix (B.3). The only additional point to watch is that the energy denominator comes in squared at the level of the horizontal bar. This concludes our discussion of the linked cluster expansion for the wavefunction normalization, $N_{v\alpha}$.

We next consider the valence contribution to the expectation value of a transition operator F,

$$\langle\alpha|F|\beta\rangle_v = N_{v\alpha}^{-\frac{1}{2}}\langle\Psi_\alpha|F|\Psi_\beta\rangle_v\, N_{v\beta}^{-\frac{1}{2}}$$

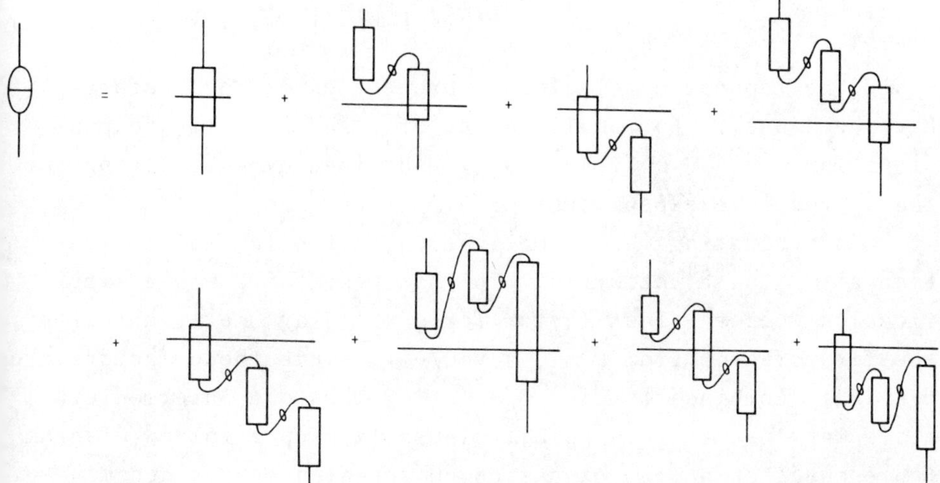

FIG. 5.11. A schematic diagrammatic representation of the lowest few terms in the fully-linked expansion for the wavefunction normalization θ, defined in eqn (5.49). The horizontal bar represents the location of the repeated energy denominator.

and consider first just the unnormalized matrix element $\langle \Psi_\alpha | F | \Psi_\beta \rangle_V$. This expectation value can be represented by the sum of all linked diagrams in which the vertex corresponding to the transition operator F appears just once. Following Brandow (1967) we introduce the symbol $\odot$ to represent this sequence of diagrams such that

$$\langle \Psi_\alpha | F | \Psi_\beta \rangle_V = \underset{\sim}{A}_\alpha^\dagger \odot \underset{\sim}{A}_\beta . \tag{5.51}$$

The procedure for finding a diagram expansion for $\odot$ follows analogously to the way that the θ-expansion was derived. Briefly, the corresponding equation to eqn (5.43) reads

$$\odot = \Omega^\dagger(E_{V\alpha}) \; F \; \Omega(E_{V\beta})$$

and by making Taylor series expansions for the wave operators Ω we obtain (cf. eqn (5.46))

$$\odot = \sum_{r,s=0}^{\infty} [-\mathscr{W}^{\dagger}P]^{r}\ \Omega^{\dagger(r)}\ F\ \Omega^{(s)}[-P\mathscr{W}]^{s}. \qquad (5.52)$$

Again the (approximate) closure properties on the vectors $\underset{\sim}{A}_{\alpha}$ have been used. For small values of r and s, series expansions for $\Omega^{\dagger(r)}\ F\ \Omega^{(s)}$ are easily obtained by multiplying out the appropriate expansions for Ω.

In evaluating the normalization, a similar multiplication of $\Omega^{\dagger(r)}\Omega^{(s)}$ occurred. In that case there was a simplification when the last factor for $\Omega^{\dagger(r)}$, say Q/e^{r_1}, met with the first factor from $\Omega^{(s)}$, say Q/e^{s_1}, since these factors are trivially combined to give $Q/e^{r_1+s_1}$. Thus one intermediate state was lost at this point. In setting up a folded diagram representation of the expansion, a repeated energy denominator always occurs at the point where the intermediate state was lost, and this point is denoted in the diagram by a horizontal bar. For the case of $\Omega^{\dagger(r)}\ F\ \Omega^{(s)}$, the last factor of $\Omega^{\dagger(r)}$ and the first factor of $\Omega^{(s)}$ are separated by the transition operator F, so that an intermediate state is not lost. However, it is fairly obvious that the results obtained for the normalization expansion θ can be taken over for $\odot$ by reinterpreting the horizontal bar, not as the level of the repeated denominator, but as the level at which the operator F acts. In the valence block notation, a dot is inserted in the diagram in place of the horizontal bar so as to distinguish the θ-series from the $\odot$-series.

On the first line in Fig. 5.12, the perturbation expansion for $\odot$ for a closed shell-plus-one system is illustrated in complete analogy with Fig. 5.11. On the next two lines the lowest few terms have been explicitly displayed with F taken to be a one-body operator, and represented by a dashed line terminating in a cross. The rules for constructing the folded diagrams in the $\odot$-expansion are exactly as stated for θ with no folded lines crossing the level at which F operates.

With the expansions derived for the normalization, N_v, and for the unnormalized matrix element $\langle \Psi_{\alpha}|F|\Psi_{\beta}\rangle_v$, we now put these pieces together to obtain the normalized matrix element $\langle \alpha|F|\beta\rangle$ of eqn (5.50). Writing

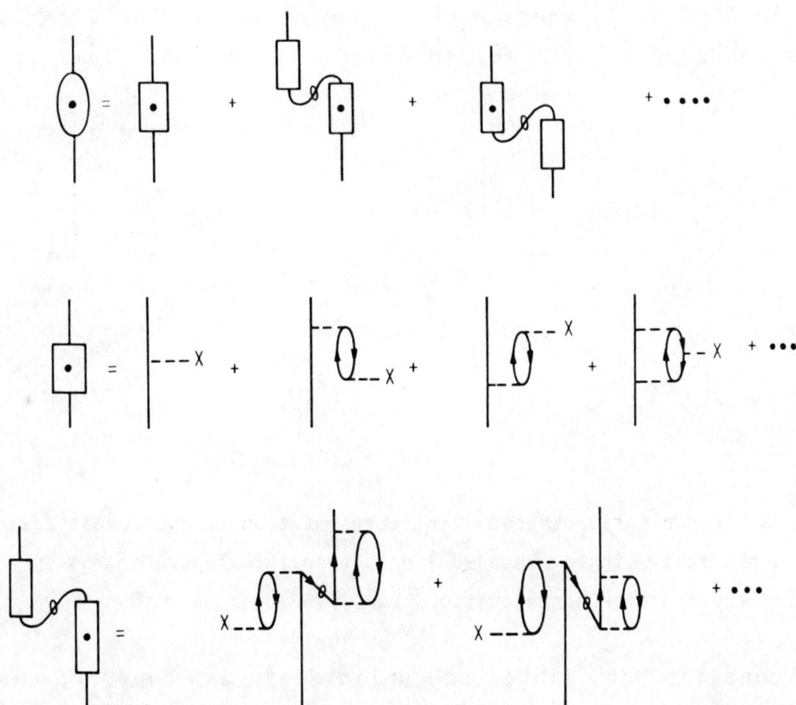

FIG. 5.12. Diagram expansion for the renormalization of a transition operator. The first line is a schematic representation of the first few terms in eqn (5.52). The second and third lines illustrate explicitly the series for a one-body transition operator in a closed-shell-plus-one nucleus.

$$\langle\alpha|F|\beta\rangle = \underset{\sim}{A}_\alpha^\dagger \mathscr{F} \underset{\sim}{A}_\beta$$

and using eqns (5.48) and (5.51), we obtain

$$\mathscr{F} = (1+\theta)^{-\frac{1}{2}} \; P \odot P \; (1+\theta)^{-\frac{1}{2}}$$

as the matrix product of three factors. Expanding $(1+\theta)^{-1/2}$ binomially, the final expression for $\mathscr{F}$ reads

$$\mathscr{F} = \sum_{r,s=0}^{\infty} \binom{-1/2}{r}\binom{-1/2}{s} [\theta P]^r \odot [P\theta]^s. \tag{5.53}$$

The corresponding folded diagram structure is shown schematically in Fig. 5.13, where each component be it θ or ⊙ is itself represented by a folded diagram expansion. This series

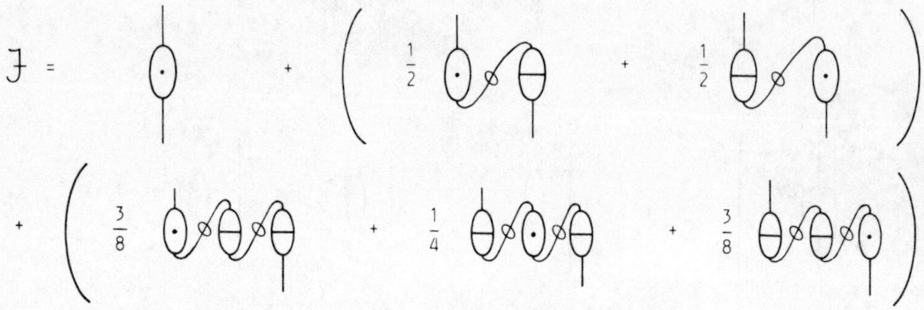

FIG. 5.13. A schematic diagrammatic representation of the lowest few terms in the fully-linked expansion for a transition operator, now including wavefunction normalization, eqn (5.53).

for $\mathscr{F}$ contains both linked and unlinked terms; however, the unlinked terms cancel with some of the folded diagrams, leaving a fully linked expansion (Brandow 1967). Note that the normalization corrections, terms such as $\frac{1}{2}\theta P\odot$, now introduce folded diagrams into second order in the perturbation series for $\mathscr{F}$.

The expansion for $\mathscr{F}$ is complicated by the need to identify the θ structure of each diagram; in particular, it is necessary to distinguish the folding lines connecting the various θ and ⊙ structures from those folding lines internal to θ and ⊙. This distinction is necessary to obtain the correct sign for the diagram. Notice in eqn (5.33) the factors $[\theta P]^r$ and $[P\theta]^s$ occur with no minus sign while the factors internal to θ in eqn (5.49), $[-P\mathscr{W}]^s$, do contain the minus sign.

The rules for evaluating the diagrams in the $\mathscr{F}$-expansion are the same as listed in paragraphs (1") to (5") in appendix (B.3), but with the following two additions:

(a) The sign factor $(-)^f$ is replaced by

$$(-)^{f-r-s} \binom{-1/2}{r} \binom{-1/2}{s}$$

where (f-r-s) is now the number of internal folds within θ and ⊙.

(b) The horizontal bar in the θ part of the diagram indicates that the energy denominator at this level is squared.

Finally, we once again mention that the derivation of eqn (5.53) for $\mathscr{F}$ is slightly incorrect, since the vectors $\underset{\sim}{A}_\alpha$ do not form an orthogonal set. The formalism can be corrected by symmetrizing the non-Hermitian $\mathscr{W}$, the details of which are given by Des Cloizeaux (1960) and Brandow (1967) and will not be discussed here.

5.8 E2-EFFECTIVE CHARGE IN CLOSED-SHELL-PLUS-ONE NUCLEI

One of the most popular tests of the theory of effective interactions and effective operators concerns the electromagnetic properties of 'single-particle' states in the closed-shell-plus-one nuclei ^{17}O and ^{17}F. A survey on the status of these calculations has been given recently by Barrett and Kirson (1973). The test involves the electric quadrupole (E2) operator defined as

$$F = \sum_i e_i r_i^2 Y_{2\mu}(\theta_i, \phi_i) \tag{5.54}$$

with $e_i = 1$ for protons and $e_i = 0$ for neutrons. Then the zeroth-order expectation for a nucleus such as ^{17}O, which is described as a closed shell plus a neutron, is that the quadrupole moments of single-particle states should be zero and that E2 decays between these states are strictly forbidden. Neither of these expectations is borne out experimentally; in fact some of the transitions are of such a magnitude as to suggest that the valence neutron has a charge not much different from that of a proton. First attempts at explaining the discrepancy were to modify the operator F in eqn (5.54) such that e_i is written as $(1+\varepsilon_p)$ for protons and $e_i = \varepsilon_n$ for neutrons, the parameter ε being known as the effective charge.

To extract a value for the effective charge from experimental data, the appropriate measurement is divided by a theoretical estimate obtained from the extreme single-particle model (with e_i=1) and evaluated with harmonic oscillator wavefunctions. For the mass 17 nuclei, three experimental values of ε have been extracted in this way (Barrett and Kirson 1973):

$$(1s_{1/2}|\varepsilon_p|0d_{5/2}) = 0.81 \pm 0.01$$

$$(1s_{1/2}|\varepsilon_n|0d_{5/2}) = 0.54 \pm 0.01$$

$$(0d_{5/2}|\varepsilon_n|0d_{5/2}) = 0.44 + 0.16.$$

The first two derive from the B(E2) value for the γ-decay of the first excited state ($1/2^+$) to the ground state ($5/2^+$) in ^{17}F and ^{17}O respectively, and the third derives from the ground-state quadrupole moment of ^{17}O. Note that the effective charge for a proton, ε_p, exceeds that for a neutron; and there is some evidence that the value of the effective charge depends on the single-particle states involved. This has been called the 'state-dependence' of the effective charge (Federman and Zamick 1969).

The description of a closed-shell-plus-one nucleus by a pure single-particle wavefunction is an extreme case of truncation in the shell-model space. Using just the bare operator, F, from eqn (5.54) to describe the transitions between these single-particle states is overly simplistic and should be modified according to the theory of effective operators discussed in the last section. A perturbation expansion for the effective operator $\mathscr{F}$ is given by eqn (5.53) and illustrated schematically in Fig. 5.13. It comprises a series © for the unnormalized matrix element of F in the valence space, eqn (5.51) and Fig. 5.12, and normalization corrections eqn (5.49) and Fig. 5.11.

Let us examine this expansion through to second order in the residual interaction for the electric quadrupole (E2) operator in closed-shell-plus-one nuclei. The relevant set o

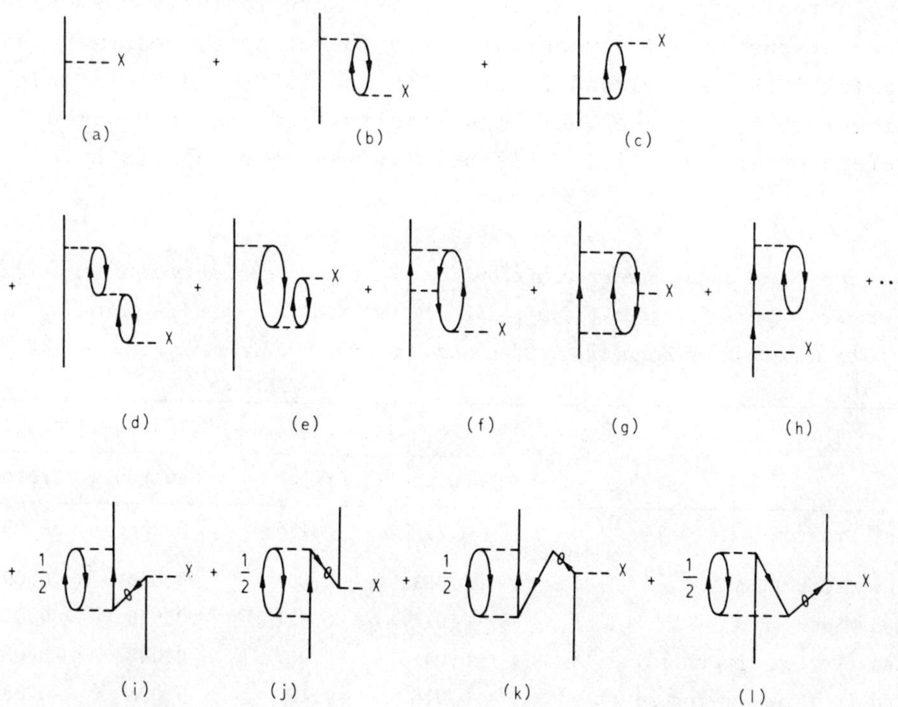

FIG. 5.14. Diagram expansion for the effective transition operator in closed-shell-plus-one nuclei. Hartree-Fock self consistency has been assumed. Diagram (a) represents the bare transition operator. Diagrams (b) and (c) are first-order corrections, and diagrams (d) to (h) are selected second-order corrections. Diagrams (i) to (ℓ) are the folded diagrams that appear in second order.

diagrams is shown in Fig. 5.14, where for brevity we have omitted all Hartree-Fock diagrams. Not all second-order diagrams have been illustrated either.

Diagram (a) is the zeroth-order term; it represents the bare operator F evaluated between single-particle oscillator states. The effective charge is given by the sum of all the other diagrams divided by this zeroth-order term. Diagrams (b) and (c) are the only first-order terms; they are known as the core-polarization graphs, and select out just the 2^+ particle-hole components in the residual interaction. These graphs are easily calculated, and historically were first evaluated by Arima and Horie (1954). More recently they have been discussed by many authors (see for example, Federman and Zamick 1969, Siegel and Zamick 1969 and 1970, and Ellis and Siegel 1971), and we quote the results from the last named reference in Table 5.1. Essentially the same results have

TABLE 5.1

First- and second-order contributions to the effective charge in A=17 nuclei (from Ellis and Siegel, 1971). Calculations were performed with the G-matrix of Barrett, Hewitt and McCarthy (1971) with $\hbar\omega$ *= 14 MeV*

	$(1s_{1/2}\|\varepsilon\|0d_{5/2})$		$(0d_{5/2}\|\varepsilon\|0d_{5/2})$	
	Neutron	Proton	Neutron	Proton
1st Order	0.219	0.115	0.258	0.099
2nd Order (TDA)	0.051	0.066	0.066	0.090
2nd order (RPA)	0.017	0.021	0.020	0.028
2nd (vertex renorm.)	-0.046	-0.057	-0.048	-0.092
2nd (no. conserving sets)	0.010	-0.105	0.008	-0.087
2nd (miscellaneous)	0.000	-0.019	0.010	0.004
Total 2nd order	0.035	-0.095	0.056	-0.057
Total 1st + 2nd	0.254	0.020	0.314	0.042
Expt.	0.54	0.81	0.44	-

been obtained by other workers with different residual interactions. It is noted that the neutron effective charge is twice as big as that for a proton (in contrast to experiment) and both are well short of the experimental value.

Consider next the second-order contributions. There are 30 diagrams appearing in the series ⊙, eqn (5.52) (some of which are illustrated in (d) to (h) in Fig. 5.14), and four folded diagrams correcting the normalization, diagrams (i) to (ℓ). These latter terms arise from $-⊙P\theta/2$ and $-\theta P⊙/2$ in the expansion, eqn (5.53), and the factor of 1/2 which has been specifically written in the diagram originates in amendment (a) to the rules for folded diagrams, discussed in appendix B.

Second-order corrections to the effective charge were

first calculated by Siegel and Zamick (1970) using the Kallio-Kolltveit (Kallio and Kolltveit 1964) potential to represent the residual interaction. They were subsequently repeated by Ellis and Siegel (1971) using the realistic G-matrix of Barrett, Hewitt and McCarthy (1971) and we shall quote from the second reference. Referring to Fig. 5.14, diagram (d) is one of the two TDA (Tamm-Dancoff) diagrams, and these give a sizeable enhancement to the effective charge (see Table 5.1). When the four RPA diagrams are evaluated, diagram (e) is one of these, the enhancement is even greater. Next consider diagram (f); this is also a core-polarization-type diagram, except that the valence particle interacts twice with the particle-hole vibration. There are four such diagrams and they have been named vertex-renormalization diagrams by Kirson. Their effect is again sizeable and cancels a large part of the TDA + RPA enhancement.

Diagram (g) is one of six topologically different diagrams in which the one-body operator acts on a particle or hole line as opposed to operating on a particle-hole pair. These six diagrams together with the four folded diagrams form what are called 'number-conserving sets'. That is, if the E2 operator is replaced by the number operator, the sum of these ten graphs would be identically zero. Thus one expects a reasonable degree of cancellation among these ten diagrams. Defining $R = \Sigma_i d_i / \Sigma_i |d_i|$ where d_i is the contribution from i^{th} diagram of the set, Ellis and Siegel find R is of order 20 per cent for neutrons and of order 40 per cent for protons. Since six of the ten neutron diagrams are identically zero in the present case, the contribution from the number-conserving sets to the neutron effective charge is small. For protons, however, the cancellation is quite poor, and the contribution is large and negative.

Of the remaining 14 diagrams, four involve single-particle intermediate states, such as diagram (h). These were not calculated by Ellis and Siegel, but they were found by Siegel and Zamick (1970) to give a very small contribution. Another four are ladder diagrams and should not be included in the perturbation expansion when a G-matrix is being used.

And finally, the last six diagrams are all individually small and mutually cancelling, such that their summed effect is negligible.

The total second-order effect (see Table 5.1) is small and positive for the neutron, but large and negative for the proton effective charge. This latter value is almost entirely due to the non-cancellation among the proton number-conserving sets. Putting the first- and second-order contributions together, the neutron results are satisfactory in that the second-order contribution is some five times smaller than the first, so that convergence order-by-order looks promising. The summed effect is still small by a factor of two compared to experiment. The proton results, however, are something of a disaster. The second-order contribution is as large as the first-order and of opposite sign, so there is no order-by-order convergence, and the summed effect is smaller than experiment by an order of magnitude.

It has been noted that both the TDA and RPA core polarization diagrams enhance the effective charge. This class of diagrams can easily be summed to all orders and it was shown by Siegel and Zamick (1970) that a very strong enhancement is obtained in this way due to the collectivity and corresponding low excitation energy of 2^+ quadrupole isoscalar phonon (particle-hole vibration). In fact there was a tendency for the RPA to collapse (i.e., to admit complex eigenvalues, see discussion in chapter 4) for ^{40}Ca. This instability of the RPA phonon can be removed, however, by including the self-screening of the particle-hole interaction. This is another class of diagrams which first appear in third-order and have the appearance of 'bubbles inside bubbles'. These diagrams have been summed to all orders by Kuo and Osnes (1973) and their results are shown in Table 5.2. It is seen that the self-screening correction produces a strong reduction in the effective charge. Note also that in all these calculations the neutron effective charge ε_n exceeds ε_p, contrary to experiment. Finally it was noted in the second-order calculations that the vertex renormalization diagrams were equally important, and that these also should be summed to all orders.

TABLE 5.2

Contributions to the effective charge in A=17 from core-polarization-type diagrams summed to all orders (from Kuo and Osnes 1973, and Barrett and Kirson, 1973). Calculations were performed with G-matrix of Kuo (1967) with ħω = 14 MeV.

	$(1s_{1/2}\|\varepsilon\|0d_{5/2})$		$(0d_{5/2}\|\varepsilon\|0d_{5/2})$	
	neutron	proton	neutron	proton
1st order	0.269	0.119	0.329	0.103
All orders (TDA)	0.404	0.277	0.501	0.315
All orders (RPA)	0.517	0.394	0.655	0.477
All orders (RPA + self-screening)	0.357	0.222	0.443	0.245
All orders (RPA + self-screening + vertex renorm. + ...)	0.274	0.141	0.325	0.126
Expt.	0.54	0.81	0.44	-

This has been carried out by Kirson (1971, 1974) and his results are shown in Table 5.2, where it is seen that the sum total of self-screening and vertex renormalizations cancels the collectiveness of the RPA, and essentially just the first-order result remains. The discrepancy with experiment remains as bad as before.

Finally, some consideration should be given to the Hartree-Fock diagrams which have been consistently omitted in this section. These have been calculated to second order by Ellis and Mavromatis (1971) and their results are shown in Table 5.3. It is clear that the Hartree-Fock and non-Hartree-Fock graphs give results of comparable magnitude in both first and second order. For neutrons, the second-order Hartree-Fock graphs are negative and over cancel the positive non-Hartree-Fock graphs, thus reducing the first-order values. For protons, the Hartree-Fock graphs are very state-dependent; however, the combined second-order contribution is negative and cancels some of the first order. For both protons and neutrons, the combined second-order effects are approximately -1/3 smaller than first-order, so that possibly the series is converging in terms of the number of interactions. Note

TABLE 5.3

Contribution to the effective charge in A=17 from Hartree-Fock diagrams to second order (from Ellis and Mavromatis 1971). Calculations were performed with the Sussex matrix elements (Elliott et al. 1968) with $\hbar\omega$ = 14.4 MeV.

	$(1s_{1/2}\|\varepsilon\|0d_{5/2})$		$(0d_{5/2}\|\varepsilon\|0d_{5/2})$	
	neutron	proton	neutron	proton
1st order non-HF	0.243	0.129	0.248	0.110
1st order HF	0	0.228	0	0.078
Total 1st order	0.243	0.357	0.248	0.188
2nd order non-HF†	0.035	-0.095	0.056	-0.057
2nd order HF	-0.096	-0.027	-0.177	-0.068
Total 2nd order	-0.061	-0.112	-0.121	-0.125
Total 1st + 2nd	0.182	0.245	0.127	0.063
Expt.	0.54	0.81	0.44	-

†from Table 5.1.

that, in contrast to the results in Table 5.1, the total first- plus second-order effective charges for the neutron and proton are now much closer together, which is favoured by experiment; but the totals are still too small by approximately a factor of three.

In summary, then, perturbation calculations have been singularly unsuccessful in explaining the effective charge in closed-shell-plus-one nuclei. This is quite disturbing. Referring back for a moment to the formal theory, the renormalization of a transition operator is expressed in eqn (5.52) in terms of the wave operator Ω. Similarly the effective interaction also depends on the same wave operator, eqn (5.12), and a consistent calculation would call for the transition operator and the effective interaction to be renormalized in the same way. Thus a failure to understand effective charges can be considered as a failure to construct an appropriate wave operator, which in turn reflects a lack of understanding on the effective interaction. This is what is disturbing.

A phenomenological approach, perhaps, is to parameterize the wave operator Ω and then to evaluate consistently the effective charge and the effective interaction. This approach has been tried with some success by Harvey and Khanna (1970), but the fundamental problem is still left unresolved.

Curiously enough, unrestricted Hartree-Fock calculations, or more correctly variation-after-projection Hartree-Fock calculations (Cusson and Lee 1973) applied to open-shell nuclei, ^{20}Ne and ^{24}Mg, have no difficulty in explaining the observed quadrupole properties of low-energy states. These calculations sample a very large configuration space involving five major oscillator shells, and as a consequence, find that no renormalization of the quadrupole operator is necessary to explain the E2 transition strengths. The obvious question then is to ask why the unrestricted Hartree-Fock calculations apparently succeed, whereas renormalization in the spherical shell model has, as yet, failed.

This question has been studied by Harvey (1974, 1975). With the aid of a simple schematic Hamiltonian, he has succeeded in identifying the terms in the perturbation series for the effective interaction and effective operators in the spherical shell model which when summed recapture the results of the unrestricted deformed Hartree-Fock model. The identification, however, requires an interesting new principle. Consider the core polarization diagram, Fig. 5.14(b), which on replacing particle-particle matrix elements by particle-hole matrix elements has the value

$$\sum_{mi} \frac{\langle a^{-1}b|V|i^{-1}m\rangle\langle i^{-1}m|F|0\rangle}{(\varepsilon_i - \varepsilon_m)}.$$

If the residual interaction V was of a separable multipole-multipole type, then the two-body matrix element can be factorized (see eqn (4.21) on schematic particle-hole models) into a part entirely within the configuration space, a_{ba}, and a part in the excluded space, a_{mi}, viz.

$$\langle a^{-1}b|V|i^{-1}m\rangle = \lambda\, a_{ba}\, a_{mi}.$$

The new principle is that the part in the model space, a_{ba}, should itself be renormalized and this now sets up an iteration scheme that has to be solved self-consistently and corresponds exactly to the self-consistency of deformed Hartree-Fock. Harvey's results are encouraging, but it remains to be seen whether their generalization to realistic residual interactions succeeds in reproducing the effective charge in closed-shell-plus-one nuclei.

6
SPECTROSCOPY IN OPEN-SHELL NUCLEI

6.1 DIAGRAM EXPANSIONS IN AN ANGULAR MOMENTUM COUPLED BASIS

The standard shell model calculation for an open-shell nucleus proceeds as follows. First, a single-particle Hamiltonian H_0 is selected, whose eigenfunctions form a basis for the calculation. This basis is of infinite dimension and must be truncated. The zeroth approximation is to place the A nucleons in the A lowest-energy orbitals. Let us assume that N of the nucleons occupy closed shells while a few (n) comprise a partially-filled shell, A = N+n. The next level of approximation is to allow the n nucleons to scatter over a few selected orbitals and to calculate all interactions between them, while constraining the N nucleons to remain undisturbed in their closed shells. Orbitals, other than those few selected, remain unfilled. This defines the 'shell-model space'.

Next, an effective interaction $\mathscr{W}$ operating in this truncated model space is constructed such that its eigenvalues match the lowest few eigenvalues of the true Hamiltonian operating in the full Hilbert space. All matrix elements of $\mathscr{W}$ between the n-particle states have to be evaluated, and the resulting matrix is diagonalized to obtain the energies and wavefunctions for the nucleus concerned. Finally these wavefunctions are used to calculate expectation values and transition rates corresponding to certain observables, and the success of the whole operation is usually based on an intercomparison of the calculated results with experimental data.

For the most part, this chapter will be concerned with the mechanics of the last step, the calculation of observables. The single-particle Hamiltonian will usually be taken to be the harmonic oscillator, and the appropriate effective interaction $\mathscr{W}$ will be assumed to be pre-calculated and available. However, we make just a few remarks here on how the diagram expansions of the last chapter are interpreted in an angular momentum coupled basis.

Consider the closed-shell-plus-two situation. Required at the outset are the two-body matrix elements $\langle cd|\mathscr{W}|ab\rangle_A$ evaluated between two-particle states, $|ab\rangle$ where

$$|ab\rangle = a_b^\dagger \; a_a^\dagger \; |C\rangle. \tag{6.1}$$

The diagram expansion for $\mathscr{W}$ has been given in Fig. 5.10. Concentrate for the moment on the core-polarization diagram which has been redrawn in Fig. 6.1a. Remember this diagram

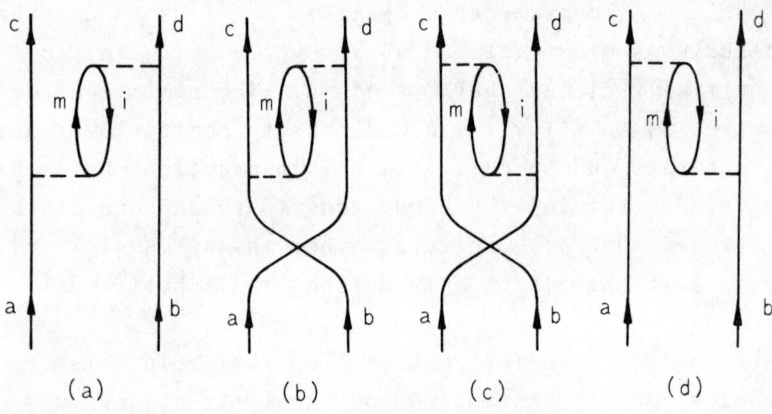

FIG. 6.1. The core polarization diagram from second-order perturbation theory for the effective interaction. Usually only diagram (a) is drawn. The exchange diagrams, (b) to (d), obtained by interchanging the external lines must, however, be included. Their presence is usually understood in the interpretation of diagram (a).

also represents the three additional exchange diagrams obtained from it by interchanging the external lines a,b and c,d. These exchange diagrams are illustrated in Figs. 6.1b, c, and d. The value of Fig. 6.1 is given by the expression

$$\sum_{mi} \frac{\langle id|V|mb\rangle_A \; \langle cm|V|ai\rangle_A}{(\varepsilon_i + \varepsilon_a - \varepsilon_m - \varepsilon_c)}$$

where the sum m spans both valence states in the model space and higher-lying particle states, while the sum i runs over the occupied core states. The single-particle energies ε in the denominator are the eigenenergies of the one-body Hamiltonian, H_0.

It is more convenient to work with two-particle states, not as expressed in eqn (6.1), but rather with states coupled to good total angular momentum. Thus the initial state is written

$$|j_a j_b;JM\rangle_A = N_{ab} \sum_{m_a m_b} \langle j_a m_a j_b m_b|JM\rangle \, a^\dagger_\beta a^\dagger_\alpha|0\rangle \qquad (6.3)$$

where

$$a^\dagger_\beta a^\dagger_\alpha|0\rangle = \frac{1}{\sqrt{2}}\{\psi_{j_a m_a}(1)\psi_{j_b m_b}(2)-\psi_{j_b m_b}(1)\psi_{j_a m_a}(2)\}$$

and where

$$N_{ab} = (1 - (-)^{2j_a - J}\delta_{j_a,j_b})^{-1/2} \qquad (6.4)$$

is a normalization constant.

There is a similar expression for the final two-particle state $|j_c j_d;JM\rangle_A$. All isospin quantum numbers have been suppressed, J ≡ J,T for example. Similarly the two-body matrix elements in eqn (6.2) are written in a coupled representation, viz.

$$\begin{aligned}\langle id|V|mb\rangle_A &\equiv \langle j_i m_i j_d m_d|V|j_m m_m j_b m_b\rangle_A \\ &= \sum_{J_1 M_1} \langle j_i m_i j_d m_d|J_1 M_1\rangle\langle j_m m_m j_b m_b|J_1 M_1\rangle \times \\ &\quad \times \langle j_i j_d;J_1|V|j_m j_b;J_1\rangle_A \qquad (6.5)\end{aligned}$$

where the coupled two-body matrix element on the right in eqn (6.5) is antisymmetrized but *not normalized*. Thus using the expressions (6.3) and (6.5), and summing over all the magnetic quantum numbers, the value of Fig. 6.1a is now given by

$$N_{ab}N_{cd}\sum_{\substack{j_m j_i\\ J_1 J_2}} \frac{\hat{J}_1\hat{J}_1\hat{J}_2}{\hat{J}\,\hat{j}_b\hat{j}_i}(-)^{J_1+J_2+j_i+j_m+J}\begin{bmatrix} j_c & j_d & J\\ j_m & J_1 & j_b\\ J_2 & j_i & j_a\end{bmatrix}\times$$

$$\times\frac{\langle j_i j_d;J_1|V|j_m j_b;J_1\rangle_A\,\langle j_c j_m;J_2|V|j_a j_i;J_2\rangle_A}{(\varepsilon_a+\varepsilon_i-\varepsilon_c-\varepsilon_m)} \tag{6.6}$$

where the array [ ] represents a recoupling coefficient, appendix (A.4), and $\hat{J} = (2J+1)^{1/2}$. The value of the three exchange diagrams can easily be deduced from eqn (6.6). For example, Fig. 6.1b, which derives from Fig. 6.1a by exchanging external lines a and b, is given by eqn (6.6) with j_b interchanged with j_a everywhere and the result multiplied by a phase factor: $-(-)^{j_a+j_b-J}$.

In certain cases, for example Fig. 5.10d, the external pair of lines both enter or both leave the same interaction vertex; this exchange has already been taken into account by the use of antisymmetrized matrix elements. In such cases the number of exchange diagrams is reduced from four to two or one as appropriate.

Following this example, it is a straightforward exercise in Racah algebra to evaluate the diagrams discussed in chapter 5 in an angular momentum coupled basis. Relevant expressions for closed-shell-plus-one and closed-shell-plus-two cases have been given by Kassis (1972) and by Barrett and Kirson (1970).

6.2 CLOSED-SHELL-PLUS-THREE NUCLEI

To appreciate some of the problems encountered with the diagram method as one moves away from the closed shell situation, we shall consider the example of closed-shell-plus-three nuclei. The first problem is to specify the three-particle states in an angular momentum coupled representation. The obvious choice is to write

$$|(j_a j_b)J_1 j_c;J\rangle = N_{abc}(J_1)\sum_{\substack{m_a m_b\\ M_1 m_c}}\langle j_a m_a j_b m_b|J_1 M_1\rangle\langle J_1 M_1 j_c m_c|JM\rangle\times$$

$$\times \frac{1}{\sqrt{6}} \begin{vmatrix} \psi_{j_a m_a}(1) & \psi_{j_a m_a}(2) & \psi_{j_a m_a}(3) \\ \psi_{j_b m_b}(1) & \psi_{j_b m_b}(2) & \psi_{j_b m_b}(3) \\ \psi_{j_c m_c}(1) & \psi_{j_c m_c}(2) & \psi_{j_c m_c}(3) \end{vmatrix} \tag{6.7}$$

with $N_{abc}(J_1)$ being the normalization constant:

$$N_{abc}(J_1) = \Big\{ 1 - (-)^{2j_a - J_1} \delta_{j_a j_b} - \\ - \frac{\hat{J}_1 \hat{J}_1}{\hat{j}_a \hat{J}} U(j_b J_1 J_1 j_b; j_a J) \delta_{j_b j_c} [1 - (-)^{2j_a - J_1} \delta_{j_a j_b}] - \\ - \frac{\hat{J}_1 \hat{J}_1}{\hat{j}_b \hat{J}} U(j_a J_1 J_1 j_a; j_b J) \delta_{j_a j_c} [1 - (-)^{2j_a - J_1} \delta_{j_a j_b}] \Big\}^{-1/2} .$$

However, this basis is over complete; states differing in the quantum number J_1 are not orthogonal. For example, consider three particles distributed over the $0d_{5/2}$, $1s_{1/2}$ and $0d_{3/2}$ orbitals. The number of states of the type given in eqn (6.7) with $J = 5/2$ $(T = 3/2)$ is sixteen, whereas only ten independent states are needed to span this model space. The difficulty is easily overcome, but more work is involved. Both the Hamiltonian matrix

$$A = \langle (j_d j_e) J_2 j_f ; J | H_0 + \mathscr{W} | (j_a j_b) J_1 j_c ; J \rangle$$

and the normalization matrix

$$N = \langle (j_d j_e) J_2 j_f ; J | (j_a j_b) J_1 j_c ; J \rangle$$

have to be constructed, and the shell-model eigenvalue problem takes the form

$$A\, X_q = E_q\, N\, X_q \tag{6.8}$$

where E_q is the energy eigenvalue and X_q the corresponding eigenvector. Both matrices A and N are symmetric with N positive definite. The first step is to find the eigenvalues of N (Wilkinson, 1969), viz.

$$R^{\dagger} N R = D^2 = \text{diag}\,(d_i{}^2)$$

where R is an orthogonal matrix ($R^{\dagger} = R^{-1}$) and D^2 a diagonal matrix containing the positive eigenvalues $d_i{}^2$ of N. The matrix D then simply comprises the square roots of these eigenvalues. In our example of the J = 5/2 states, ten of the eigenvalues are positive and six are zero. These latter six are discarded so the matrix D has dimension 10 × 10 and the matrix R,16 × 10. The generalized problem (6.8) can now be reduced to the standard eigenvalue problem

$$P\, Z_q = E_q\, Z_q$$

where P is the 10 × 10 matrix, $P = D^{-1}\, R^{\dagger} A\, RD^{-1}$ and $Z_q = DR^{\dagger} X_q$ (Wilkinson 1969). The method is numerically stable and has the advantage that the same algorithm is used twice for solving the eigenvalue problem, first for N and then for P. Note that the Z_q can be orthogonalized in the usual way, and RD^{-1} is real and non-singular so that the ten X_q corresponding to the ten non-zero E_q form a complete set. They satisfy an orthogonality relation

$$\delta_{qr} = Z_q^{\dagger}\, Z_r = (DR^{\dagger} X_q)^{\dagger}\; DR^{\dagger} X_r = X_q^{\dagger} RD\; DR^{\dagger} X_r$$

$$= X_q^{\dagger}\, N\, X_r,$$

showing that the vectors X_q are orthogonal with respect to N.

The next complication concerns the effective interaction $\mathscr{W}$, which again is determined from a perturbation expansion in the nucleon-nucleon interaction, V (or in the reaction matrix G). The relevant diagrams through to second order for the closed-shell-plus-three case are illustrated in Fig. 6.2. No Hartree-Fock diagrams are considered. Diagrams (a) and (b influence just one particle line and consequently only contribute to diagonal matrix elements. They can be considered

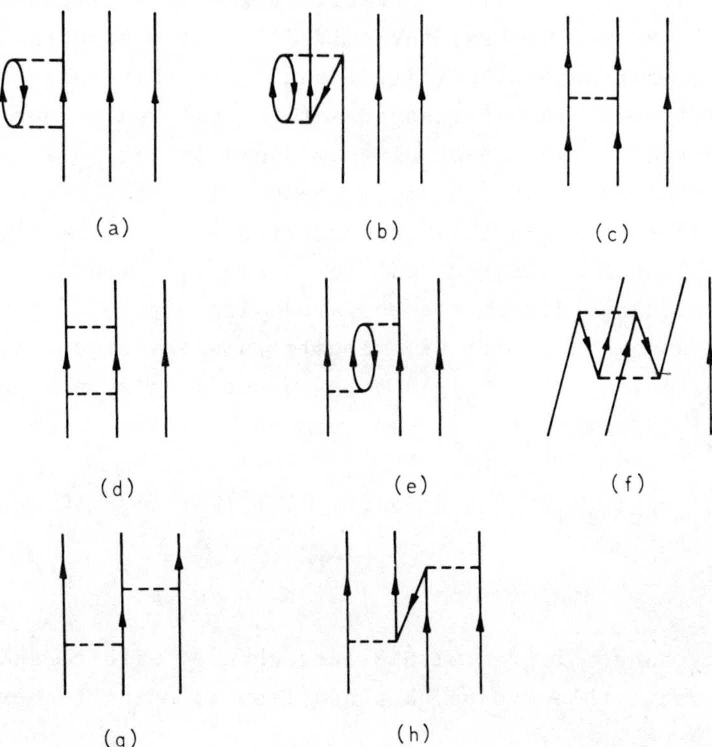

FIG. 6.2. First- and second-order diagrams for the effective interaction in closed-shell-plus-three nuclei. Diagrams (a) and (b) are effective one-body diagrams, (c) to (f) are two-body diagrams, and (g) and (h) are three-body diagrams.

as representing an effective one-body interaction. Similarly diagrams (c) to (f) represent an effective two-body and diagrams (g) and (h) an effective three-body interaction. This illustrates another property of effective interactions. Although the bare nucleon-nucleon interaction, V, contains only two-body terms, the effect of model truncations produces an effective interaction with up to m-body terms, where m is the smaller of n and (r+1). Here n is the number of nucleons outside the closed shell and r the order of perturbation theory. Thus a calculation through to second order in V for a closed-shell-plus-three system contains at most effective three-body interactions.

Evaluating the diagrams in Fig. 6.2 is a straightforward exercise in Racah algebra; the only difficulty concerns the antisymmetrization of the initial and final three-particle state. All possible interchanges of the external lines must be considered, so that each diagram shown in Fig. 6.2 represents a total of 36 direct and exchange diagrams. In certain cases where two external lines enter the same interaction vertex, this number is reduced; for example, the number of distinct exchange diagrams associated with Fig. 6.2(g) is nine. As an example, let us evaluate this three-body diagram for an initial state $|(j_a j_b)J_1 j_c;J\rangle$ and a final state $|(j_d j_e)J_2 j_f;J\rangle$, then the direct term has a value

$$\sum_{j_m K_1} U(j_d j_m J j_c;J_1 K_1)U(j_d j_e J j_f;J_2 K_1)\langle j_e j_f;K_1|V|j_m j_c;K_1\rangle \times$$
$$\times \langle j_d j_m;J_1|V|j_a j_b;J_1\rangle/(\varepsilon_a+\varepsilon_b-\varepsilon_d-\varepsilon_m) \qquad (6.9)$$

and the exchange diagram with b interchanged with c, and d interchanged with e can be obtained from it using the expression.

$$\sum_{J_1'} (-)^{J_1+J_1'-j_a-J+j_d+j_e-J_2} U(j_b j_a J j_c;J_1 J_1') \times$$
$$\times \{\text{eqn (6.9) with } b \rightleftharpoons c,\ d \rightleftharpoons e \text{ and } J_1' \rightleftharpoons J_1\}.$$

All the other exchange diagrams can be evaluated in much the same way.

It is fairly obvious that as one moves away from the closed-shell situation, the handling of the antisymmetrization of the n valence nucleons becomes unwieldy. There are two problems. First the overcompleteness of the basis states, and secondly the $(n!)^2$ exchange terms that have to be explicitly considered whenever a matrix element is being evaluated. If the n valence nucleons are placed in n distinct shell-model orbitals, then the basis states formed by straightforward angular momentum coupling do in fact form a complete set. This is because the single-particle states themselves are orthogonal. However, as soon as more than one particle is allowed in the same shell-model orbital, the basis becomes overcomplete. Similarly, there is no short cut to evaluating all $(n!)^2$ exchange terms in a matrix element when all particles are in distinct orbitals; but with more than one particle in the same orbital, many of the exchange terms become equal to each other. Thus in the next few sections we shall consider the extreme case of n nucleons in a single shell-model orbital. A complete set of antisymmetric states is constructed explicitly and the evaluation of matrix elements between such states is achieved with the aid of so-called 'coefficients of fractional parentage'. Towards the end of the chapter we shall reconsider the problem of configuration mixing with more than one shell-model orbital active in the calculation.

6.3 FRACTIONAL PARENTAGE COEFFICIENTS

Let us denote by $|j^{n-1}\, y_p J_p M_p\rangle$ a state of total angular momentum J_p and magnetic projection M_p constructed from (n-1) single-particles in just one shell model orbital j. The state is fully antisymmetric with respect to interchanges among the (n-1) nucleons. Let us further suppose that these states form a complete set, in particular they have the property of being orthogonal and normalized, i.e.

$$\langle j^{n-1}\, y_p J_p M_p | j^{n-1}\, y_p' J_p' M_p' \rangle = \delta_{y_p y_p'}\, \delta_{J_p J_p'}\, \delta_{M_p M_p'}$$

The label y_p differentiates between states with the same J_p arising from the configuration j^{n-1}.

The problem on hand is to add one more particle to the orbital j and construct a new complete set of states $|j^n\, XJM\rangle$, antisymmetric in all n nucleons, such that each member of the set is normalized and orthogonal to all other members. That is, we need to find a set of coefficients in the expansion

$$|j^n\, XJM\rangle = \sum_{M_p m\, y_p J_p} \langle j^n\, XJ\{|j^{n-1}\, y_p J_p; j\rangle \langle J_p M_p\, jm|JM\rangle |j^{n-1}\, y_p J_p M_p\rangle |jm\rangle$$

$$\equiv \sum_{y_p J_p} \langle j^n \; XJ\{|j^{n-1} \; y_p J_p;j\rangle \; |[j^{n-1} \; y_p J_p,j];JM\rangle \tag{6.10}$$

which produce an antisymmetric n-particle state satisfying

$$\langle j^n \; XJM|j^n \; X'J'M'\rangle = \delta_{XX'} \; \delta_{JJ'} \; \delta_{MM'}.$$

The orthogonality with respect to the angular momentum quantum numbers J,M is guaranteed by the algebra of Clebsch-Gordan coefficients, but the orthogonality with respect to X must be specifically imposed on the coefficients. That is

$$\sum_{y_p j_p} \langle j^n \; XJ\{|j^{n-1} \; y_p J_p;j\rangle\langle j^n \; X'J\{|j^{n-1} \; y_p J_p;j\rangle = \delta_{XX'}. \tag{6.11}$$

Notice the use of square brackets in eqn (6.10) to represent angular momentum coupling. The $\langle j^n \; XJ\{|j^{n-1} \; y_p J_p;j\rangle$ are known as the coefficients of fractional parentage (cfp).

The expansion (6.10) is not unitary; it cannot uniquely be inverted, and the symbol {| was introduced into the notation for the cfp as a reminder of this point. The number of states formed by simple vector coupling of each member of the set $|j^{n-1} \; y_p J_p M_p\rangle$ with a single-particle state $|jm\rangle$ far exceeds the number of members in the complete set $|j^n \; XJM\rangle$. The cfps, in fact, are just the coefficients for the expansion of a state $|j^n \; XJM\rangle$ in terms of the over-complete set, $|[j^{n-1} y_p J_p,j];JM\rangle$.

It is clear that the determination of cfps is intimately connected with the precise classification of the n-particle state, i.e. with the specification of the labels X. Thus the procedure pioneered by Racah (1943), Jahn and van Wieringen (1951), Edmonds and Flowers (1952), Flowers (1952), and Jahn (1954) was to examine the symmetries inherent in the angular momentum coupling and to search for a set of operators, which in addition to commuting with the angular momentum operators, had the property that the commutator of any two in the set was expressible as a linear combination of the operators in the set. That is, the set of operators formed a closed group. If such a group can be found, then it is possible to construct from it a scalar quadratic invariant, known as a Casimir operator, whose eigenvalues can be used to designate the labels X.

For example, the angular momentum operators J_x, J_y, J_z form a group, they have a closed set of commutation relations, and the corresponding Casimir operator is $J^2 = J_x^2 + J_y^2 + J_z^2$. It is a property of Casimir operators that their eigenvalues depend only on the representation labels. For example, the eigenvalue of J^2 is just $J(J+1)$.

If a Casimir operator can be found corresponding to the label X, then obtaining cfps becomes a simple eigenvalue problem. Let G be such a Casimir operator; then the expansion

$$\langle j^n\, XJ|G|j^n\, XJ\rangle$$

$$= \sum_{\substack{X_1J_1\\X_2J_2}} \langle j^n XJ\{|j^{n-1}X_1J_1;j\rangle\langle j^n XJ\{|j^{n-1}X_2J_2;j\rangle\langle[j^{n-1}X_1J_1;j]J|G|[j^{n-1}X_2J_2;j]J\rangle$$

can be cast in a matrix form

$$\sum_\beta G_{\alpha\beta}\, C_\beta = \lambda\, C_\alpha.$$

Here $G_{\alpha\beta}$ is the matrix element $\langle[j^{n-1}X_1J_1;j]J|G|[j^{n-1}X_2J_2;j]J\rangle$, C_α a cfp, and λ the eigenvalue $\langle j^n\, XJ|G|j^n XJ\rangle$. The technique is to construct the matrix $G_{\alpha\beta}$, diagonalize it, pick out the eigenvalues corresponding to λ, and the related eigenvectors are the required cfps. The real difficulty, of course, is finding the Casimir operator and knowing its eigenvalues. An example of this procedure has been given by Bayman and Lande (1966).

For many applications, the precise specification of the label X is not required. It is merely sufficient that the basis states be orthogonal, and this can be guaranteed by forcing the condition (6.11) on the cfps. The technique, then, for obtaining the cfps is essentially one of recursion.

One assumes that all cfps $\langle j^{n-1}\ y_pJ_p\{|j^{n-2}\ z_qJ_q;j\rangle$ have been obtained at a previous stage and one builds on these to find the current set. The derivation to be given here follows that of de Shalit and Talmi (1963) and is written down in j-j coupling. Analogous expressions can easily be derived for alternative coupling schemes.

Consider the state formed by coupling a single particle to an antisymmetric state of (n-1) particles

$$|[j^{n-1}(\underset{\sim}{r}_1\underset{\sim}{r}_2\cdots\underset{\sim}{r}_{n-1})y_0J_0,j(\underset{\sim}{r}_n)];JM\rangle, \tag{6.12}$$

where the antisymmetrized state can be thought of as a Slater determinant of (n-1) single-particle wavefunctions, or more correctly as a linear combination of such determinants. We have explicitly displayed the particle coordinates $\underset{\sim}{r}_1,\underset{\sim}{r}_2\cdots\underset{\sim}{r}_{n-1}$. The first step is to antisymmetrize fully the state (6.12) with respect to interchanges among all n nucleons by operating with the antisymmetrizer

$$\mathscr{A} = \frac{1}{n}\ [1 - \sum_{k=1}^{n-1} P_{kn}]. \tag{6.13}$$

Here the operator P_{kn} simply interchanges the coordinates $\underset{\sim}{r}_k$ and $\underset{\sim}{r}_n$. The antisymmetrizer has been normalized such that $\mathscr{A}^2 = \mathscr{A}$. Next, overlap this antisymmetric state with another state of type (6.12); the result to within a normalization constant will be a fractional parentage coefficient. That is

$$\langle[j^{n-1}(\underset{\sim}{r}_1\underset{\sim}{r}_2\cdots\underset{\sim}{r}_{n-1})y_pJ_p,j(\underset{\sim}{r}_n)];JM|\ \times$$

$$\times\mathscr{A}[j^{n-1}(\underset{\sim}{r}_1\underset{\sim}{r}_2\cdots\underset{\sim}{r}_{n-1})y_0J_0,j(\underset{\sim}{r}_n)];JM\rangle$$

$$= N_{y_0J_0}\ \langle j^n\ XJ\{|j^{n-1}\ y_pJ_p;j\rangle \tag{6.14}$$

The bra and ket notation indicates an integration over the coordinates $\underset{\sim}{r}_1,\ \underset{\sim}{r}_2\cdots\ \underset{\sim}{r}_n$. The first step is to evaluate $\mathscr{A}$ operating on eqn (6.12):

$$\mathscr{A}[j^{n-1}(\underset{\sim}{r}_1,\underset{\sim}{r}_2\cdots\underset{\sim}{r}_{n-1})y_0J_0,j(\underset{\sim}{r}_n)];JM\rangle$$

$$= \frac{1}{n} |[j^{n-1}(\underset{\sim}{r}_1, \underset{\sim}{r}_2 \cdots \underset{\sim}{r}_{n-1}) y_0 J_0, j(\underset{\sim}{r}_n)]; JM\rangle -$$

$$- \frac{1}{n} \sum_{k=1}^{n-1} |[j^{n-1}(\underset{\sim}{r}_1 \cdots \underset{\sim}{r}_{k-1}\ \underset{\sim}{r}_n\ \underset{\sim}{r}_{k+1} \cdots \underset{\sim}{r}_{n-1}) y_0 J_0, j(\underset{\sim}{r}_k)]; JM\rangle$$

$$= D + E, \text{ say}, \tag{6.15}$$

where D represents the direct term and E the term in which coordinates $\underset{\sim}{r}_k$ and $\underset{\sim}{r}_n$ have been interchanged. Next the particles in the group $|j^{n-1}\rangle$ are reordered so that the nth particle is last in the list. This is achieved by successively interchanging columns in the Slater determinant, the result being

$$E = - \frac{1}{n} \sum_{k=1}^{n-1} (-)^{p_k} |[j^{n-1}(\underset{\sim}{r}_1 \cdots \underset{\sim}{r}_{k-1} \underset{\sim}{r}_{k+1} \cdots \underset{\sim}{r}_{n-1} \underset{\sim}{r}_n) y_0 J_0, j(\underset{\sim}{r}_k)]; JM\rangle$$

where $(-)^{p_k}$ is -1 or +1, depending on whether an odd number or an even number of columns were interchanged in the determinant. Next the fractional parentage expansion, eqn (6.10), is made on the group $|j^{n-1}\rangle$ to separate out the nth particle:

$$E = - \frac{1}{n} \sum_{k=1}^{n-1} (-)^{p_k} \sum_{z_q J_q} \langle j^{n-1}\ y_0 J_0 \{| j^{n-2}\ z_q J_q; j\rangle \times$$

$$\times |[[j^{n-2}(\underset{\sim}{r}_1 \cdots \underset{\sim}{r}_{k-1} \underset{\sim}{r}_{k+1} \cdots \underset{\sim}{r}_{n-1}) z_q J_q, j(\underset{\sim}{r}_n)]^{J_0}, j(\underset{\sim}{r}_k)]; JM\rangle$$

$$= - \frac{1}{n} \sum_{k=1}^{n-1} (-)^{p_k} \sum_{z_q J_q J_0'} \langle j^{n-1} y_0 J_0 \{| j^{n-2} z_q J_q; j\rangle U(J_q jjJ; J_0 J_0') \times$$

$$\times (-)^{J_0 + J_0' - J - J_q} |[[j^{n-2}(\underset{\sim}{r}_1 \cdots \underset{\sim}{r}_{n-1}) z_q J_q, j(\underset{\sim}{r}_k)]^{J_0'}, j(\underset{\sim}{r}_n)]; JM\rangle. \tag{6.16}$$

The last step just reorders the angular momentum coupling, introducing a recoupling coefficient (appendix A.4). Inserting eqns (6.15) and (6.16) back into the overlap integral, eqn (6.14), produces the expression

$$n\, N_{y_0 J_0} \langle j^n\ xJ \{| j^{n-1}\ y_p J_p; j\rangle = \delta_{y_0 y_p}\ \delta_{J_0 J_p} -$$

$$- \sum_{k=1}^{n-1} (-)^{p_k} \sum_{z_q J_q J_0'} \langle j^{n-1} y_0 J_0 \{| j^{n-2} z_q J_q ; j \rangle \, U(J_q jJ; J_0 J_0') (-)^{J_0 + J_0' - J - J_q} \times$$

$$\times \langle [j^{n-1}(\underset{\sim}{r}_1 \ldots \underset{\sim}{r}_{n-1}) y_p J_p, \, j(\underset{\sim}{r}_n); JM \Big| [[j^{n-2}(\underset{\sim}{r}_1 .. \underset{\sim}{r}_{k-1} \underset{\sim}{r}_{k+1} \ldots \underset{\sim}{r}_{n-1}) z_q J_q, \times$$

$$\times j(\underset{\sim}{r}_k)]^{J_0'}, j(\underset{\sim}{r}_n)] JM \rangle. \tag{6.17}$$

Once again the fractional parentage expansion is used on the group $|j^{n-1}\rangle$ to separate out the kth particle, and as before p_k permutations of the columns in the Slater determinant must first be executed to place the kth particle in the last column. The resulting expression for the overlap integral shows no distinction between the kth particle and any other of the (n-1) particles in the group $|j^{n-1}\rangle$, so the sum over k gives (n-1) equivalent terms. With these manipulations eqn (6.17) reduces to

$$n\, N_{y_0 J_0} \langle j^n \, XJ \{| j^{n-1} \, y_p J_p ; j \rangle = \delta_{y_0 y_p} \, \delta_{J_0 J_p} -$$

$$-(n-1) \sum_{z_q J_q} \langle j^{n-1} \, y_0 J_0 \{| j^{n-2} z_q J_q ; j \rangle \langle j^{n-1} y_p J_p \{| j^{n-2} z_q J_q ; j \rangle \times$$

$$\times \, U(J_q jJ; J_0 J_p) \, (-)^{J_0 + J_p - J - J_q}. \tag{6.18}$$

Notice that the right-hand side of this equation remains unchanged when $y_0 J_0$ is interchanged with $y_p J_p$. Therefore, the left-hand side should also satisfy this symmetry property, suggesting that

$$n\, N_{y_0 J_0} \langle j^n \, XJ \{| j^{n-1} \, y_p J_p ; j \rangle = n\, N_{y_p J_p} \langle j^n XJ \{| j^{n-1} y_0 J_0 ; j \rangle$$

or that

$$N_{y_0 J_0} = \langle j^n \, XJ \{| j^{n-1} \, y_0 J_0 ; j \rangle.$$

With this identification, the final expression for the recursion relation between cfps emerges:

$$n \langle j^n\ XJ\{|j^{n-1} y_0 J_0; j\rangle \langle j^n\ XJ\{|j^{n-1} y_p J_p; j\rangle$$

$$= \delta_{y_0 y_p}\ \delta_{J_0 J_p} - (n-1) \sum_{z_q J_q} \langle j^{n-1} y_0 J_0\{|j^{n-2} z_q J_q; j\rangle \langle j^{n-1} y_p J_p\{|j^{n-2} z_q J_q; j\rangle \times$$

$$\times\ U(J_q jjJ; J_0 J_p)\ (-)^{J_0+J_p-J-J_q}. \tag{6.19}$$

Use of the recursion formula begins with the following observations. There is only one state with zero particles in the orbital j, namely the vacuum state, and only one state $|j^1\ J\rangle$ with one particle in the orbital j, which necessarily has angular momentum J = j. Thus from the definition (6.10) one trivially identifies

$$\langle j^1\ J{=}j\{|j^0\ 0; j\rangle = +1. \tag{6.20}$$

For states $|j^2\rangle$, eqn (6.20) is substituted in eqn (6.19) to obtain

$$\langle j^2\ J\{|j^1\ j; j\rangle^2 = \tfrac{1}{2}(1 - (-)^{2j-J}). \tag{6.21}$$

Again the label X is redundant, since corrsponding to a given angular momentum J, and isospin T, there is only one state with configuration j^2. Note, however, that only even values of J are allowed (for T = 1 states, odd J for T = 0 states). The positive square root can be chosen in eqn (6.21).

The first non-trivial case then concerns states $|j^3\ XJ\rangle$, for which eqn (6.19) can be simplified to

$$3\ \langle j^3\ XJ\{|j^2\ J_0; j\rangle \langle j^3\ XJ\{|j^2\ J_p; j\rangle = \delta_{J_0 J_p} - 2\ U(jjjJ; J_0 J_p)(-)^{J+j}.$$

This is evaluated first for $J_0 = J_p$, where J_0 is any of the even values allowed by the triangle relation $\Delta(jjJ_0)$ and $\Delta(JjJ_0)$ to give

$$3\,|\langle j^3\ XJ\{|j^2\ J_0;j\rangle|^2 = 1 - 2(-)^{J+j}\ U(jjjJ;J_0J_0);$$

then choosing the positive square root one finds for the remaining cfps

$$\langle j^3\ XJ\{|j^2J_p;j\rangle = -2(-)^{J+j}\ U(jjjJ;J_0J_p)\{3-6(-)^{J+j}\ U(jjjJ;J_0J_0)\}^{-\frac{1}{2}}, J_0 \neq J_p.$$

This typifies the general procedure.

A terminology frequently associated with fractional parentage coefficients is to name the states $|j^n\ XJ\rangle$ the daughter states, the states $|j^{n-1}\ y_pJ_p\rangle$ the parent states, and the states $|j^{n-2}\ z_qJ_q\rangle$ the grandparent states. In particular the state $|j^{n-1}\ y_0J_0\rangle$, which plays a significant role in the recursion formula (6.19), is known as the 'principal parent'. Any one of the states with configuration j^{n-1} for which the cfp $\langle j^n\ XJ\{|j^{n-1}\ y_pJ_p;j\rangle$ is non-zero can be chosen as a principal parent. Let us denote the vector $\underset{\sim}{C}$ with components C_p, the cfps connecting the daughter state $|j^n\ XJ\rangle$ to all N possible parent states, that is

$$C_p = \langle j^n\ XJ\{|j^{n-1}\ y_pJ_p;j\rangle \quad p = 0,1\ldots(N-1).$$

The procedure then is to choose one parent state, for example $p = 0$ as the principal parent, and to evaluate eqn (6.19) repeatedly for $p = 0, 1, 2\ldots(N-1)$. In this way values for $|C_0|^2$, C_0C_1, $C_0C_2\ldots$, C_0C_{N-1} are obtained. If the sign of C_0 is chosen positive, then the signs of all the remaining components C_1, C_2,... are determined with respect to this choice. Thus, there is some degree of arbitrariness concerning the sign of a cfp, as there should be since the overall phase of a wavefunction is arbitrary.

This procedure gives a unique solution for the cfps providing that the daughter state $|j^n\ XJ\rangle$ is the only state in the j^n configuration with angular momentum J, that is if X has only one value. However, there may be d states of the type $|j^n\ XJ\rangle$; then it is necessary to make several choices of principal parents until d independent vectors of the type $\underset{\sim}{C}$ have been generated. These vectors will not necessarily be orthogonal, but they can be made so by taking linear combinations and using the method of Schmidt orthogonalization. This leads to a further arbitrariness in the cfps, since any orthogonal combination of vectors satisfying eqn (6.11) is sufficient for the shell model.

As the number of nucleons, n, in the shell model orbital increases, so the number of cfps escalates enormously. For example, by the time one has reached the middle of the j = 5/2 orbital (in an isospin formalism), the number of cfps required is 1728, while for the j = 7/2 orbital, the number if 122070 at the mid-point (see Table 6.1). This represents a practical limit even for present-day computers.

TABLE 6.1

The number of states N(n) with configuration $|j^n\ XJT\rangle$ *from Flowers (1952) and the number of* cfps $\langle j^n\ XJT\{|j^{n-1}\ y_pJ_pT_p;j\rangle = N(n)\times N(n-1)$, *for the* j = 3/2, 5/2 *and* 7/2 *orbitals, in the isospin formalism.*

n	1	2	3	4	5	6	7	8
				j = 3/2				
N(n)	1	4	5	8	5	4	1	1
no. of cfps	1	4	20	40	40	20	4	1
				j = 5/2				
N(n)	1	6	12	29	36	48	36	29
no. of cfps	1	6	72	348	1044	1728	1728	1044
				j = 7/2				
N(n)	1	8	22	72	133	256	313	390
no. of cfps	1	8	176	1584	9576	34048	80128	122070

Note that cfps only have to be evaluated for the first half of the shell. There is a one-one correspondence between states $|j^n\ XJ\rangle$ and states $|j^{n_{max}-n}\ XJ\rangle$ where n_{max} is the maximum number of particles accommodated in the shell-model orbital. In the isospin formalism $n_{max} = 4j+2$ and in the proton-neutron formalism, $n_{max} = 2j+1$. The corresponding relation between cfps in the first half and the second half of the shell is (Macfarlane and French 1960):

$$\langle j^{n_{max}-n+1} XJ\{|j^{n_{max}-n} yJ';j\rangle$$

$$= (-)^{n}(-)^{j+J'-J}\left[\frac{n(2J'+1)}{(n_{max}-n+1)(2J+1)}\right]^{\frac{1}{2}}\langle j^{n} yJ'\{|j^{n-1} XJ;j\rangle$$

with a possible ambiguity in the sign.

Remember that the overall sign of a cfp vector $\underset{\sim}{C}$ is arbitrary. Thus in consulting a table of cfps, the user is perfectly free to replace the cfp $\langle j^n XJ\{|j^{n-1} y_pJ_p;j\rangle$ by $-\langle j^n XJ\{|j^{n-1} y_pJ_p;j\rangle$ providing it is done for all parent states p and providing all cfps $\langle j^{n+1} W_qJ_q\{|j^n XJ;j\rangle$ are also reversed in sign, since the construction of the latter cfp is based on the former through the recurrence relation.

Tables of j-j coupled cfps in the isospin formalism have been given by Edmonds and Flowers (1952), Glaudemans, Wiechers, and Brussaard (1964), Balashov (1959), Towner, and Hardy (1969), and Hubbard (1971), and in proton-neutron formalism by Sato (as quoted in de Shalit and Talmi 1963) and by Bayman and Lande (1966). Tables of L-S coupled cfps have been given by Jahn and van Wieringen (1951).

6.4 TWO-PARTICLE FRACTIONAL PARENTAGE COEFFICIENTS

In analogy with the one-particle cfp discussed in the last section, it is sometimes of interest to construct states $|j^n XJM\rangle$, antisymmetric in the coordinates of the n nucleons, by coupling an antisymmetric two-particle state $|j^2 J_2M_2\rangle$ to the grandparent states $|j^{n-2} z_qJ_qM_q\rangle$. The relevant expansion is written (Elliott, Hope, and Jahn 1953) as

$$|j^n XJM\rangle = \sum_{z_qJ_qJ_2} \langle j^n XJ\{|j^{n-2} z_qJ_q;j^2J_2\rangle|[j^{n-2} z_qJ_q,j^2J_2];JM\rangle \quad (6.22$$

with the coefficient $\langle j^n XJ\{|j^{n-2} z_qJ_q;j^2J_2\rangle$ being known as a two-particle cfp. The requirement that the states $|j^n XJM\rangle$ be orthogonal leads, as before, to an orthogonality condition on the cfps:

$$\sum_{z_q J_q J_2} \langle j^n\, XJ\{|j^{n-2}\, z_q J_q; j^2 J_2\rangle \langle j^n\, X'J\{|j^{n-2}\, z_q J_q; j^2 J_2\rangle = \delta_{XX'} \tag{6.23}$$

For the evaluation of these two-particle cfps it is sufficient to find an expression relating them with one-particle cfps. The first step is to break up the two-particle state $|j^2\, J_2\rangle$ on the right-hand side of eqn (6.22) using the cfp expansion

$$|j^n(\underset{\sim}{r}_1 \underset{\sim}{r}_2 \cdots \underset{\sim}{r}_{n-1} \underset{\sim}{r}_n) XJM\rangle$$

$$= \sum_{z_q J_q J_2} \langle j^n\, XJ\{|j^{n-2}\, z_q J_q; j^2 J_2\rangle \langle j^2\, J_2\{|j^1\, j; j\rangle \times$$

$$\times\, |[j^{n-2}(\underset{\sim}{r}_1 \cdots \underset{\sim}{r}_{n-2}) z_q J_q,\, [j(\underset{\sim}{r}_{n-1}),\, j(\underset{\sim}{r}_n)]^{J_2}]; JM\rangle. \tag{6.24}$$

The second step is to start once again with the state $|j^n\, XJ\rangle$ and make two successive cfp expansions

$$|j^n(\underset{\sim}{r}_1 \cdots \underset{\sim}{r}_n) XJM\rangle = \sum_{y_p J_p z'_q J'_q} \langle j^n\, XJ\{|j^{n-1}\, y_p J_p; j\rangle \langle j^{n-1} y_p J_p\{|j^{n-2} z'_q J'_q; j\rangle \times$$

$$\times\, |[[j^{n-2}(\underset{\sim}{r}_1 \cdots \underset{\sim}{r}_{n-2}) z'_q J'_q, j(\underset{\sim}{r}_{n-1})]^{J_p}, j(\underset{\sim}{r}_n)]; JM\rangle$$

$$= \sum_{y_p J_p z'_q J'_q J'_2} \langle j^n\, XJ\{|j^{n-1} y_p J_p; j\rangle \langle j^{n-1} y_p J_p\{|j^{n-2} z'_q J'_q; j\rangle\, U(J'_q jJj; J_p J'_2) \times$$

$$\times\, |[j^{n-2}(\underset{\sim}{r}_1 \cdots \underset{\sim}{r}_{n-2}) z'_q J'_q, [j(\underset{\sim}{r}_{n-1}), j(\underset{\sim}{r}_n)]^{J'_2}]; JM\rangle,$$

where the last step is just a simple recoupling of three angular momenta. Equating, in eqns (6.24) and (6.25), the coefficients of the state $|[j^{n-2}(\underset{\sim}{r}_1 \cdots \underset{\sim}{r}_{n-2}) z_q J_q, [j(\underset{\sim}{r}_{n-1}), j(\underset{\sim}{r}_n)]^{J_2}]; JM\rangle$, gives the required result

$$\langle j^n\, XJ\{|j^{n-2}\, z_q J_q; j^2 J_2\rangle \langle j^2 J_2\{|j^1 j; j\rangle$$

$$= \sum_{y_p J_p} \langle j^n\, XJ\{|j^{n-1}\, y_p J_p; j\rangle \langle j^{n-1} y_p J_p\{|j^{n-2} z_q J_q; j\rangle U(J_q jJj; J_p J_2)$$

or

$$\langle j^n \chi J\{|j^{n-2} z_q J_q ; j^2 J_2\rangle = \tfrac{1}{2}(1-(-)^{2j-J_2}) \sum_{y_p J_p} \langle j^n \chi J\{|j^{n-1} y_p J_p ; j\rangle \times$$

$$\times \langle j^{n-1} y_p J_p \{|j^{n-2} z_q J_q ; j\rangle \, U(J_q jJj; J_p J_2) \qquad (6.26)$$

where from eqn (6.21) the cfp $\langle j^2 J_2\{|j^1 j;j\rangle = +1$ when J_2 is even. It is straightforward to show that two-particle cfps constructed from eqn (6.26) satisfy the orthogonality condition eqn (6.23), providing the one-particle cfps themselves satisfy eqn (6.11). Thus, given a complete set of one-particle cfps, it is trivial to obtain the required two-particle cfps.

In an analogous way, one can introduce r-particle cfps defined by the expansion

$$|j^n \chi_n J_n M_n\rangle = \sum_{\substack{\chi_{n-r} J_{n-r} \\ \chi_r J_r}} \langle j^n \chi_n J_n \{|j^{n-r} \chi_{n-r} J_{n-r} ; j^r \chi_r J_r\rangle \times$$

$$\times \, |[j^{n-r} \chi_{n-r} J_{n-r}, j^r \chi_r J_r] ; J_n M_n\rangle,$$

and evaluated in terms of one-particle cfps with the following formula

$$\langle j^n \chi_n J_n \{|j^{n-r} \chi_{n-r} J_{n-r} ; j^r \chi_r J_r\rangle$$

$$= \sum_{\substack{\chi_{n-k} J_{n-k} \chi_k J_k \\ k=1\ldots r-1}} \langle j^n \chi_n J_n \{|j^{n-1} \chi_{n-1} J_{n-1} ; j\rangle \langle j^{n-1} \chi_{n-1} J_{n-1} \{|j^{n-2} \chi_{n-2} J_{n-2} ; j\rangle \ldots$$

$$\ldots \langle j^{n-r+1} \chi_{n-r+1} J_{n-r+1} \{|j^{n-r} \chi_{n-r} J_{n-r} ; j\rangle \times$$

$$\times \langle j^r \chi_r J_r \{|j^{r-1} \chi_{r-1} J_{r-1} ; j\rangle \langle j^{r-1} \chi_{r-1} J_{r-1} \{|j^{r-2} \chi_{r-2} J_{r-2} ; j\rangle \ldots$$

$$\ldots \langle j^2 X_2 J_2 \{| j^1 X_1 J_1 ; j \rangle \times$$

$$\times\, U(J_{n-r} J_1 J_{n-r+2} j ; J_{n-r+1} J_2) U(J_{n-r} J_2 J_{n-r+3} j ; J_{n-r+2} J_3) \ldots$$

$$\ldots U(J_{n-r} J_{r-1} J_n j ; J_{n-1} J_r).$$

These coefficients would only be required in the shell model if one was evaluating a matrix element of an r-body operator between n-particle states, $n \geqslant r$. This is not very common for $r > 2$.

6.5 ONE-BODY OPERATORS

As a first example in the use of fractional parentage coefficients, let us evaluate the expectation value of a one-body operator

$$F_q^{(k)} = \sum_{i=1}^{n} f_q^{(k)}(\underset{\sim}{r}_i)$$

of tensorial rank k, and magnetic projection q between states of configuration j^n. The operator $f_q^{(k)}(\underset{\sim}{r}_i)$ acts only on the coordinates of the ith nucleon. The first step is to note that the n nucleons comprising the wavefunction $|j^n\, XJM\rangle$ are indistinguishable, so that the reduced matrix element of the one-body operator $F_q^{(k)}$ is just n times that of $f_q^{(k)}(\underset{\sim}{r}_n)$, namely

$$\begin{aligned}\langle F^{(k)} \rangle &\equiv \langle j^n(\underset{\sim}{r}_1, \underset{\sim}{r}_2, \ldots, \underset{\sim}{r}_n) XJ \| F^{(k)} \| j^n(\underset{\sim}{r}_1, \underset{\sim}{r}_2, \ldots, \underset{\sim}{r}_n) X'J' \rangle \\ &= n \langle j^n(\underset{\sim}{r}_1, \underset{\sim}{r}_2, \ldots, \underset{\sim}{r}_n) XJ \| f^{(k)}(\underset{\sim}{r}_n) \| j^n(\underset{\sim}{r}_1, \underset{\sim}{r}_2, \ldots, \underset{\sim}{r}_n) X'J' \rangle.\end{aligned} \tag{6.27}$$

Next, the cfp expansion is made on both the bra and ket wavefunction separating the nth nucleon from the rest.

$$\langle F^{(k)} \rangle = n \sum_{\substack{y_p J_p \\ y'_p J'_p}} \langle j^n XJ \{| j^{n-1} y_p J_p ; j \rangle \langle j^n X'J' \{| j^{n-1} y'_p J'_p ; j \rangle \times$$

$$\times\langle[j^{n-1}(\underset{\sim}{r}_1,..,\underset{\sim}{r}_{n-1})y_pJ_p;j(\underset{\sim}{r}_n)]XJ\| f^{(k)}(\underset{\sim}{r}_n)\|[j^{n-1}(\underset{\sim}{r}_1,..,\underset{\sim}{r}_{n-1})y_p'J_p';j(\underset{\sim}{r}_n)]X'J'\rangle$$

Finally note that the reduced matrix element on the right is exactly of the form (A.38b), with the bra and ket wavefunctions being vector couplings of two parts, and the operator acting only on the second part. Integrating over the n-1 coordinates leads to the final result

$$\langle F^{(k)}\rangle = n \sum_{y_pJ_p} \langle j^nXJ\{|j^{n-1}y_pJ_p;j\rangle\langle j^nX'J'\{|j^{n-1}y_pJ_p;j\rangle \times$$

$$\times\ U(jkJ_pJ;jJ')\ \langle j\| f^{(k)}\| j\rangle. \tag{6.28}$$

For the special case in which $F^{(k)}$ is a scalar operator, $k = 0$, the U-coefficient reduces to $\delta_{JJ'}$, and the sum over the cfps becomes just the orthogonality condition, eqn (6.11). In this case we have

$$\langle F^{(0)}\rangle = n\ \langle j\| f^{(0)}\| j\rangle\ \delta_{XX'}\ \delta_{JJ'} \tag{6.29}$$

In general, eqn (6.28) can be cast in the form

$$\langle j^nXJ\| F^{(k)}\| j^n\ X'J'\rangle = C(njXJX'J'k)\ \langle j\| f^{(k)}\| j\rangle \tag{6.30}$$

where the coefficient C(njXJX'J'k) depends on the configurations of the nuclear states in question, but not on the details of the operator, other than its rank, k. This form shows that for these equivalent particle configurations, the ratio of two matrix elements of two one-body operators $F^{(k)}$ and $U^{(k)}$ with the same rank is equal to the corresponding ratio of their single particle matrix elements, i.e.

$$\frac{\langle j^nXJ\| F^{(k)}\| j^nX'J'\rangle}{\langle j^nXJ\| U^{(k)}\| j^nX'J'\rangle} = \frac{\langle j\| f^{(k)}\| j\rangle}{\langle j\| u^{(k)}\| j\rangle}\ . \tag{6.31}$$

As an example let us consider an operator of rank one in orbital-spin space (but a scalar in isospin space, that is the implied reference to isospin quantum nubmers is dropped for the rest of this section). Another operator of the same character is $\underset{\sim}{J} = \Sigma_i\ \underset{\sim}{j}(\underset{\sim}{r}_i)$, so that

$$\langle j^n XJ \| F^{(1)} \| j^n X'J' \rangle = \frac{\langle j \| f^{(1)} \| j \rangle}{\langle j \| \underset{\sim}{j} \| j \rangle} \langle j^n XJ \| \underset{\sim}{J} \| j^n X'J' \rangle$$

$$= \delta_{XX'} \delta_{JJ'} \left[\frac{J(J+1)}{j(j+1)}\right]^{1/2} \langle j \| f^{(1)} \| j \rangle \quad (6.32)$$

This result leads to the interesting theorem: within a j^n configuration of identical particles, all off-diagonal matrix elements involving one-body operators of rank one are zero. For example, there can be no magnetic dipole radiation between such configurations. In practice this theorem can never be experimentally verified, since configuration mixing is always present to some degree; however, one can at least observe that such transitions are strongly hindered. In Table 6.2 some experimental data on M1 decays in ^{19}O, ^{19}F, ^{43}Ca, and ^{43}Sc have been listed. Naively these decays can be interpreted in the shell model as simple $j^3 \rightarrow j^3$ transitions between

TABLE 6.2

M1 transitions in closed-shell-plus-three nuclei. Hindrance factor is the ratio of the Weisskopf estimate to the experimental width. Data were taken from the compilations of Ajzenberg-Selove (1972) and Endt and Van der Leun (1973).

Nucleus	J_i	J_f	E_γ	Width (eV)	Weisskopf estimate (eV)	Hindrance factor
^{19}O	$3/2^+$	$5/2^+$	0.096	$(3.3\pm0.1)\times10^{-7}$	1.8×10^{-5}	55
^{19}F	$3/2^+$	$5/2^+$	1.357	$(1.4\pm0.7)\times10^{-1}$	5.2×10^{-2}	0.4
^{43}Ca	$5/2^-$	$7/2^-$	0.373	$(1.3\pm0.1)\times10^{-5}$	1.1×10^{-3}	85
^{43}Sc	$5/2^-$	$7/2^-$	0.845	$(2.0\pm0.7)\times10^{-3}$	1.3×10^{-2}	6

states with angular momentum $J = j-1$ and $J = j$ with j being either the $d_{5/2}$ or $f_{7/2}$ orbital. When the experimental widths are compared with single-particle estimates (Weisskopf estimates), a tremendous hindrance, between 50 and 100, is observed for ^{19}O and ^{43}Ca, nuclei for which the theorem applies, but little or no hindrance is observed in ^{19}F and ^{43}Sc for which the theorem does not apply. (These latter nuclei involve two neutrons and one proton in the orbital j; that is, they are not three identical particles.)

In the isospin formalism, a one-body operator typically has the structure

$$F^{(k_J,k_T)} = \Sigma_i \, f^{(k_J)}(\underset{\sim}{r}_i) \, g^{(k_T)}(\underset{\sim}{\tau}_i)$$

where k_J and k_T are the ranks of the operator in spin space and isospin space respectively, and $\underset{\sim}{r}_i$ and $\underset{\sim}{\tau}_i$ are the spin and isospin coordinates of the ith nucleon. Equations (6.28) to (6.31) still apply with the notation reinterpreted as discussed in appendix A. Now the rank k_T can only take the value zero or one; the operators are therefore known as either isoscalar or isovector operators respectively. The theorem (6.32) applies only to isoscalar operators since the isospin dependence trivially drops out, but does not hold for isovector operators. This is because $\Sigma_i \, \underset{\sim}{j}(\underset{\sim}{r}_i)\underset{\sim}{t}(\underset{\sim}{\tau}_i)$ does *not* equal $\underset{\sim}{J}\,\underset{\sim}{T}$, so that the operator cannot be trivially evaluated between states of configuration j^n. There is now no reason why an off-diagonal matrix element of an isovector operator of rank one in spin space should vanish. The M1 operator is in this category; it comprises an isoscalar and an isovector term, the first term obeying the theorem, the second term not.

Another application of the formula is to insert into eqn (6.28) a one-body operator whose reduced matrix elements are trivially evaluated in both single-particle and j^n configurations. This enables sum rules on the cfps to be derived. For example, suppose the operator $\underset{\sim}{J}$ is inserted once again in eqn (6.28), then the expression

$$[J(J+1)]^{1/2} = n \sum_{y_p J_p} \langle j^n x J\{|j^{n-1} y_p J_p; j\rangle^2 \, U(j1J_pJ;jJ)\,[j(j+1)]^{1/2}$$

is obtained, and explicitly evaluating the U-coefficient leads to the sum rule

$$\sum_{y_pJ_p} \langle j^n XJ\{|j^{n-1} y_pJ_p;j\rangle^2 J_p(J_p+1) = j(j+1) + \frac{n-2}{n} J(J+1).$$

Such relations provide useful checks on cfp tables.

6.6 TWO-BODY OPERATORS

In exact analogy to the last section, let us evaluate the expectation value of a two-body operator

$$G_q^{(k)} = \sum_{i<j} g_q^{(k)} (\underset{\sim}{r}_i,\underset{\sim}{r}_j)$$

of tensorial rank k, magnetic projection q between states of configuration j^n. Again the n nucleons are indistinguishable so that the value of the matrix element is given by the value for any one pair of nucleons multiplied by the number of pairs, viz.

$$\langle G^{(k)}\rangle \equiv \langle j^n(\underset{\sim}{r}_1\ldots\underset{\sim}{r}_n)XJ\| G^{(k)}\| j^n(\underset{\sim}{r}_1\ldots\underset{\sim}{r}_n)X'J'\rangle$$

$$= \frac{1}{2}\, n(n-1)\langle j^n(\underset{\sim}{r}_1\ldots\underset{\sim}{r}_n)XJ\| g^{(k)}(\underset{\sim}{r}_{n-1},\underset{\sim}{r}_n)\| j^n(\underset{\sim}{r}_1\ldots\underset{\sim}{r}_n)X'J'\rangle.$$

Next the two-particle cfp expansion, eqn (6.22), is used to break the j^n configurations into two parts

$$\langle G^{(k)}\rangle = \frac{1}{2}\, n(n-1) \sum_{\substack{z_pJ_pJ_2\\ z_p'J_p'J_2'}} \langle j^n XJ\{|j^{n-2}z_pJ_p;j^2J_2\rangle\langle j^n X'J'\{|j^{n-2}z_p'J_p';j^2J_2'\rangle \times$$

$$\times \langle [j^{n-2}(\underset{\sim}{r}_1\ldots\underset{\sim}{r}_{n-2})z_pJ_p;j^2(\underset{\sim}{r}_{n-1},\underset{\sim}{r}_n)J_2]XJ\| g^{(k)}(\underset{\sim}{r}_{n-1},\underset{\sim}{r}_n)\| \times$$

$$\times \quad [j^{n-2}(\underset{\sim}{r}_1\ldots\underset{\sim}{r}_{n-2})z_p'J_p';j^2(\underset{\sim}{r}_{n-1},\underset{\sim}{r}_n)J_2']X'J'\rangle$$

and using the reduction formula (A.38b) to simplify the reduced matrix element, the final result is obtained:

$$\langle G^{(k)}\rangle = \frac{1}{2}\, n(n-1) \sum_{\substack{z_pJ_p\\ J_2J_2'}} \langle j^n XJ\{|j^{n-2}z_pJ_p;j^2J_2\rangle\langle j^n X'J'\{|j^{n-2}z_pJ_p;j^2J_2'\rangle \times$$

$$\times\, U(J_2'kJ_pJ;J_2J')\langle j^2J_2\|g^{(k)}\|j^2J_2'\rangle_A. \tag{6.33}$$

The two-body matrix element on the right in eqn (6.33) is both antisymmetrized *and* normalized. The normalization is guaranteed from the way in which the two-particle cfps were constructed. By examining eqn (6.28) for the one-body operator and eqn (6.33) for the two-body operator, it is fairly straightforward to see how the matrix element for an r-body operator between states j^n can be evaluated with the aid of r-particle cfps.

For the most part we shall be concerned only with scalar two-body operators, such as potential energy $V = \sum_{i<j} V_{ij}$ for which eqn (6.33) simplifies to

$$\begin{aligned}\langle j^nXJ|V|j^nX'J'\rangle &= \tfrac{1}{2}\,n(n-1)\delta_{JJ'} \sum_{z_pJ_pJ_2} \langle j^nXJ\{|j^{n-2}z_pJ_p;j^2J_2\rangle \times \\ &\quad\times \langle j^nX'J\{|j^{n-2}z_pJ_p;j^2J_2\rangle\langle j^2J_2|V|j^2J_2\rangle_A \\ &= \sum_{J_2} A(njXX'JJ_2)\,\langle j^2J_2|V|j^2J_2\rangle_A.\end{aligned} \tag{6.34}$$

The coefficients $A(njXX'JJ_2)$ depend on the nuclear configurations of the basis states but not on the details of the interaction. They satisfy a simple sum rule

$$\sum_{J_2} A(njXX'JJ_2) = \tfrac{1}{2}\,n(n-1)\delta_{XX'}$$

which follows immediately from the orthogonality of the cfps. These coefficients have been tabulated by Band and Kharitonov (1971) for use in the proton-neutron formalism.

The usefulness of eqn (6.34) is that the matrix elements of n-particle configurations can be expressed in terms of relatively few two-body matrix elements. In fact, one approach has been to treat these two-body matrix elements as parameters, whose values are fixed by some least squares fitting procedure to a selection of experimental data. In this way the details of the nuclear interaction never have to be specified; it only has to be assumed that the interaction is two-body in character and that the states have a simple configuration structure. This procedure was first adopted by Goudsmit and Bacher (1934) in atomic spectroscopy, and was later applied to the calculation of nuclear binding energies and nuclear spectra by Talmi (Talmi and Unna 1960, Talmi 1962).

Let us illustrate the method by calculating the energy spectra in a closed-shell-plus-three nucleus, using a severely truncated model space, namely confining the three valence particles to a single j orbital. To calculate the effective interaction for this model space starting from some realistic G-matrix would be a horrendous task, so this approach is dropped immediately. Instead an effective interaction containing only zero-body, one-body and two-body terms is assumed. Any three-body or higher-body terms are neglected. Furthermore only the relative level spacings are calculated so that the zero-body and one-body terms can be dropped as they only add a constant to the absolute energies.

It remains to fix the two-body matrix elements

$$V_J = \langle j^2 J | V | j^2 J \rangle_A,$$

or rather the differences $V_J - V_0$. These differences are given by the energy spacings between the J-state and the ground state of the j^2 configuration, and can, for example, be taken from the experimental energy differences in the closed-shell-plus-two nucleus. Thus for two neutrons in the $d_{5/2}$ orbit, $V_2 - V_0$ is fixed at 1.982 MeV, the excitation energy of the first 2^+ state in ^{18}O, and $V_4 - V_0$ at 3.553 MeV. The prediction for the spectra of ^{19}O, assuming a $5/2^3$ configuration for the three neutrons, is given in Table 6.3. We have denoted the excitation energy of the state of spin J by E(J) which in terms of the model is given by

$$E(J) = \langle j^3 J | V | j^3 J \rangle - \langle j^3\ J{=}j | V | j^3\ J{=}j \rangle$$

$$E(J) = \sum_{J_2,\text{even}} a^J_{J_2} \; (V_{J_2} - V_0) \qquad (6.35)$$

TABLE 6.3

Prediction for the energy levels of ^{19}O *and* ^{43}Ca*, assuming a* j^3 *neutron configuration. The relative energy* E(J) *is given by eqn (6.35) with the coefficients* $a^J_{J_2}$ *extracted from the tables of Band and Kharitonov (1971). Experimental data from compilations of Ajzenberg-Selove (1972) and Poletti et al. (1976).*

Nucleus	J^π	a^J_2	a^J_4	a^J_6	E(J) MeV	Expt. MeV
^{19}O	$3/2^+$	55/42	-9/14	-	0.312	0.096
^{19}O	$9/2^+$	-4/21	6/7	-	2.668	2.371
^{43}Ca	$5/2^-$	17/12	-25/44	-13/132	0.282	0.373
^{43}Ca	$3/2^-$	19/84	45/28	-13/12	1.310	0.593
^{43}Ca	$11/2^-$	5/12	-7/44	65/132	1.768	1.678
^{43}Ca	$9/2^-$	-3/28	369/308	-15/44	2.045	(2.094)
^{43}Ca	$15/2^-$	-5/12	-3/44	163/132	3.117	2.754

where the coefficients $a^J_{J_2}$ can easily be expressed in terms of the coefficients $A(njXX'JJ_2)$, tabulated by Band and Kharitanov (1971). Note that in this simple illustration the labels X are redundant, since there is only one state of spin J in the configuration j^3 with $j \leq 7/2$; thus evaluating the expectation value of the interaction V is equivalent (to within a constant) to evaluating the energy of the state. In more complicated cases one would evaluate the matrix elements of V and diagonalize the Hamiltonian matrix to obtain the energy.

Returning then to Table 6.3, the agreement between the predictions and the experimental excitation energies for the $3/2^+$ and $9/2^+$ state in ^{19}O is seen to be good - the discrepancies being of order 300 keV. Similarly for ^{43}Ca, with the parameters $V_2 - V_0 = 1.524$ MeV, $V_4 - V_0 = 2.751$ MeV, and $V_6 - V_0 = 3.190$ MeV taken from the ^{42}Ca spectrum, the agreement is again of order 300 keV, but there is one glaring exception. The exception is the $3/2^-$ state where the discrepancy is twice as large, 700 keV. This points out the limitation, or possibly even the usefulness of the method, since the breakdown of the model signifies the presence of a nearby state, a so-called 'intruder' state, whose basic configuration is not contained in the model space. That a $3/2^-$ intruder state should exist comes as no surprise. The next unfilled orbital for the calcium isotopes is a $p_{3/2}$ orbital, therefore a state of configuration $f_{7/2}{}^2 p_{3/2}$ may well mix strongly with the state $f_{7/2}{}^3$.

This conjecture is borne out by the calculations of Federman and Pittel (1970). In their work, the model space for the description of the calcium isotopes spans the $f_{7/2}$ and $p_{3/2}$ orbitals. The effective interaction is then specified by two one-body matrix elements, ten diagonal, and five off-diagonal two-body matrix elements. The off-diagonal matrix elements were taken from the G-matrix of Kuo and Brown (1968), while the diagonal matrix elements were determined in a least square fit to 38 selected energy levels in the calcium isotopes, following the spirit of the Talmi method. The calculated spectrum is very close to the values E(J) listed in Table 6.3 with the exception of the $3/2^-$ state which shifts down by 430 keV to lie at an excitation energy of 0.875 MeV. This is still high compared with the experimental position of 0.593 MeV but the discrepancy, of order of 300 keV, is now compatible with the level of agreement obtained for the other J-states. Furthermore, the wavefunctions for all states but one are still better than 95 per cent pure $f_{7/2}{}^3$, the exception being the $3/2^-$ state, which is found to have approximately 78 per cent $f_{7/2}{}^3$ and 22 per cent $f_{7/2}{}^2\ p_{3/2}$ configurations.

6.7 SENIORITY, AVERAGE ENERGIES

We now return to the problem of classification of many-body states $|j^n\ XJM\rangle$ and the precise specification of the label X. We confine our discussion to the example of j-j coupling of n identical particles (all protons or all neutrons) in a single j shell.

Consider the set of basis states $|j^n\, XJM\rangle$. The labels n, J and M characterize the irreducible transformation properties under the groups SU_{2j+1}, R_3, and R_2. Each of these groups is a subgroup of the previous one, and the group R_2, being one-dimensional, is clearly the end of the chain. The problem now is to try and find a subgroup of SU_{2j+1} which itself contains R_3 as a subgroup.

We start by introducing Racah's unit tensor operator, u^k_q, (Racah 1943) defined for single-particle states by its matrix elements

$$\langle j'm'|u^k_q|jm\rangle = \langle kqjm|j'm'\rangle\, \delta_{jj'}$$

or equivalently

$$\sum_{qm} \langle kqjm|j'm'\rangle\, u^k_q|jm\rangle = \delta_{jj'}|j'm'\rangle.$$

Applying this result twice, it is straightforward to show that

$$\sum_{q_1q_2} \langle k_1q_1k_2q_2|kq\rangle\, u^{k_1}_{q_1}\, u^{k_2}_{q_2} = U(k_1k_2jj;kj)u^k_q.$$

That is the set of $(2j+1)\times(2j+1)$ operators u^k_q, $k = 0,1,\ldots,2j$, $q = -k,\ldots, +k$ form a closed group. They are known as the infinitesmal generators of the group U_{2j+1}. The corresponding many-particle operators

$$U^k_q = \sum_{i=1}^{n} u^k_q(i)$$

are the infinitesimal generators of the group of unitary substitutions U_{2j+1} on n-particle wavefunctions. These many-body operators satisfy the same commutation relations as the single-particle operators, that is

$$U^{k_1}_{q_1}\, U^{k_2}_{q_2} = \sum_{kq} (k_1q_1k_2q_2|kq)\, U(k_1k_2jj;kj)U^k_q$$

$$U^{k_2}_{q_2}\, U^{k_1}_{q_1} = \sum_{kq} (-)^{k_1+k_2-k}(k_1q_1k_2q_2|kq)U(k_1k_2jj;kj)U^k_q,$$

which on subtracting give the result

$$[U^{k_1}_{q_1}, U^{k_2}_{q_2}]_- = \sum_{kq} (1-(-)^{k_1+k_2-k})(k_1q_1k_2q_2|kq)U(k_1k_2jj;kj)U^k_q.$$

We now harness the following theorem (Bayman 1957): if a sub-set $U^{k'}_{q'}$ of the U^k_q can be found such that the commutator of any pair in the sub-set is expressible as a linear combination of only the $U^{k'}_{q'}$ in the sub-set, then the members of this sub-set are the generators of a sub-group of the full unitary group U_{2j+1}. This is important, since corresponding to each sub-group, we can associate a quantum number which assists in the classification of the many-body states $|j^n\ XJM\rangle$.

First, note that if U^0_0 is to occur on the right-hand side of the commutation relations, the Clebsch-Gordan coefficient requires that $k_1 = k_2$. But then the factor $(1 - (-)^{k_1+k_2-k})$ vanishes. Thus we can be certain that U^0_0 will never occur in such an expansion. Therefore the set of all U^k_q except U^0_0 will generate a sub-group of U_{2j+1}. This sub-group is known as the unimodular SU_{2j+1} group. The operator U^0_0 is simply the unit operator, and its eigenvalue for many-particle states is just the number of particles, n.

Next, note that if $k_1 = k_2 = 1$, the triangle conditions allow k to equal 0, 1 or 2. But for k = 0 or 2 the factor $[1 - (-)^{k_1+k_2-k}]$ vanishes. Thus we have that the three operators U^1_{-1}, U^1_0, U^1_{+1} generate a sub-group. Their commutation relations, in fact, are just those of the angular momentum operators, $\underset{\sim}{J}_q$, so we identify

$$U^1_q = -[j(j+1)]^{-1/2}\ \underset{\sim}{J}_q$$

and assert that the three operators U^1_q are the generators of the rotation group R_3. The eigenvalues associated with this group are of course the total angular momentum J and the magnetic projection M.

Next, consider the situation in which the k_1 and k_2 are odd. Then the factor $[1 - (-)^{k_1+k_2-k}]$ ensures that only odd k occur in the expansion. Hence the set of all U^k_q, odd k are the generators of a group, known as the sympletic group,

Sp_{2j+1}, and the associated eigenvalue for this group is known as the seniority quantum number, v. We therefore identify the label X with the seniority v.

The classification problem has not, however, been completely solved. For the higher spin orbitals, $j \geqslant 9/2$, there exists more than one state with a given set of quantum numbers, n, v, J, and M, so an additional label or labels are still required. However, it has not been possible to find any further sub-groups of U_{2j+1}, so the additional labels have to remain arbitrary and an orthogonal set of states $|j^n\ XJM\rangle$ is generated by imposing orthogonality on the cfp expansions, eqn (6.11).

Consider, for the moment, the two-particle states $|j^2\ JM\rangle$ with $J = 0,2,4,\ldots,2j-1$. The antisymmetry requirement limits J to even values only. Then for odd k we have that

$$\langle j^2\ JM|U^k_q|j^2\ 00\rangle = 0 \qquad \text{odd, } k$$

since the triangle relations require $J = k$ for a non-vanishing matrix element and we have already noted that J has to be even. Therefore the set of operators U^k_q, odd k, the generators of the sympletic group, leave the two-particle state $|j^2\ 00\rangle$ invariant. This particular state is called a seniority zero, $v = 0$, state. All other two-particle states with $J \neq 0$ are defined to have seniority two.

We can generalize these ideas by considering the state $|j^n\ XJM\rangle$, where the label X characterizes the transformation properties of the state under the group of operators Sp_{2j+1}. Consider the two-particle fractional parentage expansion (eqn (6.22)):

$$|j^n\ XJ\rangle = \sum_{\substack{X_1 J_1\\ X_2 J_2}} \langle j^n\ XJ\{|j^{n-2}X_1J_1;j^2X_2J_2\rangle[|j^{n-2}X_1J_1\rangle|j^2X_2J_2\rangle]^J_M.$$

We note that if $\langle j^n\ XJ\{|j^{n-2}X_1J_1;j^2\ 00\rangle \neq 0$, then the state $|j^n\ XJM\rangle$ contains a term $[|j^{n-2}\ XJ\rangle\ |j^2\ 00\rangle]^J_M$. That is the j^{n-2} configuration has a term with the irreducible transforma tion properties XJ. Similarly if $\langle j^{n-2}\ XJ\{|j^{n-4}\ XJ;j^2 00\rangle \neq 0$, there will be a term in the configuration j^{n-4} with these labels XJ. Continuing in this way, a point is reached for which $\langle j^{v+2}\ XJ\{|j^v\ XJ;j^2\ 00\rangle \neq 0$ but $\langle j^v\ XJ\{|j^{v-2}\ XJ;j^2\ 00\rangle = 0$; then the states $|j^v\ XJ\rangle$, $|j^{v+2}\ XJ\rangle,\ldots,|j^n\ XJ\rangle$ are all said to have a seniority v. That is, a state $|j^v\ XJ\rangle$ has seniority v if all cfps $\langle j^v\ XJ\{|j^{v-2}\ X_1J;j^2\ 00\rangle = 0$ for every X_1. Equivalently we can say that a state $|j^n\ vJ\rangle$ has seniority v if it can be obtained by adding successively pairs coupled to $J = 0$ to the parent state $|j^v\ vJ\rangle$ and antisymmetrizing. The number v essentially counts the number of particles not contained in $J = 0$ coupled pairs.

An alternative way of discussing this is to define a scalar two-body operator $Q = \sum_{i<j} q_{ij}$ where

$$\langle j^2\ JM|q_{12}|j^2\ JM\rangle = (2j+1)\delta_{J,0}.$$

hen to every state $|j^n\ vJM\rangle$ with a non-vanishing expectation alue of Q, there is a state j^{n-2} with the same irreducible ransformation properties under Sp_{2j+1}. Evaluating the expecation value of Q for an n-particle state follows straightorwardly from eqn (6.34), viz.

$$\langle j^n v'J|Q|j^n vJ\rangle = \frac{n(n-1)}{2} \sum_{v_1J_1J_2} \langle j^n v'J\{|j^{n-2}v_1J_1;j^2J_2\rangle\langle j^n vJ\{|j^{n-2}v_1J_1;j^2J_2\rangle \times$$

$$\times \langle j^2J_2|Q|j^2J_2\rangle$$

$$= (2j+1)\ \frac{n(n-1)}{2} \langle j^n vJ\{|j^{n-2}vJ;j^2 0\rangle^2\ \delta_{vv'},$$

hat is Q is diagonal in the seniority scheme, and its expecation value is zero for a state with $n = v$.

To simplify this expression further requires evaluating he two-particle cfp. This is a little tedious and has been pelled out in de Shalit and Talmi's book (1963, Chap. 27), so t will not be repeated here. The final result is

$$\langle j^n v'J|Q|j^n vJ\rangle = \tfrac{1}{2}(n-v)(2j+3-n-v)\delta_{vv'}.$$

Note the eigenvalue is independent of J, as of course it should be.

Let us now apply these ideas to the calculation of the energy spectrum for nuclei describable by j^n configurations. As in section (6.6), we again consider the j = 5/2 shell. The number of two-body matrix elements $V_{J_2} = \langle \frac{5}{2}^2 \; J_2 | V | \frac{5}{2}^2 \; J_2 \rangle$ necessary to span this space is just three, V_0, V_2, and V_4. Thus without any loss of generality, the interaction V can be written in terms of any three independent two-body operators, viz.

$$V_{ij} = a + 2b \; \underset{\sim}{j}(\underset{\sim}{r}_i)\cdot\underset{\sim}{j}(\underset{\sim}{r}_j) + c \; q_{ij} \tag{6.36}$$

where parameters a, b, c now replace the parameters V_0, V_2, and V_4.

This choice of two-body operators was not arbitrary. The unit operator has the number of particles as its eigenvalue,

$$\langle j^n \; XJ | 1 | j^n \; X'J' \rangle = \frac{n(n-1)}{2} \; \delta_{XX'}\delta_{JJ'},$$

the angular momentum operator has J as its eigenvalue,

$$\langle j^n \; XJ | \; 2 \sum_{i<j} \; \underset{\sim}{j}(\underset{\sim}{r}_i)\cdot\underset{\sim}{j}(\underset{\sim}{r}_j) | j^n \; X'J' \rangle$$

$$= \langle j^n \; XJ | \underset{\sim}{J}^2 - \Sigma_i \; \underset{\sim}{j}(\underset{\sim}{r}_i)^2 | j^n \; X'J' \rangle$$

$$= \delta_{XX'}\delta_{JJ'}[J(J+1)-nj(j+1)],$$

and the expectation value of the seniority operator, Q, has already been spelled out.

These three operators are chosen because they can be represented in terms of the generators, U^k_q, of the group U_{2j+1}, or more precisely in terms of the Casimir operators for the subgroups SU_{2j+1}, R_3, and Sp_{2j+1} respectively. It is a property of Casimir operators that their eigenvalues depend only on the representation labels of the basis states $|j^n\, vJ\rangle$.

Thus returning to the problem of calculating the spectra in a model space $5/2^n$ using the general interaction (6.36), the matrix elements in the n-particle states can immediately be written down as

$$\langle \tfrac{5}{2}^n\ vJ|V|\tfrac{5}{2}^n\ vJ\rangle = a\,\frac{n(n-1)}{2} + b[J(J+1) - \frac{35n}{4}] + \\ + \frac{c}{2}(n-v)(8-n-v). \qquad (6.37)$$

If, as before, the parameters are fixed to the spectra of the closed-shell-plus-two nucleus, then inserting in eqn (6.37) $n = 2$, $v = 0$ for the state with $J = 0$, and $v = 2$ for the states with $J = 2$ and 4, the following relations are obtained:

$$V_0 = a - \frac{35}{2}b + 6c$$

$$V_2 = a - \frac{23}{2}b$$

$$V_4 = a + \frac{5}{2}b.$$

Inverting these equations gives

$$a = \frac{1}{28}(5V_2 + 23V_4)$$

$$b = \frac{1}{14}(V_4 - V_2)$$

$$c = \frac{1}{6}V_0 - \frac{1}{42}(10V_2 - 3V_4). \qquad (6.38)$$

Thus, if eqn (6.38) is substituted back into eqn (6.37), one obtains expressions for the coefficients $A(njvv'JJ_2)$ of eqn (6.34) for the case of the $j = 5/2$ orbital. These coefficients have been found now without resorting to the use of fractional parentage coefficients.

The method cannot be extended to the $j = 7/2$ shell, since four parameters are now required to represent a two-body interaction in that model space, and it is not possible to find four scalar two-body operators whose matrix elements are trivially evaluated in j^n-states. However, if one is satisfied with a calculation of *average energies* over groups of states with the same seniority in the j^n configuration, then the above method can be applied after a slight modification.

Consider first the averages of states with the same seniority in the j^2 configuration. Since there are only two possible values for the seniority v, namely v = 0 and v = 2, these averages can be represented by the matrix elements of an arbitrary interaction V' containing just two parameters. Thus in analogy with eqn (6.36), V' is written as

$$V'_{ij} = a + c\, q_{ij} \tag{6.39}$$

and the requirement is imposed that the interaction eqn (6.39) will have the same averages as the original interaction V, i.e.

$$\begin{aligned} a + (2j+1)c &= V_0 \\ a &= \overline{V}_2 \end{aligned} \tag{6.40}$$

where

$$V_0 = \langle j^2\ v{=}0\ J{=}0|V|j^2\ v{=}0\ J{=}0\rangle$$

and

$$\begin{aligned} \overline{V}_2 &= \sum_J (2J+1)\,\langle j^2\ v{=}2\ J|V|j^2\ v{=}2\ J\rangle / \sum_J (2J+1) \\ &= \frac{1}{(j+1)(2j-1)} \sum_{\substack{J>0\\ \text{even}}} (2J+1)V_J . \end{aligned}$$

Inverting eqn (6.40) shows that the parameters a and c must satisfy the relationship

$$\begin{aligned} a &= \overline{V}_2 \\ c &= \frac{1}{2j+1}\,(V_0 - \overline{V}_2) . \end{aligned} \tag{6.41}$$

Next we would like to evaluate the average energies in the j^n configuration, and for this the following theorem is required:

The averages of *any* given interaction V over all states with the same seniority v in the j^n configuration are linear combinations of V_0 and $\overline{V}_2$ with coefficients which are *independent* of the interaction.

The proof of this statement is given in de Shalit and Talmi (1963), and we shall not give the details here. The argument is essentially one of induction. If the theorem holds for all configurations j^{n-1} and for all seniorities, then it can be shown to hold for all configurations j^n. The theorem is trivially true for n = 2.

The theorem is exploited by evaluating the coefficients for V_0 and $\overline{V}_2$ using the very simple interaction V' (eqn (6.39)). Thus

$$\sum_{XJ} (2J+1) \langle j^n XvJ|V|j^n XvJ\rangle / \sum_{XJ} (2J+1)$$

$$= \langle j^n vJ|V'|j^n vJ\rangle$$

$$= \frac{n(n-1)}{2}\overline{V}_2 + \left[\frac{n-v}{2}(2j+2) - \frac{n(n-1)}{2} + \frac{v(v-1)}{2}\right]\left(\frac{V_0-\overline{V}_2}{2j+1}\right), \tag{6.42}$$

producing a result depending only on n and v, and being independent of J as should be expected.

There are two very simple cases for which there is only one state contained in the average, namely the J=0 state when v=0 and the J=j state when v=1. In these specific cases, eqn (6.42) can be simplified to read

$$\langle j^n vJ|V|j^n vJ\rangle = \frac{n(n-1)}{2(2j+1)}[(2j+2)\overline{V}_2-V_0] + \frac{n-v}{2}\frac{2j+2}{2j+1}(V_0-\overline{V}_2), \text{ v=0 or 1.} \tag{6.43}$$

The ground states of even-even nuclei have spin J=0 and to a quite good approximation can be considered to be seniority zero states. Likewise, odd nuclei have a spin equal to that of the last unpaired nucleon J=j and similarly can be considered seniority one states. Thus for a sequence of nuclei whose dominant configurations are j^n, eqn (6.43) can be used to estimate the contribution from the two-body part of the effective Hamiltonian to their ground-state binding energies.

Let us illustrate the procedure with reference to the calcium isotopes. In Table 6.4 are listed the ground-state binding energies of these isotopes expressed relative to that of ^{40}Ca. The effective interaction is assumed to contain a

TABLE 6.4

Binding energies of the ground states of the calcium isotopes and the calculated two-body contributions obtained from eqn (6.43)

Calcium Isotope	Binding Energy[a] (MeV)	Binding Energy[b] (2-body part)	Calculated Energy (MeV) From ^{42}Ca[c]	Calculated Energy (MeV) Fitted[d]
41	-8.364			
42	-19.837	-3.109	-3.109	-3.129
43	-27.769	-2.677	-3.173	-2.679
44	-38.906	-5.450	-6.347	-5.358
45	-46.320	-4.500	-6.475	-4.459
46	-56.724	-6.540	-9.713	-6.688
47	-64.000	-5.452	-9.905	-5.339
48	-73.951	-7.039	-13.207	-7.118

a Binding energy relative to that for ^{40}Ca taken from mass tables of Wapstra and Gove (1971).

b BE($^{40+n}$Ca) - n BE(^{41}Ca).

c From eqn (6.43) with V_0 = -3.109, $\overline{V}_2$ = -0.374 MeV.

d From eqn (6.43) with V_0 = -3.129, $\overline{V}_2$ = -0.148 MeV.

one-body and a two-body part, and the one-body contribution is determined from the binding energy of ^{41}Ca. Thus in column three of Table 6.4 are listed the binding energies of the calcium isotopes, $^{40+n}$Ca, minus n times the binding energy of ^{41}Ca. This represents just the two-body contribution to the ground-state energy and is the quantity to be fitted by eqn (6.43).

Next, the parameters V_0 and $\overline{V}_2$ have to be chosen. There are two approaches. One is that they can be determined from the available data on the ^{42}Ca spectrum, in which case V_0 is fixed from the ground-state energy V_0 = -3.109 MeV and $\overline{V}_2$ is an average of the excited-state energies V_2 = -1.585, V_4 = -0.358 and V_6 = +0.081 MeV, weighted according to eqn (6.40) to give $\overline{V}_2$ = -0.374 MeV. The second approach is to determine V_0 and $\overline{V}_2$ by a least squares fit to the data in column three of Table 6.4. The results of both methods are given in columns four and five.

First it is noted that fitting the parameters V_0 and $\overline{V}_2$ to the spectrum of ^{42}Ca gives a rather poor description of the binding energies of the remaining calcium isotopes. This is an indication of the inadequacy of our assumptions; either the configurations are not pure $f^n_{7/2}$ or the effective interaction contains a significant three-body and possibly higher-body contributions. This latter possibility has received some attention recently (Quesne 1970, Eisenstein and Kirson 1973, Yariv 1974, Koltun 1973), and the consensus is that for neutrons in the $f_{7/2}$ shell the average strength of an effective three-body force is about 120 keV per triplet of particles (Koltun 1973), and is repulsive.

However, just because the parameters V_0 and $\overline{V}_2$ deduced from ^{42}Ca do not fit the binding energies of the calcium isotopes is not in itself a sufficient justification to invoke three-body forces. The data should also be tested to see whether they are compatible with eqn (6.43). In column five of Table 6.4 are the results of a least squares fit; the resulting parameters are V_0 = -3.13 ± 0.02 MeV and $\overline{V}_2$ = -0.148 ± 0.005 MeV. One notes that V_0 is essentially the same as that deduced from ^{42}Ca, while $\overline{V}_2$ has been reduced by half. However, it is the combination of parameters $9\overline{V}_2 - V_0$ contributing to the first term of eqn (6.43) that is relevant to the binding energies. This combination changes sign, being -0.284 MeV using the ^{42}Ca parameters and +1.787 MeV from the least squares fit. The quality of the fit can be gauged from the table, the mean deviation being of order 100 keV. This gives some limit on the magnitude of three-body forces.

So far, our remarks have been confined to systems of identical particles (all protons or all neutrons); however, most of the techniques illustrated above can be used in the isospin formalism. A seniority operator Q can be defined, with eigenvalues (see eqn (34.26) of de Shalit and Talmi (1963))

$$\begin{aligned}\langle j^n\, vt\; JT|Q|j^n\, vt\; JT\rangle \\ = \frac{n-v}{4}(4j+8-n-v) - T(T+1) + t(t+1),\end{aligned} \tag{6.44}$$

which vanish when n=v. The state $|j^n\, vt\; JT\rangle$ is said to have seniority v and *reduced isospin* t, if it can be obtained by adding successively pairs of particles coupled to J=0 and T=1 to the principal parent $|j^v\, vt\; Jt\rangle$ and antisymmetrizing. Note that states with configuration j^n have the same spin J as the principal parent j^v; however, the value of T need not be the same. Thus the states j^n are characterized by an additional label t, being the isospin of the principal parent (Flowers 1952).

For the problem of calculating spectra, matrix elements of scalar two-body operators are still given by eqn (6.34) in terms of two-particle cfps in the isospin formalism. The number of two-body matrix elements necessary to characterize the interaction in a single j-shell is now (2j+1), twice as many as before, since both T=0 and T=1 matrix elements are required. For $j \leqslant 3/2$, the use of fractional parentage coefficients can be avoided by writing the interaction in terms of four simple two-body operators, viz.

$$V_{ij} = a + 2b\; \underset{\sim}{j}(r_i)\cdot\underset{\sim}{j}(\underset{\sim}{r}_j) + 2c\; \underset{\sim}{t}_i\cdot\underset{\sim}{t}_j + d\; q_{ij}.$$

Then the matrix elements are given by

$$\langle j^n\, vt\; JT|V|j^n\, vt\; JT\rangle = a\,\frac{n(n-1)}{2} + b[J(J+1)-nj(j+1)]+c[T(T+1)-\frac{3}{4}n] +$$

$$+ \; d[\tfrac{n-v}{4}(4j+8-n-v)-T(T+1)+t(t+1)]. \qquad (6.45)$$

As before, the parameters a, b, c, and d can be fixed from the spectra of the closed-shell-plus-two nuclei.

Similarly for shells with $j > 3/2$, an expression can be found for the average energy of a group of states with the same seniority and reduced isospin in the j^n configuration. For two particles there are three such groups to consider, viz.

$$V_0 = \langle j^2 \; v=0 \; t=1 \; J=0 \; T=1|V|j^2 \; v=0 \; t=1 \; J=0 \; T=1\rangle,$$

$$\bar{V}_{\text{even}} = \sum_{\substack{J>0\\ \text{even}}} \frac{(2J+1)\langle j^2 \; v=2 \; t=1 \; J \; T=1|V|j^2 \; v=2 \; t=1 \; J \; T=1\rangle}{(j+1)(2j-1)},$$

$$\bar{V}_{\text{odd}} = \sum_{\substack{J\\ \text{odd}}} \frac{(2J+1)\langle j^2 \; v=2 \; t=0 \; J \; T=0|V|j^2 \; v=2 \; t=0 \; J \; T=0\rangle}{(j+1)(2j+1)}, \qquad (6.46)$$

therefore the average energy can be represented by an interaction V' containing just three parameters. Thus V' is written

$$V'_{ij} = a + 2c \; \underset{\sim}{t}_i \cdot \underset{\sim}{t}_j + d \; q_{ij}$$

and the requirement is imposed that this interaction produces the same averages, eqn (6.46), as the original interaction. This fixes the parameters as

$$a = \frac{1}{4}(3\,\bar{V}_{\text{even}} + \bar{V}_{\text{odd}})$$

$$c = \frac{1}{2}(\bar{V}_{\text{even}} - \bar{V}_{\text{odd}}) \qquad (6.47)$$

$$d = \frac{1}{2j+1}(V_0 - \bar{V}_{\text{odd}})$$

and leads to an average energy in the j^n configuration given by

$$\sum_{XJ} (2J+1) \langle j^n \, Xvt \, JT|V|j^n \, Xvt \, JT\rangle / \sum_{XJ} (2J+1)$$

$$= \frac{n(n-1)}{8} [3 \bar{V}_{even} + \bar{V}_{odd}] + \frac{1}{2} [T(T+1) - \frac{3}{4} n](\bar{V}_{even} - \bar{V}_{odd}) +$$

$$+ \frac{1}{(2j+1)} [\frac{n-v}{4} (4j+8-n-v) - T(T+1) + t(t+1)](V_0 - \bar{V}_{odd}). \quad (6.48)$$

For v=0 and v=1, there is just one state in the average for any given T, so that eqn (6.48) gives directly the value of that one matrix element. Furthermore, the ground states of even-even and odd-even nuclei are to a good approximation described by the lowest seniority state with configuration j^n. Thus eqn (6.48) can be harnessed in the same way that eqn (6.43) was used to estimate the contribution from the two-body part of the effective Hamiltonian to ground-state binding energies. One must remember, however, in using experimental data, to subtract out the Coulomb contribution. The isospin formalism is based on a charge-symmetric Hamiltonian. The Coulomb force, of course, breaks this symmetry.

6.8 HOLES AND PARTICLES

There is a one-one correspondence between states of configuration j^n and states of configuration $j^{n_{max}-n}$ (which we shall abbreviate to j^{-n}) where n_{max} is the number of nucleons that completely fills the single j-shell. In the proton-neutron formalism n_{max} equals 2j+1, and in the isospin formalism n_{max} = 2(2j+1). We shall call j^n the n-particle configuration and j^{-n} the conjugate n-hole configuration.

The first statement to be proved is that: 'the relative energy spacings in a nucleus described by an n-particle configuration j^n are idential to the spacings in the conjugate nucleus j^{-n} under the assumption of two-body effective interactions'. To see this, we go back to the m-scheme and write the states in a second-quantization notation

$$|j^n \, m_1 m_2 \ldots m_n\rangle = a^\dagger_{m_n} \ldots . a^\dagger_{m_2} \, a^\dagger_{m_1} \, |0\rangle$$

where only the magnetic substate needs be specified in the subscript on the creation operators. Note that each m_i must necessarily be different (Pauli principle). The expectation value of a two-body operator V between these states is given by eqn (1.32) and is

$$\langle j^n\, m_1 \ldots m_n | V | j^n\, m_1 \ldots m_n \rangle = \sum_{i<j=1}^{n} \langle m_i m_j | V | m_i m_j \rangle_A . \quad (6.49)$$

Similarly an off-diagonal matrix element must have two m-values different (remember that total $M = \sum_i m_i$ is conserved and all m_i are different), so that from eqn (1.33)

$$\langle j^n \ldots m_i \ldots m_j \ldots | V | j^n \ldots m_i' \ldots m_j' \ldots \rangle = \langle m_i m_j | V | m_i'\, m_j' \rangle_A . \quad (6.50)$$

Next consider the conjugate hole states, and remember the definition of a single-hole state (eqn (1.65))

$$|m_i^{-1}\rangle = b_{m_i}^{\dagger} |C\rangle = (-)^{j+m_i} a_{-m_i} |C\rangle = a_{\tilde{m}_i} |C\rangle$$

where $|\tilde{m}_i\rangle$ is the time-reversed state. In this case $|C\rangle$ represents the closed single j shell as defined in eqn (1.63),

$$|C\rangle = a_j^{\dagger}\, a_{j-1}^{\dagger} \ldots a_{-j+1}^{\dagger}\, a_{-j}^{\dagger}\, |0\rangle .$$

The conjugate n-hole state is now written

$$|j^{-n}\, m_1^{-1}\, m_2^{-1} \ldots m_n^{-1}\rangle = b_{m_n}^{\dagger} \ldots b_{m_2}^{\dagger}\, b_{m_1}^{\dagger}\, |C\rangle , \quad (6.51)$$

this state having total magnetic projection $M = \sum_i m_i$.

First let us calculate the diagonal matrix element between states (eqn (6.51)). The two-body interaction V is written in terms of hole creation operators in eqn (1.90) and there are three terms giving a contribution, eqns (1.90i), (1.90m) and (1.90n), thus

$$\langle j^{-n}\, m_1^{-1}\, m_2^{-1} \ldots m_n^{-1} | V | j^{-n}\, m_1^{-1}\, m_2^{-1} \ldots m_n^{-1}\rangle$$

$$= \sum_{i<j=1}^{n} \langle \tilde{m}_i\tilde{m}_j|V|\tilde{m}_i\tilde{m}_j\rangle_A -$$

$$- \sum_{m=-j}^{+j} \sum_{i=1}^{n} \langle m\ \tilde{m}_i|V|m\ \tilde{m}_i\rangle_A +$$

$$+ \frac{1}{2} \sum_{m,m'=-j}^{+j} \langle mm'|V|mm'\rangle_A. \tag{6.52}$$

Consider just the last two terms and denote their contribution to eqn (6.52) by E_0. Then we shall show that E_0 is just a constant, independent of m_i, by expressing the two-body matrix elements appearing in eqn (6.52) in terms of angular momentum coupled matrix elements as in eqn (1.39). Thus

$$E_0 = -2\sum_{i=1}^{n} \sum_{mJ_0M_0} \langle jm\ j\text{-}m_i|J_0M_0\rangle\langle jm\ j\text{-}m_i|J_0M_0\rangle\langle j^2J_0|V|j^2J_0\rangle_A +$$

$$+ \sum_{mm'J_0M_0} \langle jm\ jm'|J_0M_0\rangle\langle jm\ jm'|J_0M_0\rangle\langle j^2J_0|V|j^2J_0\rangle_A$$

$$= -2\sum_{i=1}^{n} \sum_{J_0} \frac{(2J_0+1)}{(2j+1)}\langle j^2J_0|V|j^2J_0\rangle_A + \sum_{J_0M_0} \langle j^2J_0|V|j^2J_0\rangle_A$$

$$= \left(1 - \frac{2n}{2j+1}\right) \sum_{J_0} (2J_0+1)\rangle\langle j^2J_0|V|j^2J_0\rangle_A. \tag{6.53}$$

The factor of two originates in the definition of a normalized, antisymmetrized, coupled two-particle state $|j^2\ J_0\rangle$. Combining eqns (6.49), (6.52), and (6.53), and noting that $\langle \tilde{m}_i\tilde{m}_j|V|\tilde{m}_i\tilde{m}_j\rangle_A$ equals $\langle m_im_j|V|m_im_j\rangle_A$, gives

$$\langle j^{-n}\ m_1^{-1}\ldots m_n^{-1}|V|j^{-n}\ m_1^{-1}\ldots m_n^{-1}\rangle$$

$$= E_0 + \langle j^n m_1\ldots m_n|V|j^n m_1\ldots m_n\rangle. \tag{6.54}$$

Similarly the off-diagonal matrix element is given by

$$\langle j^{-n}\ldots m_i^{-1}\ldots m_j^{-1}\ldots|V|j^{-n}\ldots m_1'^{-1}\ldots m_j'^{-1}\ldots\rangle$$

$$= \langle \tilde{m}_i' \ \tilde{m}_j' | V | \tilde{m}_i \tilde{m}_j \rangle. \qquad (6.55)$$

So by examining the structure of eqns (6.50), (6.54), and (6.55), it is seen that the whole matrix of the interaction V in the n-hole configuration taken in the m-schemes, is equal to a constant E_0 plus the transposed matrix in the same scheme of the conjugate n-particle configuration. The eigenvalues of V taken in the j^{-n} configuration are therefore equal to E_0 plus the eigenvalues of V in the j^n configuration. The level spacings in the two configurations are therefore equal.

This proves the statement made at the beginning of the section. Although the proof resorts to the m-scheme, it can be used for angular momentum coupled states, since the allowed angular momenta in the n-hole configuration are identical to those allowed in the n-particle configuration. Thus states which have the same energy in the two configurations (apart from a constant E_0) have the same total angular momentum.

Let us examine this statement with reference to the calcium isotopes. In Table 6.5 are listed the experimental

TABLE 6.5

Experimental energy level spacings in ^{42}Ca *and* ^{43}Ca *and in their respective conjugate nuclei* ^{46}Ca *and* ^{45}Ca, *for states whose configuration is predominantly* $f_{7/2}{}^n$.

J^π	$^{42}Ca^a$	$^{46}Ca^b$	J^π	$^{43}Ca^c$	$^{45}Ca^d$
0^+	0.0	0.0	$7/2^-$	0.0	0.0
2^+	1.524	1.346	$5/2^-$	0.373	0.174
4^+	2.751	2.575	$3/2^-$	0.593	1.435
6^+	3.190	2.974	$11/2^-$	1.678	1.562
			$9/2^-$	(2.094)	1.895
			$15/2^-$	2.754	2.877

Experimental data from

a) Endt and van der Leun (1973)

b) Kutschera *et al.* (1975)

c) Poletti *et al.* (1976)

d) Nann *et al.* (1976).

level spacings in ^{42}Ca (configuration $f_{7/2}{}^2$) and ^{43}Ca, together with their respective conjugate nuclei ^{46}Ca and ^{45}Ca. The spacings in the hole nuclei are on the average 200 keV smaller than their partners in the conjugate nuclei. This can be interpreted as an indication for the need of an effective three-body interaction (Koltun 1973). Note, however, one glaring exception. This is the $3/2^-$ level, for which the breakdown in the particle-hole relation (eqn (6.54)) is 800 keV, four times larger than the average. This is further evidence for $p_{3/2}$ intruder states, as discussed at the end of section (6.6).

Next, let us consider matrix elements of a one-body operator of rank k in n-particle and conjugate hole configurations. Again the results are derived in the m-scheme. First the expectation value of $F_q^{(k)}$ between n-particle states is given by eqn (1.24) and is

$$\langle j^n\ m_1 \ldots m_n | F_q^{(k)} | j^n\ m_1 \ldots m_n \rangle = \sum_{i=1}^{n} \langle m_i | F_q^{(k)} | m_i \rangle, \qquad (6.56)$$

where for these diagonal matrix elements, the Wigner-Eckhart theorem dictates that q = 0. For the off-diagonal matrix element we have from eqn (1.25)

$$\langle j^n \ldots m_i \ldots | F_q^{(k)} | j^n \ldots m_i' \ldots \rangle = \langle m_i | F_q^{(k)} | m_i' \rangle \qquad (6.57)$$

with $q = m_i - m_i'$.

For the hole configurations, the one-body operator is written in terms of hole creation operators in eqn (1.87). Then for diagonal matrix elements there are two terms, eqns (1.87d) and (1.87e), giving a contribution, resulting in

$$\langle j^{-n}\ m_1^{-1}\ m_2^{-1} \ldots m_n^{-1} | F_q^{(k)} | j^{-n}\ m_1^{-1}\ m_2^{-1} \ldots m_n^{-1} \rangle$$
$$= - \sum_{i=1}^{n} \langle \tilde{m}_i | F_q^{(k)} | \tilde{m}_i \rangle + \sum_{m=-j}^{+j} \langle m | F_q^{(k)} | m \rangle. \qquad (6.58)$$

The second term just reduces to a constant F_0 independent of m_i, as can be seen from using the Wigner-Eckhart theorem:

$$F_0 = \sum_m \langle m | F_q^{(k)} | m \rangle$$

$$= \sum_m \langle jmkq | jm \rangle \langle j \| F^{(k)} \| j \rangle$$

$$= \sum_m (-)^{j-m} \frac{\hat{j}}{\hat{k}} (jm\ j{-m} | kq)\ (j \| F^{(k)} \| j)$$

$$= \frac{(2j+1)}{\hat{k}} \sum_m (jm\ j{-m} | 00)(jm\ j{-m} | kq)(j \| F^{(k)} \| j)$$

$$= (2j+1)\ \delta_{k,0}\ (j \| F^{(k)} \| j). \tag{6.59}$$

This term, therefore, only contributes for scalar one-body operators. The sign on the magnetic quantum numbers in the first term in eqn (6.58) can be reversed, introducing a sign factor $(-)^k$ (standard symmetry of Clebsch-Gordan coefficients), so the final result for diagonal matrix elements reads

$$\langle j^{-n}\ m_1^{-1} \ldots m_n^{-1} | F_q^{(k)} | j^{-n}\ m_1^{-1} \ldots m_n^{-1} \rangle$$

$$= (-)^{k+1} \langle j^n\ m_1 \ldots m_n | F_q^{(k)} | j^n\ m_1 \ldots m_n \rangle + \delta_{k,0}\ F_0. \tag{6.60}$$

For an off-diagonal matrix element, we have

$$\langle j^{-n} \ldots m_i^{-1} \ldots | F_q^{(k)} | j^{-n} \ldots m_i'^{-1} \ldots \rangle = -\langle \tilde{m}_i' | F_q^{(k)} | \tilde{m}_i \rangle$$

$$= (-)^{k-q+1} \langle m_i' | F_{-q}^{(k)} | m_i \rangle$$

$$= (-)^{k-q+1} \langle m_i | F_{-q}^{(k)\dagger} | m_i' \rangle$$

$$= (-)^{p+k+1} \langle m_i | F_q^{(k)} | m_i' \rangle \tag{6.61}$$

$$= (-)^{p+k+1} \langle j^n \ldots m_i \ldots | F_q^{(k)} | j^n \ldots m_i' \ldots \rangle$$

where the matrix element was transposed using the relation

(A.26a) from the appendix, and the phase factor $(-)^{p+q}$ was introduced from the Hermitian property (eqn (A.27)) of the one-body operator.

Eqns (6.60) and (6.61) show how matrix elements in the n-hole configurations relate to matrix elements in the conjugate n-particle configurations, in the m-scheme. The same relations hold in any other scheme, and in particular in the scheme with definite total angular momentum J. Thus eqns (6.60) and (6.61) also hold when written in terms of reduced matrix elements, providing some care is taken over identifying the correct conjugate states (de Shalit and Talmi 1963).

Let us consider multipole moments of nuclei, which are described in the shell model in terms of diagonal matrix elements of the appropriate multipole operator. For the odd multipoles, such as dipole and octopole, the same moment is obtained in an n-particle configuration as in the conjugate n-hole configuration, while for even multipoles the magnitudes of the moments are the same, but the signs are reversed. In particular at the half-filled shell, the hole configuration and particle configuration are identical - thus the even multipole moments all vanish for states of a half-filled shell.

These observations are borne out experimentally. Magnetic moments of nuclei do not show any systematic changes from the beginning to the end of a shell. For example, ^{51}V (three protons in the $f_{7/2}$ shell) has a magnetic moment 5.149 nm (Fuller and Cohen 1969), whereas the conjugate nucleus ^{53}Mn has a magnetic moment ± 5.01 nm. The magnitudes, therefore, are in very good agreement, and from the considerations discussed here a plus sign is strongly suggested for the ^{53}Mn moment. Similarly, quadrupole moments are generally negative at the start of a j shell and positive at the end of the shell in accordance with expectations.

6.9 CONFIGURATION MIXING

In the last few sections, we have considered simple j^n configurations. Unfortunately, only in very few selected regions of the periodic table can one hope to make the assumption that valence nucleons are filling just one j shell. The more typical situation is that several j shells are being filled concurrently.

Let us by way of an example consider configuration mixing with just two shell-model orbitals. Then a typical basis state has the structure

$$\phi = |j_1^{\,n_1}\, j_2^{\,n_2}(\underset{\sim}{r}_1,\underset{\sim}{r}_2,\ldots,\underset{\sim}{r}_{n_1},\underset{\sim}{r}_{n_1+1},\ldots,\underset{\sim}{r}_{n_1+n_2})X_1J_1,X_2J_2;JM\rangle \tag{6.62}$$

in which the n_1 nucleons in the j_1 shell are coupled to angular momentum J_1, the n_2 nucleons in the j_2 shell are coupled to J_2, and J_1 is coupled to J_2 to form a resultant J. The state is fully antisymmetric with respect to interchanges among the n_1+n_2 coordinates.

A general wavefunction for a system of n nucleons (outside closed shells) will be a linear combination of states of type (6.62), viz.

$$\psi_q = \sum_{\alpha=1}^{d} a_\alpha^{(q)}\ \phi_\alpha \tag{6.63}$$

where the summation extends over all possible values of J_1 and J_2 that can couple to J, and over all distributions of n_1 and n_2 with n_1+n_2 = n. The construction of the states ϕ_α is such that they form an orthonormal set.

To calculate the energy of the nuclear system, the eigenvalue problem

$$H\psi_q = E_q\ \psi_q \tag{6.64}$$

has to be solved, where H is the effective Hamiltonian appropriate to this truncated model space. Substituting eqn (6.63) into eqn (6.64), multiplying both sides with ϕ_β^* and integrating, one obtains d eigenvalues, E_q, and d eigenvectors $a_\alpha^{(q)}$, q = 1,...,d from the set of equations

$$\sum_{\alpha=1}^{d} a_\alpha^{(q)}\ \langle\phi_\alpha|H|\phi_\beta\rangle = E_q\ a_\beta^{(q)} \qquad \beta,q = 1,\ldots,d. \tag{6.65}$$

Thus the essence of calculating the energy spectrum is to

evaluate all the matrix elements $\langle \phi_\alpha | H | \phi_\beta \rangle$ and then to diagonalize the resulting matrix.

Similarly the expectation value of a transition operator F (the effective operator appropriate to the truncated model space) between eigensolutions ψ_q and ψ_r is given by

$$\langle \psi_r | F | \psi_q \rangle = \sum_{\alpha,\beta} a_\alpha^{(r)^*} \langle \phi_\alpha | F | \phi_\beta \rangle \, a_\beta^{(q)} \tag{6.66}$$

so again all matrix elements of F between the basis states have to be calculated.

In this section, then, we shall indicate how these matrix elements are evaluated for configuration mixed basis states. We shall work with the example of just two shell-model orbitals; the extension to many orbitals is fairly straightforward, but results in more complicated angular momentum coupling. For a discussion of the more general case, the reader is referred to the article by French *et al.* (1969).

The first problem is to write the basis states (6.62) in terms of the states χ:

$$\chi = | j_1^{n_1}(\underset{\sim}{r}_1,\ldots,\underset{\sim}{r}_{n_1}) X_1 J_1,\, j_2^{n_2}(\underset{\sim}{r}_{n_1+1},\ldots,\underset{\sim}{r}_{n_1+n_2}) X_2 J_2; JM\rangle \tag{6.67}$$

in which the n_1 nucleons in the j_1-shell are fully antisymmetric with respect to interchanges among their coordinates $\underset{\sim}{r}_1,\ldots,\underset{\sim}{r}_{n_1}$, the n_2 nucleons in the j_2-shell are similarly antisymmetrized, and the two groups are then just vector coupled to form a state of angular momentum J. These states χ, therefore, are *not* antisymmetric in all n_1+n_2 particles. It is more convenient to work with the states χ, since the fractional parentage expansions can now be readily exploited to simplify the evaluation of many-body matrix elements.

Note that we are discussing systems comprising nucleons of one type (isospin formalism) so that antisymmetrization over all nucleons is required. In the proton-neutron formalism, however, the proton configurations are antisymmetrized amongst themselves, and the neutron configurations amongst themselves, and then the proton and neutron groups are just vector coupled to form states of good total angular momentum.

Thus in our example, if j_1 was a proton orbit, and j_2 a neutron orbit, the basis states ϕ would already be in the form of χ.

However, in the isospin formalism, an antisymmetrizer $\mathfrak{a}_{n_1+n_2}$ is introduced to relate the two bases (Macfarlane and French, 1960) such that

$$\phi = \mathfrak{a}_{n_1+n_2}\,\chi \tag{6.68}$$

where

$$\mathfrak{a}_{n_1+n_2} = \binom{n_1+n_2}{n_1}^{-\frac{1}{2}} \sum{}' \,(-)^r\, P_r.$$

Here the binomial coefficient $\binom{n_1+n_2}{n_1}$ is just a normalization constant, and the summation Σ' is over all permutations between the groups $j_1^{\,n_1}$ and $j_2^{\,n_2}$ that are order-preserving. The coordinates of the particles in the j_1 shell have been labelled with subscripts $1,2,\ldots,n_1$ in eqn (6.67) and those in the j_2 shell with $n_1+1,\ldots,n_1+n_2$. Then an order-preserving permutation is any permutation of the labels $1,\ldots,n_1$ with the labels $n_1+1,\ldots,n_1+n_2$ such that the resulting set of labels is in ascending order in each shell. For example, consider a system of five nucleons, three in shell j_1, and two in shell j_2; then the particles can be labelled

1 2 3 ; 4 5.

Examples of order-preserving permutations are

1 2 4 ; 3 5	2 3 4 ; 1 5
1 2 5 ; 3 4	2 3 5 ; 1 4
1 3 4 ; 2 5	1 4 5 ; 2 3
1 3 5 ; 2 4	2 4 5 ; 1 3
	3 4 5 ; 1 2

hat is ten permutations in total, one in which no particles

were transferred between the shells, six in which one particle was transferred, and three in which two were transferred.

That ten is the correct total can be seen from the following argument. The basis states ϕ are represented by a single Slater determinant of dimension n_1+n_2, which when written out in full produces $(n_1+n_2)!$ terms. Similarly the basis states χ are a product of two Slater determinants of dimension n_1 and n_2 respectively producing $n_1!n_2!$ terms. Thus the number of terms in the antisymmetrizer $\mathfrak{a}_{n_1+n_2}$ which will equate both sides in eqn (6.68) is

$$\frac{(n_1+n_2)!}{n_1!n_2!} = \binom{n_1+n_2}{n_1},$$

vindicating the normalization constant in eqn (6.69) and producing the value ten for our example of $n_1=3$, $n_2=2$.

The antisymmetrizer, eqn (6.69), can be explicitly written out in terms of elementary exchange operators as

$$\mathfrak{a}_{n_1+n_2} = \binom{n_1+n_2}{n_1}^{-\frac{1}{2}} \Big[1 - \sum_{i=1}^{n_1} \sum_{j=n_1+1}^{n_1+n_2} P_{ij} + \\ + \sum_{i<k=1}^{n_1} \sum_{j<\ell=n_1+1}^{n_1+n_2} P_{ij}P_{k\ell} - \ldots\Big] \tag{6.70}$$

where P_{ij} exchanges the particles i and j.

The next step is to rewrite a many-particle matrix element with basis states ϕ in terms of matrix elements with states χ. As an illustration we evaluate $\langle \phi_\alpha | V | \phi_\beta \rangle$ where

$$V = \sum_{i<j=1}^{n_1+n_2} V_{ij}$$

is a scalar two-body operator (the two-body part of the effective Hamiltonian), then

$$\langle \phi_\alpha | V | \phi_\beta \rangle = \binom{n_1+n_2}{n_1}^{-\frac{1}{2}} \langle \phi_\alpha | V \sum_r{}' (-)^r P_r | \chi_\beta \rangle$$

where ϕ_β is a basis state, eqn (6.62), with n_1 particles in shell j_1 and n_2 particles in shell j_2. The operator V is symmetric in the particle labels, so the permutation operator commutes with V and now operates on ϕ_α. Irrespective of how the n nucleons are distributed among the shells j_1 and j_2, the state ϕ_α is fully antisymmetric in all n particle coordinates, so that each and every order-preserving permutation leaves ϕ_α unaffected and gives an equal contribution to the matrix element. Thus we obtain the result

$$\langle\phi_\alpha|V|\phi_\beta\rangle = \binom{n_1+n_2}{n_1}^{+\frac{1}{2}} \langle\phi_\alpha|V|\chi_\beta\rangle. \qquad (6.71)$$

This step is the first stage in the evaluation of all matrix elements.

To continue, it is necessary to specify exactly how the n nucleons are distributed in the shells j_1 and j_2 in the basis state ϕ_α. There are three cases to consider: (a) the distribution is the same as in ϕ_β, namely $j_1^{n_1} j_2^{n_2}$, (b) the distribution is $j_1^{n_1-1} j_2^{n_2+1}$, (c) the distribution is $j_1^{n_1-2} j_2^{n_2+2}$. We shall follow the treatment of Glaudemans *et al.* (1964) and consider each in turn.

Case (a). The expansion eqn (6.69) is applied once again, so that eqn (6.71) becomes

$$\langle\phi_\alpha|V|\phi_\beta\rangle = \langle\chi_\alpha|V \sum_r{}' (-)^r P_r|\chi_\beta\rangle.$$

Next V is explicitly written out as a sum of three terms

$$V = \sum_{i<j=1}^{n_1} V_{ij} + \sum_{i<j=n_1+1}^{n_1+n_2} V_{ij} + \sum_{i=1}^{n_1} \sum_{j=n_1+1}^{n_1+n_2} V_{ij}.$$

In the first term, the interaction V has no influence on the nucleons in the second shell j_2. Therefore the structure of this part of the wavefunction must be identical in χ_α and χ_β and the only permutation in the sum $\sum_r{}'$ which will give a non-vanishing matrix element is the unit operator. A similar result holds for the second term. For the third term in V, the order-preserving permutations which give non-vanishing contributions are $P_r=1$ and $P_r=P_{ij}$. Thus we have

$$\langle\phi_\alpha|V|\phi_\beta\rangle \equiv \langle j_1^{n_1}X_1'J_1',j_2^{n_2}X_2'J_2';J|\sum_{i<j=1}^{n_1} V_{ij}|j_1^{n_1}X_1J_1,j_2^{n_2}X_2J_2;J\rangle +$$

$$+\langle j_1^{n_1}X_1'J_1',j_2^{n_2}X_2'J_2';J|\sum_{i<j=n_1+1}^{n_1+n_2} V_{ij}|j_1^{n_1}X_1J_1,j_2^{n_2}X_2J_2;J\rangle +$$

$$+\langle j_1^{n_1}X_1'J_1',j_2^{n_2}X_2'J_2';J|\sum_{i=1}^{n_1}\sum_{j=n_1+1}^{n_1+n_2} V_{ij}(1-P_{ij})| \times$$

$$\times\ j_1^{n_1}X_1J_1,j_2^{n_2}X_2J_2;J\rangle.$$

It remains to make the angular momentum reduction of these many-particle matrix elements in terms of a linear combination of two-body matrix elements. The first two terms essentially follow from the single-shell results, eqn (6.34), while the third term is evaluated as follows:

$$\text{Third term} = n_1n_2\langle j_1^{n_1}X_1'J_1',j_2^{n_2}X_2'J_2';J|V_{n_1,n_1+n_2}(1-P_{n_1,n_1+n_2})| \times$$

$$\times\ j_1^{n_1}X_1J_1,j_2^{n_2}X_2J_2;J\rangle$$

$$= n_1n_2 \sum_{\substack{y_1'K_1'y_2'K_2'\\ y_1K_1y_2K_2}} \langle j_1^{n_1}X_1'J_1'\{|j_1^{n_1-1}\ y_1'K_1';j_1\rangle\langle j_2^{n_2}X_2'J_2'\{|j_2^{n_2-1}\ y_2'K_2';j_2\rangle \times$$

$$\times\langle j_1^{n_1}X_1J_1\{|j_1^{n_1-1}\ y_1K_1;j_1\rangle\langle j_2^{n_2}X_2J_2\{|j_2^{n_2-1}\ y_2K_2;j_2\rangle \times$$

$$\times\langle [j_1^{n_1-1}\ j_1]^{J_1'}\ [j_2^{n_2-1}\ j_2]^{J_2'};J|V(1-P)|[j_1^{n_1-1}\ j_1]^{J_1}\ [j_2^{n_2-1}\ j_2]^{J_2};J\rangle$$

$$= n_1 n_2 \sum_{\substack{y_1'K_1'y_2'K_2' \\ y_1K_1y_2K_2 \\ L'K'LK}} (4 \text{ cfps}) \begin{bmatrix} K_1' & j_1 & J_1' \\ K_2' & j_2 & J_2' \\ K' & L' & J \end{bmatrix} \begin{bmatrix} K_1 & j_1 & J_1 \\ K_2 & j_2 & J_2 \\ K & L & J \end{bmatrix} \times$$

$$\times \langle [j_1^{n_1-1} \; j_2^{n_2-1}]^{K'} \; [j_1 j_2]^{L'}; J | V(1-P) | [j_1^{n_1-1} \; j_2^{n_2-1}]^{K} [j_1 j_2]^{L}; J \rangle$$

$$= n_1 n_2 \sum_{\substack{y_1K_1y_2K_2 \\ LK}} (4 \text{ cfps}) \; (2 \text{ recoupling coeffs.}) \; \langle j_1 j_2; L | V(1-P) | j_1 j_2; L \rangle .$$

The various steps involved were: (i) Of the terms in the sum $\sum_{i=1}^{n_1}$ we pick the n_1 term; all others gave equivalent contributions. Similarly from $\sum_j$ the (n_1+n_2) term is picked. (ii) Fractional parentage expansions separate the n_1 particle from the group $j_1^{n_1}$ and the (n_1+n_2) particle from the group $j_2^{n_2}$. (iii) Angular momentum recoupling is required (eqn (A.17)), to couple the n_1 and the (n_1+n_2) particles together. (iv) The operator V(1-P) depends only on $\underset{\sim}{r}_{n_1}$ and $\underset{\sim}{r}_{n_1+n_2}$; integration over the other n-2 coordinates gives the orthogonality conditions δ_{K_1',K_1}, δ_{K_2',K_2}, and $\delta_{K',K}$. Lastly note that $\langle j_1 j_2; L | V(1-P) | j_1 j_2; L \rangle$ equals $\langle j_1 j_2; L | V | j_1 j_2; L \rangle_A$, where the subscript A has been inserted to remind us that the two particle states have been antisymmetrized.

Putting all these pieces together, the final expression for the matrix element reads

$$\langle j_1^{n_1} j_2^{n_2} \; X_1'J_1', X_2'J_2'; J | V | j_1^{n_1} j_2^{n_2} \; X_1 J_1, X_2 J_2; J \rangle$$

$$= \tfrac{1}{2} n_1(n_1-1) \sum_{y_1K_1L} \langle j_1^{n_1} X_1'J_1' \{ | j_1^{n_1-2} \; y_1 K_1; j_1^2 L \rangle \langle j_1^{n_1} X_1 J_1 \{ | j_1^{n_1-2} \; y_1 K_1; j_1^2 L \rangle \times$$

$$\times \langle j_1^2; L | V | j_1^2; L \rangle_A \; \delta_{J_1,J_1'} \; \delta_{X_2,X_2'} \; \delta_{J_2,J_2'} +$$

$$+ \tfrac{1}{2} n_2(n_2-1) \sum_{y_2K_2L} \langle j_2^{n_2} X_2'J_2' \{ | j_2^{n_2-2} \; y_2 K_2; j_2^2 L \rangle \langle j_2^{n_2} X_2 J_2 \{ | j_2^{n_2-2} \; y_2 K_2; j_2^2 L \rangle \times$$

$$\times \langle j_2{}^2;L|V|j_2{}^2;L\rangle_A \; \delta_{J_2,J_2'} \; \delta_{X_1,X_1'} \; \delta_{J_1,J_1'} +$$

$$+ n_1n_2 \sum_{y_1K_1y_2K_2} \langle j_1{}^{n_1} X_1'J_1'\{|j_1{}^{n_1-1} y_1K_1;j_1\rangle\langle j_2{}^{n_2} X_2'J_2'\{|j_2{}^{n_2-1} y_2K_2;j_2\rangle \times$$

$$\times \langle j_1{}^{n_1} X_1J_1\{|j_1{}^{n_1-1} y_1K_1;j_1\rangle\langle j_2{}^{n_2} X_2J_2\{|j_2{}^{n_2-1} y_2K_2;j_2\rangle \times$$

$$\times \sum_{KL} \begin{bmatrix} K_1 & j_1 & J_1' \\ K_2 & j_2 & J_2' \\ K & L & J \end{bmatrix} \begin{bmatrix} K_1 & j_1 & J_1 \\ K_2 & j_2 & J_2 \\ K & L & J \end{bmatrix} \langle j_1j_2;L|V|j_1j_2;L\rangle_A. \qquad (6.72)$$

In the last line of eqn (6.72), the sum over K of two recoupling coefficients can be written in terms of a 12j-symbol of the second kind (Yutsis *et al.* 1962), which in turn can be written

$$\sum_K \begin{bmatrix} K_1 & j_1 & J_1' \\ K_2 & j_2 & J_2' \\ K & L & J \end{bmatrix} \begin{bmatrix} K_1 & j_1 & J_1 \\ K_2 & j_2 & J_2 \\ K & L & J \end{bmatrix} = \sum_z \frac{\hat{L}\,\hat{J}_1}{\hat{j}_2\hat{K}_1} (-)^{j_2+J_2+K_1-L-J_2'-J_1} \times$$

$$\times\; U(j_1j_1j_2j_2;zL)\; U(j_2zK_2J_2;j_2J_2')\; U(J_2'zJJ_1;J_2J_1')\; U(J_1'J_1j_1j_1;zK_1).$$

This relation is of considerable help in the numerical evaluation of eqn (6.72).

Case (b). With the distribution of particles in the state ϕ_α being $j_1{}^{n_1-1}\; j_2{}^{n_2+1}$, we start, as before, with eqn (6.71) and apply the antisymmetrizer eqn (6.69) to ϕ_α to obtain

$$\langle\phi_\alpha|V|\phi_\beta\rangle = \begin{pmatrix} n_1+n_2 \\ n_1 \end{pmatrix}^{\frac{1}{2}} \begin{pmatrix} n_1+n_2 \\ n_1-1 \end{pmatrix}^{-\frac{1}{2}} \times$$

$$\times \langle j_1{}^{n_1-1}(\underset{\sim}{r}_1 \cdot\cdot \underset{\sim}{r}_{n_1-1})X_1'J_1', j_2{}^{n_2+1}(\underset{\sim}{r}_{n_1} \cdot\cdot \underset{\sim}{r}_{n_1+n_2})X_2'J_2';J|V \sum_r{}'(-)^r P_r|$$

$$\times\; j_1{}^{n_1}(\underset{\sim}{r}_1 \cdot\cdot \underset{\sim}{r}_{n_1})X_1J_1, j_2{}^{n_2}(\underset{\sim}{r}_{n_1+1} \cdot\cdot \underset{\sim}{r}_{n_1+n_2})X_2J_2;J\rangle,$$

where the permutation operator written out explicitly takes the form

$$\sum_r{}' (-)^r P_r = 1 - \sum_{i=1}^{n_1-1} \sum_{j=n_1}^{n_1+n_2} P_{ij} + \ldots.$$

$$= 1 - \sum_{i=1}^{n_1-1} P_{in_1} - \sum_{i=1}^{n_1-1} \sum_{j=n_1+1}^{n_1+n_2} P_{ij} + \ldots..$$

Note that nucleons $1,\ldots,n_1-1$ in shell j_1 and nucleons $n_1+1,\ldots,n_1+n_2$ are common to both sides of the matrix element, so that the only order-preserving permutations which will give a non-vanishing matrix element are 1 and $\sum_{i=j}^{n_1-1} P_{in_1}$. These permutations acting on the group $j_1^{\,n_1}$ in the bra wavefunction give n_1 identical terms, so the summation $\sum_r{}'$ can be replaced by unity and the result multiplied by n_1. For this one term the particle with coordinate $\underset{\sim}{r}_{n_1}$ is the one transferred from the j_1-shell to the j_2-shell so that the only terms in V contributing to the matrix element are those involving $\underset{\sim}{r}_{n_1}$, viz.

$$\sum_{i=1}^{n_1-1} V_{in_1} + \sum_{j=n_1+1}^{n_1+n_2} V_{n_1j}.$$

Putting these pieces altogether, we obtain

$$\langle\phi_\alpha|V|\phi_\beta\rangle = [n_1(n_2+1)]^{\frac{1}{2}} \langle j_1^{\,n_1-1} X_1'J_1', j_2^{\,n_2+1} X_2'J_2'; J| \sum_{i=1}^{n_1-1} V_{in_1} |j_1^{\,n_1} X_1J_1, j_2^{\,n_2} X_2J_2; J\rangle$$

$$+ [n_1(n_2+1)]^{\frac{1}{2}} \langle j_1^{\,n_1-1} X_1'J_1', j_2^{\,n_2+1} X_2'J_2'; J| \sum_{j=n_1+1}^{n_1+n_2} V_{n_1j} |j_1^{\,n_1} X_1J_1, j_2^{\,n_2} X_2J_2; J\rangle. \tag{6.73}$$

It now remains to make the angular momentum reductions. Consider just the first term:

$$=[n_1(n_2+1)]^{\frac{1}{2}}(n_1-1)\sum_{\substack{y_1'K_1'\\ y_2'K_2'\\ y_1K_1L}} \langle j_1^{n_1-1} X_1'J_1'\{|j_1^{n_1-2} y_1'K_1';j_1\rangle\langle j_2^{n_2+1} X_2'J_2'\{|j_2^{n_2} y_2'K_2';j_2\rangle$$

$$\times\langle j_1^{n_1}X_1J_1\{|j_1^{n_1-2} y_1K_1;j_1^2L\rangle (-)^{n_2}\times$$

$$\times\langle [j_1^{n_1-2}\ j_1]^{J_1'}\ [j_2^{n_2}\ j_2]^{J_2'};J|V_{n_1-1\ n_1}|[j_1^{n_1-2}\ j_1^2]^{J_1}\ j_2^{n_2}J_2;J\rangle$$

$$=[n_1(n_2+1)]^{\frac{1}{2}}(n_1-1)\sum_{y_1K_1LK} \langle j_1^{n_1-1} X_1'J_1'\{|j_1^{n_1-2} y_1K_1;j_1\rangle\langle j_2^{n_2+1} X_2'J_2'\{|j_2^{n_2}X_2J_2;j_2\rangle\times$$

$$\times\langle j_1^{n_1}X_1J_1\{|j_1^{n_1-2} y_1K_1;j_1^2L\rangle \begin{bmatrix} K_1 & j_1 & J_1' \\ J_2 & j_2 & J_2' \\ K & L & J \end{bmatrix} (-)^{n_2}\times$$

$$\times\ U(K_1LJ_2J;J_1K)\ (-)^{J_1+K-J-K_1}\langle j_1j_2;L|V|j_1^2;L\rangle$$

$$=[\tfrac{1}{2}n_1(n_2+1)]^{\frac{1}{2}}(n_1-1)(-)^{n_2}(-)^{J_2+j_2-J_2'}\langle j_2^{n_2+1} X_2'J_2'\{|j_2^{n_2}X_2J_2;j_2\rangle U(J_1'j_2JJ_2;J_1J_2')\times$$

$$\times\sum_{y_1K_1L}\langle j_1^{n_1}X_1J_1\{|j_1^{n_1-2} y_1K_1;j_1^2L\rangle\langle j_1^{n_1-1} X_1'J_1'\{|j_1^{n_1-2} y_1K_1;j_1\rangle\times$$

$$\times\ U(K_1j_1J_1j_2;J_1'L)\langle j_1j_2;L|V|j_1^2;L\rangle_A. \qquad (6.74a)$$

The various steps involved were: (i) Of the terms in $\sum_i V_{in_1}$, choose $i = n_1-1$ and multiply the result by the number of terms, viz. n_1-1. (ii) Make fractional parentage expansions to separate the (n_1-1) and the n_1 particles from the group $j_1^{n_1}$, the (n_1-1) particle from the group $j_1^{n_1-1}$, and the n_1 particle from the group $j_2^{n_2+1}$. Notice that in the latter case, the particle n_1 occupies the first column of the Slater determinant and has to be permuted to the last column before the cfp expansion can be performed. Hence the introduction of the phase factor $(-)^{n_2}$, the so-called Pauli phase (French

1965a). (iii) Recouple the angular momenta in the bra and ket wavefunctions so that an integration over all coordinates except $\underset{\sim}{r}_{n_1-1}$ and $\underset{\sim}{r}_{n_1}$ can be performed. (iv) Sum the [ ]-coefficient and the U-coefficient over K to give two U-coefficients. (v) Finally, replace the two-body matrix element $\langle j_1 j_2;L|V|j_1{}^2;L\rangle$ by $(2)^{-\frac{1}{2}}\langle j_1 j_2;L|V|j_1{}^2;L\rangle_A$ with the two-particle state $|j_1 j_2\rangle$ now being antisymmetrized.

A similar sequence of steps can be used to reduce the second term in eqn (6.73), giving

$$=[\tfrac{1}{2}n_1(n_2+1)]^{\frac{1}{2}} n_2 (-)^{n_2} (-)^{J_2+j_2-J_2'} \langle j_1{}^{n_1} X_1 J_1\{|j_1{}^{n_1-1} X_1'J_1';j_1\rangle U(J_1'j_1JJ_2;J_1J_2') \times$$

$$\times \sum_{y_2K_2L} \langle j_2{}^{n_2} X_2J_2\{|j_2{}^{n_2-1} y_2K_2;j_2\rangle \langle j_2{}^{n_2+1} X_2'J_2'\{|j_2{}^{n_2-1} y_2K_2;j_2{}^2L\rangle \times$$

$$\times U(K_2j_2J_2'j_1;J_2L)\langle j_1j_2;L|V|j_2{}^2;L\rangle_A. \tag{6.74b}$$

Eqns (6.74a) and (6.74b) taken together make up the expression for the off-diagonal matrix element for case (b).

Case (c). As with case (b), we start once again with eqn (6.71) and apply the antisymmetrizer eqn (6.69) to ϕ_α, obtaining

$$\langle\phi_\alpha|V|\phi_\beta\rangle = \begin{pmatrix} n_1+n_2 \\ n_1 \end{pmatrix}^{\frac{1}{2}} \begin{pmatrix} n_1+n_2 \\ n_1-2 \end{pmatrix}^{-\frac{1}{2}} \times$$

$$\times\langle j_1{}^{n_1-2}(\underset{\sim}{r}_1..\underset{\sim}{r}_{n_1-2})X_1'J_1', j_2{}^{n_2+2}(\underset{\sim}{r}_{n_1-1}..\underset{\sim}{r}_{n_1+n_2})X_2'J_2';J|V \sum_r{}' (-)^r P_r|$$

$$\times j_1{}^{n_1}(\underset{\sim}{r}_1..\underset{\sim}{r}_{n_1})X_1J_1, j_2{}^{n_2}(\underset{\sim}{r}_{n_1+1}..\underset{\sim}{r}_{n_1+n_2})X_2J_2;J\rangle.$$

Of all the order-preserving permutations, just those leaving nucleons $1,\ldots,n_1-2$ in shell j_1 and nucleons $n_1+1,\ldots,n_1+n_2$ in shell j_2 produce a non-vanishing result. There are $n_1(n_1-1)/2$ such permutations coming from

$$1 - \sum_{i=1}^{n_1-2} \sum_{j=n_1-1}^{n_1} P_{ij} + \sum_{i<k=1}^{n_1-2} P_{in_1-1} P_{kn_1}$$

and each term gives an identical contribution. Thus the summation $\sum_r{}'$ can be replaced by unity and the result multiplied by $n_1(n_1-1)/2$. For this one term, nucleons n_1-1 and n_1 are explicitly transferred from the j_1 shell to the j_2 shell, so the only term in V contributing to the matrix element is $V_{n_1-1\, n_1}$. Thus we obtain

$$\langle \phi_\alpha | V | \phi_\beta \rangle = \frac{1}{2} [n_1(n_1-1)(n_2+1)(n_2+2)]^{1/2} \times$$

$$\times \langle j_1^{n_1-2} X_1'J_1', j_2^{n_2+2} X_2'J_2'; J | V_{n_1-1\, n_1} | j_1^{n_1} X_1J_1, j_2^{n_2} X_2J_2; J\rangle.$$

It remains to separate out nucleons n_1-1 and n_1 using fractional parentage expansions and to integrate over the other (n-2) coordinates. The final result is

$$\langle \phi_\alpha | V | \phi_\beta \rangle = \frac{1}{2}[n_1(n_1-1)(n_2+1)(n_2+2)]^{\frac{1}{2}} \sum_L (-)^{L+J_2-J_2'} U(J_1'LJJ_2; J_1J_2') \times$$

$$\times \langle j_1^{n_1} X_1J_1 \{| j_1^{n_1-2} X_1'J_1'; j_1^2 L\rangle \langle j_2^{n_2+2} X_2'J_2' \{| j_2^{n_2} X_2J_2; j_2^2 L\rangle \times$$

$$\times \langle j_2^2; L | V | j_1^2; L\rangle_A. \qquad (6.75)$$

Equations (6.72), (6.74), and (6.75) give expressions for the diagonal and off-diagonal matrix elements $\langle \phi_\alpha | V | \phi_\beta \rangle$ arising from a two-body effective interaction. To the diagonal matrix elements are added the one-body terms (single-particle energies) and the resulting Hamiltonian matrix is diagonalized, eqn (6.65), to obtain energies and wavefunctions. A comparison with experiment gives some measure of success for the shell model calculation.

The wavefunctions can be further tested by evaluating

expectation values of one-body operators, eqn (6.66). Required are expressions for the reduced matrix elements $\langle\phi_\alpha\|F^{(K)}\|\phi_\beta\rangle$ evaluated with the configuration mixed basis states, ϕ of eqn (6.62). Again the states ϕ have to be replaced by χ, eqn (6.67) and the first stage is the result (eqn (6.71))

$$\langle\phi_\alpha\|F^{(K)}\|\phi_\beta\rangle = \begin{pmatrix} n_1+n_2 \\ n_1 \end{pmatrix}^{\frac{1}{2}} \langle\phi_\alpha\|F^{(K)}\|\chi_\beta\rangle.$$

Next the states ϕ_α must be specified more explicitly. There are two cases to consider: (a) the particle distribution between the shells is the same as in ϕ_β, namely $j_1^{n_1} j_2^{n_2}$, and (b) the distribution is $j_1^{n_1-1} j_2^{n_2+1}$.

Case (a). The state ϕ_α is expanded as in eqn (6.69), giving

$$\langle\phi_\alpha\|F^{(K)}\|\phi_\beta\rangle = \langle\chi_\alpha\|F^{(K)}\sum_r{}'(-)^r P_r\|\chi_\beta\rangle.$$

The one-body operator is written as a sum of two terms,

$$F^{(K)} = \sum_{i=1}^{n_1} f^{(K)}(\underset{\sim}{r}_i) + \sum_{i=n_1+1}^{n_1+n_2} f^{(K)}(\underset{\sim}{r}_i).$$

For the first term, F has no influence on the nucleons in the second shell j_2. Therefore the structure of this part of the wavefunction must be identical in χ_α and χ_β and the only permutation in the sum $\sum_r{}'$ which will give a non-vanishing matrix element is just the unit operator. A similar argument holds for the second term, resulting with

$$\langle\phi_\alpha\|F^{(K)}\|\phi_\beta\rangle = \langle j_1^{n_1}X_1'J_1', j_2^{n_2}X_2'J_2';J'\|\sum_{i=1}^{n_1} f^{(K)}\|j_1^{n_1}X_1J_1, j_2^{n_2}X_2J_2;J\rangle +$$

$$+ \langle j_1^{n_1}X_1'J_1', j_2^{n_2}X_2'J_2';J'\|\sum_{i=n_1+1}^{n_1+n_2} f^{(K)}\|j_1^{n_1}X_1J_1, j_2^{n_2}X_2J_2;J\rangle.$$

Consider just the first term. It remains to integrate over the coordinates of the nucleons in the second shell, using the factorization theorem, eqn (A.38a). Then one is left with a single-shell matrix element of the type evaluated in eqn (6.28). With a similar argument for the second term, we put all these pieces together to obtain the final result

$$
\begin{aligned}
\langle \phi_\alpha \| F^{(K)} \| \phi_\beta \rangle = {} & n_1 \, (-)^{J_1'+J-J'-J_1} \, U(J'J_2KJ_1;J_1'J) \times \\
& \times \sum_{y_1K_1} \langle j_1^{\,n_1} X_1'J_1' \{| j_1^{\,n_1-1} y_1K_1; j_1 \rangle \langle j_1^{\,n_1} X_1J_1 \{| j_1^{\,n_1-1} y_1K_1; j_1 \rangle \times \\
& \times U(J_1'K_1Kj_1;j_1J_1) \langle j_1 \| f^{(K)} \| j_1 \rangle \, \delta_{X_2,X_2'} \, \delta_{J_2,J_2'} + \\
& + n_2 \, U(J'J_1KJ_2;J_2'J) \times \\
& \times \sum_{y_2K_2} \langle j_2^{\,n_2} X_2'J_2' \{| j_2^{\,n_2-1} y_2K_2; j_2 \rangle \langle j_2^{\,n_2} X_2J_2 \{| j_2^{\,n_2-1} y_2K_2; j_2 \rangle \times \\
& \times U(J_2'K_2Kj_2;j_2J_2) \, \langle j_2 \| f^{(K)} \| j_2 \rangle \, \delta_{X_1,X_1'} \, \delta_{J_1,J_1'}.
\end{aligned}
\tag{6.76}
$$

Notice that for scalar one-body operators (K=0) the expression considerably reduces to

$$
n_1 \, \langle j_1 \| f^{(K)} \| j_1 \rangle + n_2 \, \langle j_2 \| f^{(K)} \| j_2 \rangle
$$

and only diagonal matrix elements are non-zero. Thus the effective one-body interaction gives a contribution to the diagonal elements of the Hamiltonian matrix that is particularly simple to evaluate.

Case (b). With the particles in ϕ_α distributed as $j_1^{\,n_1-1} \; j_2^{\,n_2+1}$, applying the antisymmetrizer, eqn (6.69), to ϕ_α gives

$$
\langle \phi_\alpha \| F^{(K)} \| \phi_\beta \rangle = \begin{pmatrix} n_1+n_2 \\ n_1 \end{pmatrix}^{\frac{1}{2}} \begin{pmatrix} n_1+n_2 \\ n_1-1 \end{pmatrix}^{-\frac{1}{2}} \times
$$

$$\times \langle j_1^{n_1-1}(\underset{\sim}{r}_1 .. \underset{\sim}{r}_{n_1-1})X_1'J_1', j_2^{n_2+1}(\underset{\sim}{r}_{n_1} .. \underset{\sim}{r}_{n_1+n_2})X_2'J_2'; J' \| F^{(K)} \sum_r{}' (-)^r P_r \| \times$$

$$\times\ j_1^{n_1}(\underset{\sim}{r}_1 .. \underset{\sim}{r}_{n_1})X_1J_1, j_2^{n_2}(\underset{\sim}{r}_{n_1+1} .. \underset{\sim}{r}_{n_1+n_2})X_2J_2; J\rangle.$$

Note that nucleons $1,\ldots,n_1-1$ in shell j_1 and nucleons $n_1+1,\ldots,n_1+n_2$ are common to both sides of the matrix element, so that the only order-preserving permutations which give a non-vanishing matrix element are 1 and $\sum_{i=1}^{n_1-1} P_{in_1}$. These permutations acting on the group $j_1^{n_1}$ in the bra wavefunction give n_1 identical terms, so the summation $\sum_r{}'$ can be replaced by unity and the result multiplied by n_1. For this one term the particle with coordinate $\underset{\sim}{r}_{n_1}$ is transferred from the j_1 shell to the j_2 shell, so the only term in F contributing to the matrix element is $f^{(K)}(\underset{\sim}{r}_{n_1})$. Thus one obtains

$$\langle \phi_\alpha \| F^{(K)} \| \phi_\beta \rangle = [n_1(n_2+1)]^{\frac{1}{2}} \langle j_1^{n_1-1} X_1'J_1', j_2^{n_2+1} X_2'J_2'; J' \| f^{(K)}(\underset{\sim}{r}_n) \| \times$$

$$\times\ j_1^{n_1} X_1J_1, j_2^{n_2} X_2J_2; J\rangle$$

$$= [n_1(n_2+1)]^{\frac{1}{2}} (-)^{n_2} (-)^{j_1+j_2-K+J_1'+j_1-J_1} \langle j_1^{n_1} X_1J_1 \{| j_1^{n_1-1} X_1'J_1'; j_1\rangle \times$$

$$\times \langle j_2^{n_2+1} X_2'J_2' \{| j_2^{n_2} X_2J_2; j_2\rangle \left[\frac{(2J_1+1)(2J_2'+1)}{(2J'+1)(2j_1+1)}\right]^{\frac{1}{2}} \begin{bmatrix} J_1' & J_2' & J' \\ J_1 & J_2 & J \\ j_1 & j_2 & K \end{bmatrix} \langle j_2 \| f^{(K)} \| j_1 \rangle. \qquad (6.77)$$

Equations (6.72) to (6.77) are the life-blood of a configuration-mixing shell model calculations based on fractional parentage expansions and angular momentum couplings. In the next section we discuss a small selection of shell model calculations and comment on the state of the art.

6.10 SOME SHELL MODEL RESULTS

The earliest shell model attempts at explaining the spacings between nuclear energy levels did not involve the diagonalization of a Hamiltonian matrix. Instead, only a very restrictive class of effective interactions was considered, namely those that are diagonal in the chosen basis. For example, Feenberg and Phillips (1937) described the p-shell nuclei (^{4}He to ^{16}O) using a simple central interaction with Wigner and Majorana exchange terms, this interaction being diagonal in the L-S coupling scheme. The influence of spin exchange terms and Coulomb forces were then estimated from first-order perturbation theory. Kurath (1952) took an opposing viewpoint. He assumed there was a very strong one-body spin-orbit force in the residual interaction such that the nuclei between ^{4}He and ^{12}C could be described in j-j coupling as pure $(p_{3/2})^n$ configurations.

It was observed that L-S coupling was more successful at the beginning of the p-shell, and j-j coupling preferred at the end. For example, a ground-state spin of 3^+ is favoured for ^{6}Li in j-j coupling, yet the observed value is 1^+, indicating that the high orbital symmetry of the triplet S-state, as suggested by L-S coupling, is dominant. For ^{10}B the situation is reversed. L-S coupling predicts a 1^+ ground-state spin, whereas j-j coupling reproduces the observed value of 3^+. Similarly the splitting between the $3/2^-$ and $1/2^-$ states in odd nuclei, the so-called doublet splitting, increases through the shell and changes sign in going from A = 11 to A = 13, suggesting the closure of the $p_{3/2}$ subshell at ^{12}C.

The first configuration-mixing shell model calculation was performed by Inglis (1953) and was dubbed at that time an 'intermediate coupling' calculation. The name arose because the calculation was mounted in order to investigate the range of validity of the two coupling schemes. Inglis used a Hamiltonian that was the sum of two-body central interactions (with general exchange force mixtures) and a one-body spin-orbit term. The complete intermediate coupling calculation was performed for two particles (A = 6) and two holes (A = 14 but for the more complicated cases further removed from a shell closure, Inglis estimated the spectra by computing the two extremes (L-S and j-j) and interpolating between them. Inglis plotted the calculated spectra as a function of the spin-orbit strength and looked for the value that gave the best fit to the experimental spectrum. Subsequently, Kurath (1956) repeated the Inglis calculation, avoiding now the need for interpolation in the more complicated cases, by setting up what was probably the first computer shell-model calculation. Kurath further calculated the magnetic dipole moments of the ground states and found agreement with experiment for substantially the same value of the spin-orbit strength as was needed for the spectrum.

The success of the intermediate coupling calculations in interpreting the low-lying states in the 1p-shell led Elliott and Flowers (1955) to perform similar investigations for two and three nucleons in the 2s,1d shell. Although the j-j model generally seemed to work quite well beyond ^{16}O, the ground state of ^{19}F is a glaring exception, the last nucleon apparently occupying the $2s_{1/2}$ orbit while a $1d_{5/2}$ state is predicted. This suggested that the tendency towards maximum orbital symmetry is still playing a dominant role in determining the coupling scheme for this nuclide. Elliott and Flowers confirmed that this was indeed the case. A decomposition of their calculated ^{19}F wavefunctions indicated that the T = 1/2 states contained greater than 85 per cent maximum orbital symmetry configurations, whereas the T = 3/2 states, curiously enough, were better described in j-j coupling. This they believed was a general result: that L-S coupling was a better approximation for low isotopic spin states and j-j coupling for high isotopic spin.

Leaping ahead of the historical progression for just a moment, this conclusion has been borne out by more recent calculations. The lowest 0^+, 2^+, 4^+, 6^+, and 8^+ T = 0 states in ^{20}Ne are found to contain between 80 and 90 per cent the state of maximum orbital symmetry, and this fact has been used by Akiyama, Arima, and Sebe (1969) as a guide for truncating the number of configurations to be retained in a calculation of the spectra of ^{21}Ne, ^{22}Ne, ^{22}Na, and ^{24}Mg. As with the 1p-shell, the preference for L-S coupling is confined to the first half of the s-d shell, j-j coupling being a better approximation in the latter half. Curiously enough, L-S coupling does poorly at the start of the p-f shell. This is because the f single-particle orbital lies lower than the p orbital; the reverse order is preferred if states of maximum orbital symmetry are to be low lying in the spectrum.

The choice between using L-S and j-j coupled basis functions is ultimately just a matter of expediency for those calculations in which a complete set of states belonging to a chosen configuration are retained. Only when truncations within the configuration of n particles in a major oscillator shell are being contemplated does the choice of basis functions become important.

The advent of high-speed computers during the 1960s also saw the advent of large-scale shell model computations. Most of the computer codes written preferred to use j-j coupled bases, since there is less Racah algebra involved, and the codes were more versatile in being adapted for different regions of the nuclear periodic table.

One of the first of the new regime of computer calculations appeared in 1964, when Glaudemans, Wiechers, and Brussaard (1964) computed the spectra of the nuclei A = 29 to A = 39 using a model space based on a closed-shell core at ^{28}Si, and valence nucleons distributed in the shell-model orbitals $2s_{1/2}$ and $1d_{3/2}$. The effective interaction is specified by 15 two-body matrix elements and 2 one-body matrix elements (single-particle energies). These matrix elements were treated as parameters and adjusted in a least squares fitting procedure in which computed energies were compared with energies of 50 states with experimentally well-known spin and isospin. The sizes of the matrices that had to be constructed and diagonalized were fairly modest, the largest being 14×14 for the 2^+ states in ^{34}S.

The same sized model space was used by Cohen and Kurath (1965) to describe nuclei in the 1p-shell, A = 5 to A = 15. The core is taken at ^{4}He and the valence nucleons distributed this time in the orbitals $1p_{3/2}$ and $1p_{1/2}$. In one fit, the two-body matrix elements were expressed in an L-S coupling representation, where again there are ten diagonal and five off-diagonal elements. However, four of the five off-diagonal elements connect states of opposite spatial symmetry. Realistic potentials deduced from nucleon-nucleon scattering data are symmetric under permutation of the spatial coordinates and lead to vanishing off-diagonal matrix elements between such states. Thus in one fit, the number of parameters varied was reduced by four, but the resulting spectra were very similar to the full fit, the r.m.s. energy deviation being 430 keV for the former and 400 keV for the latter. The resulting wavefunctions were tested by calculating magnetic dipole moments, and probabilities for M1 gamma transitions and beta decay (Cohen and Kurath 1965), spectroscopic amplitudes for one-nucleon transfer reactions (Cohen and Kurath 1967), for two-nucleon transfer reactions (Cohen and Kurath 1970), and alpha-cluster transfer reactions (Kurath 1973). The quality of agreement with experiment is most impressive.

The two-body matrix elements derived from such forced fits are not determined unambiguously. For example, the matrix elements can be multiplied by any of the following classes of sign changes and the calculated spectra would be unaffected, although relative signs on certain components in the wavefunction will change. The ambiguity arises because a single-particle wavefunction, eqn (A.1), never has to be defined. Thus all matrix elements $\langle j_1 j_2;J|V|j_3 j_4;J\rangle_A$ (and ere j represents the n, ℓ, j-value of the orbital) can be ultiplied by: (a) $(-)^{1/2(\ell_3+\ell_4-\ell_1-\ell_2)}$ to compensate for ncluding the time-reversal phase, or by (b) $(-)^{n_1+n_2+n_3+n_4}$ o compensate for defining the radial wavefunctions positive symptotically rather than at the origin. Any calculated bservable should be insensitive to these sign changes, and ould be if the two-body matrix elements were derived from irst principles. But in calculating transition rates, a pecific choice of sign convention for the single-particle unction has to be made (e.g. for the radial integrals in 2 γ-decay, or the form factor for a reaction cross-section calculation), so that a user of wavefunctions from a forced fit has complete freedom to make sign changes (a) or (b) or both, to see which gives the best description of the experimental data. This phase consistency problem does not arise in the p-shell calculations of Cohen and Kurath.

A more serious difficulty is that some matrix elements are poorly determined in the least squares fitting. For example, Glaudemans *et al.* (1964) found that an equally acceptable fit to the energy spectra could be obtained irrespective of whether the matrix element $\langle s_{1/2}{}^2; J=0|V|d_{3/2}{}^2; J=0\rangle_A$ was positive or negative. The dilemma, however, can be resolved by appealing to data on transition rates or reaction cross-sections. This particular matrix element has a crucial effect on the predicted cross-sections for two-nucleon transfer reactions, and its sign could be determined unambiguously from data on the ^{28}Si(t,p)^{30}Si intensities to the ground and excited 0^+ states (Towner and Hardy 1967).

In a similar study on the oxygen, fluorine and neon isotopes, using an ^{16}O core and a valence space comprising only the $1d_{5/2}$ and $2s_{1/2}$ orbitals, Arima *et al.* (1968) found by examining the error matrix of the forced fit that of the 16 two-body matrix elements, only 11 were well determined.

Another problem with the forced fit is deciding which data should be included in the least squares analysis and which data intrude – that is, represent states outside the model space. This was also discussed by Arima *et al.* (1968) and MacFarlane (1967) with reference to the well-known deformed 1^+ state at an excitation energy of 1.7 MeV in the spectrum of ^{18}F. When this state was not included in the fitting procedure, the model predicted an excitation energy of 8.75 MeV for the first excited 1^+ state in ^{18}F. This spectacular discrepancy confirms the conclusion that the 1^+ state at 1.7 MeV in ^{18}F belongs predominantly to configurations outside the model space. However, what happens when this state is forced into the fit? The result is that large changes are induced in some two-body matrix elements. The 1^+ state now appears at 2.34 MeV, but the quality of agreement between theory and experiment for all the other states deteriorates uniformly. ^{18}F acquires a model ground-state spin of 3^+ instead of 1^+. This gives some clue as to a procedure for identifying intruders. The states with the largest errors in the original fit are removed one at a time and the fit repeated. If the new fit shows a markedly different character and a distinctly improved chi-square, then the singled-out state may well be an intruder. Of course considerations of moments, transition rates, and reaction cross-sections can all help in the identification.

Another feature of the phenomenological shell model, that is one with the effective interaction fitted to experimental spectra, is its ability to absorb large amounts of 'configuration impurity'. This is illustrated by a series of studies on 'pseudonium' by Cohen, Lawson, and Soper (1966). Here a theoretical spectrum was generated from a two-shell configuration-mixing calculation, and then fitted by a single-shell calculation. The fitted interaction gave an excellent description of the pseudo-experimental data: excitation energies, M1 and E2 γ-decay rates, and spectroscopic factors. Out of an enormous collection of data, only a tiny handful were in gross disagreement. Engel and Unna (1968) point out that this agreement was to some extent fortunate, since the two orbitals chosen in pseudonium were of opposite parity. Mixing was then confined to configurations of type $(j_1)^{n-r}(j_2)^r$ with r even only. If the two orbitals were of the same parity, then both r even and r odd terms would be present and Engel and Unna show that the latter type of mixing cannot be concealed in a shell-model calculation.

So far we have only discussed configuration mixing among two shell-model orbitals, resulting in fairly modest sized matrices having to be diagonalized for the energy eigenvalues. The extension to larger matrices requires extensive book-keeping and sophisticated computer programming. Two very powerful computer codes were developed during the 1960s. One was the Rochester-Oak Ridge code (French, Halbert, McGrory, and Wong 1969) which relies primarily on fractional parentage expansions and angular momentum algebra as discussed in this chapter. The second was the Argonne code (Cohen, Lawson,

MacFarlane, and Soga 1966) in which basis states of good total angular momentum are specifically constructed in terms of linear combinations of Slater determinants following the methods used in atomic spectroscopy (see for example, Condon and Shortley 1951). Of the two codes, the former has received the wider use, a version of the code being available in many nuclear physics centres.

Popular for the testing of these large scale calculations are the (2s,1d)-shell nuclei, A = 17 to 39. The shell-model space here comprises an ^{16}O core with the valence nucleons distributed over the three 2s,1d orbitals subject only to the restrictions imposed by the Pauli principle. The effective interaction is specified by 63 two-body matrix elements and three single-particle energies. This is too many parameters for a meaningful least squares fitting procedure to be adopted. Calculations have therefore used either realistic interactions (see Kuo and Brown 1966 or Kuo 1967), suitably renormalized for the model space, or a modelistic interaction, such as the modified surface delta interaction (MSDI), with parameters determined by a minimization procedure. Results for nuclei with up to six particles in the s-d shell (Halbert, McGrory, Wildenthal, and Pandya 1971, Preedom and Wildenthal 1972) or six holes in the shell (Wildenthal, Halbert, McGrory, and Kuo 1971) show an impressive agreement with experiment for energy spectra, electromagnetic and beta decay transition rates, and spectroscopic factors for one-nucleon transfer reactions.

For the nuclei in the middle of the shell, A = 23 to 33, the size of the matrices to be handled is just too great (the largest matrix is 6706 × 6706 for 3^+ T=1 state in ^{28}Si), and further model space truncations are introduced. Only modelistic interactions have been used (Wildenthal, McGrory, Halbert, and Graber 1971, Wildenthal and McGrory 1973) and the calculated energies and spectroscopic factors are nevertheless still in good agreement with experiment. Similarly, the model wavefunctions satisfactorily reproduce experimental quadrupole moments and E2 transition rates provided effective charges of e_p = 1.6e and e_n = 0.6e (see eqn (5.54)) are used (de Voigt *et al.* 1972, 1973). However, it is in the area of magnetic moments, M1 transition rates, and Gamow-Teller beta decay probabilities (Lanford and Wildenthal 1973) that the calculations have the most difficulty. The deficiency involves a failure to predict adequate inhibition, a failure which can be traced to the fact that the operator $\underset{\sim}{\sigma}$ is very sensitive to the degree of mixing between the $d_{5/2}$ and $d_{3/2}$ orbitals. The calculations involving a truncation within the s,d-model space fail to produce sufficient mixing between these two orbitals.

The Oak Ridge code was probably pushed to its limit by Soyeur and Zuker (1972) in a recent calculation for the 0^+ T=0 (dimension of matrix, 839 × 839) and the 8^+ T=0 (dimension 1205) states in ^{28}Si using the full s-d shell model space. The results were compared with a 'small space' calculation in which at least 8 nucleons were constrained to occupy the $d_{5/2}$ orbital. The enormous difference between the small space and the exact one (the same effective interaction was used in both cases) can be appreciated by comparing the schematic representation of the ^{28}Si ground-state wavefunction

$$|0^+ \text{ small space}\rangle = 12\% \; |0p\text{-}0h\rangle + 47\% \; |2p\text{-}2h\rangle + 14\% \; |3p\text{-}3h\rangle + 27\% \; |4p\text{-}4h\rangle$$

$$|0^+ \text{ exact}\rangle = 0.1\% \; |0p\text{-}0h\rangle + 1\% \; |2p\text{-}2h\rangle + 3\% \; |3p\text{-}3h\rangle + 17\% \; |4p\text{-}4h\rangle$$

$$+ 26\% \; |5p\text{-}5h\rangle + 29\% \; |6p\text{-}6h\rangle + 14\% \; |7p\text{-}7h\rangle + 4\% \; |8p\text{-}8h\rangle + \ldots .$$

Here $|0p\text{-}0h\rangle$ implies a closed $d_{5/2}$ sub-shell, $|2p\text{-}2h\rangle$ implies two particles in the $s_{1/2}$, $d_{3/2}$ orbitals and two holes in the $d_{5/2}$ sub-shell and so on. For the other spin states, a 'large space' calculation containing excitations of up to 8p-8h and containing most of the important configurations of the exact calculation was performed. The calculated spectra in the small space and in the large space gave an equally good fit to experiment even though the corresponding wavefunctions in the two spaces were quite different. In such circumstances one expects the electromagnetic transition rates to distinguish the calculations, but this proved not to be the case;

both spaces yielded similar results in pleasing agreement with experiment. This phenomenon is reminiscent of the pseudonium results discussed above. The result is not understood and further work on these large shell-model spaces is required.

A different numerical approach to shell model calculations has been advocated by Whitehead (1972). His approach breaks away from the traditional shell model formalism of group theory, fractional parentage, and angular momentum coupling, and replaces these with the more elementary operations of second quantization. The basis states are taken as Slater determinants and are represented in the computer by assigning a single-particle state to each position in the computer word, a 1-bit representing an occupied state and a 0-bit an unoccupied one. (Similar ideas are used in the Argonne code.) Matrix elements of the Hamiltonian are evaluated using the commutation relations of the creation and annihilation operators and can be performed very efficiently using normal computer logic and bit-handling techniques. The entire Hamiltonian matrix is not calculated (this would be a huge $d \times d$ matrix in the m-scheme), but rather the Lanczos altorithm (Wilkinson 1969) is used which iteratively constructs an $r \times r$ tri-diagonal matrix whose eigenvalues converge monotonically to the eigenvalues of the full $d \times d$ Hamiltonian matrix. At each iteration the dimension of the tri-diagonal matrix is increased by one, and after d iterations a tri-diagonal representation of the Hamiltonian matrix would be obtained. However, the beauty of the method is that the iteration is stopped with r much less than d. For a typical case with $d \sim 1000$, 25 iterations will produce the ground state, and 50 iterations the lowest few eigenvalues, so that only a tiny fraction of the total computation need be done. The method is suitable for handling really huge calculations, since the computation time depends linearly on the size of the basis, rather than the cubic dependence of the conventional approach.

The code has been used (Whitehead and Watt 1971, 1972, Cole, Watt, and Whitehead 1973) to tackle the nuclei A = 23 to 33 in the middle of the s-d shell. The calculated spectra, using the Kuo realistic interaction (Kuo 1967), show clearly-identifiable rotation bands with energy spacings between the various members of the band being in good agreement with experiment. However, the separation of one band relative to another is poorly reproduced by the Kuo interaction. In an extreme example, ^{26}Aℓ, three rotation bands are evident, the lowest member of each band, the 5^+, 3^+ and 1^+ states, being the ground, first, and second excited states in that order. The calculated spectrum produces a 1^+ ground state, the 3^+ is the fourth and the 5^+ the eleventh excited state. It is clear that large shell-model calculations magnify any defects in the effective interaction. In fact, it is rather ironic that ^{28}Si teaches us more about the effective two-body force than does ^{18}O, but that appears to be the way it is.

7
SPECTROSCOPIC FACTORS

7.1 DIRECT REACTIONS

In the last chapter, fractional parentage expansions were introduced and extensively used within the framework of the shell model. These expansions represent the breakup of one shell model configuration j^n into two others j^{n-x} and j^x. The coefficients were explicitly calculated for nucleons in a single j shell, but as we shall see the concept can be extended to mixed configurations and a generalized parentage coefficient defined.

Experimental information on the overlap of one nucleus with another can be obtained from so-called 'direct' nuclear reactions. In the analysis of these scattering experiments, the assumption is made that the reaction proceeds with a minimum of rearrangement. Then the total cross-section is essentially proportional to the degree of overlap between the nuclear states, and under favourable conditions a parentage factor can be deduced and compared with the shell model prediction.

Let us consider the reaction schematically pictured in Fig. 7.1. A projectile nucleus, a, impinges on a target

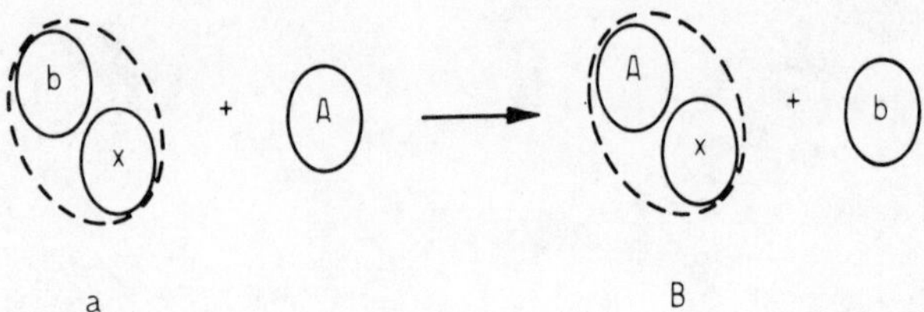

FIG. 7.1. A schematic representation of a direct reaction, in which a projectile, a, being a bound state of b plus x, impinges on a target A. The cluster x is transferred from the projectile to the target resulting in a final nucleus B, being a bound state of A plus x, and an outgoing projectile, b.

nucleus, A, and a cluster of x nucleons is transferred from the projectile to the target. If during this transfer the constituent nuclei, A, b, and x remain unchanged, then the reaction is said to have occurred with a minimum of rearrangement and the process is called 'direct stripping'. The inverse reaction in which nucleons are transferred from the target to the projectile is called 'direct pick-up'. The cross-section for such a process clearly depends on two parentage factors: that relating to the breakup of B into A+x, and that relating to the breakup of a into b+x.

The kinematics of these reactions have traditionally been handled using the distorted wave Born approximation (DWBA). In the so-called 'post-form' of the DWBA, the transition amplitude is written (Satchler 1964, Austern, Drisko, Halbert, and Satchler 1964):

$$T_{ba} = \int d\underset{\sim}{r}_{aA}\, d\underset{\sim}{r}_{bB}\, \chi_b^{*(-)}(\underset{\sim}{k}_b, \underset{\sim}{r}_{bB}) F(\underset{\sim}{r}_{aA}, \underset{\sim}{r}_{bB}) \chi_a^{(+)}(\underset{\sim}{k}_a, \underset{\sim}{r}_{aA}) \quad (7.1)$$

where the $\chi_a^{(+)}$ and $\chi_b^{(-)}$ are the distorted waves describing the relative motion of the pair A,a before collision and the pair B,b after collision. These functions are generated from optical model potentials and they describe the observed elastic scattering in the entrance and exit channel.

The kernel of the integral, $F(\underset{\sim}{r}_{aA}, \underset{\sim}{r}_{bB})$, known as the form factor, is an integral over the internal coordinates ξ of the unchanged constituents A, b and x and is written

$$F(\underset{\sim}{r}_{aA}, \underset{\sim}{r}_{bB}) = \langle \Phi_{B,b} | V | \Phi_{A,a} \rangle$$

$$= \int d\xi_x [\tbinom{B}{x}^{\frac{1}{2}} \int \Phi_B^*(\xi_B)\Phi_A(\xi_A)d\xi_A]\ V\ [\tbinom{a}{x}^{\frac{1}{2}} \int \Phi_b^*(\xi_b)\Phi_a(\xi_a)d\xi_b] \quad (7.2)$$

The combinational factors $\binom{B}{x}$ and $\binom{a}{x}$ count the number of equivalent ways in which x particles can be transferred from projectiles to nuclei without exciting the cores A and b. This can be seen from the following argument.

The matrix element written in the first line of eqn (7.2) is evaluated between states fully antisymmetrized with respect to interchanges among the A+a = B+b coordinates. It is convenient to rewrite the state $\Phi_{A,a}$ as a linear combination of product states $\Phi_A\Phi_a$ in which nucleons in A are fully antisymmetrized amongst themselves and similarly the nucleons in a, but the product function is not antisymmetrized in all a+A nucleons. The antisymmetrizer relating the two states was discussed in chapter 6, and using eqn (6.69), the form factor is expressed as

$$\begin{aligned}\langle\Phi_{B,b}|V|\Phi_{A,a}\rangle &= \binom{A+a}{a}^{-\frac{1}{2}} \langle\Phi_{B,b}|V \sum_r{}' (-)^r P_r|\Phi_A\Phi_a\rangle \\ &= \binom{A+a}{a}^{+\frac{1}{2}} \langle\Phi_{B,b}|V|\Phi_A\Phi_a\rangle.\end{aligned} \tag{7.3}$$

The summation $\sum_r{}'$ is over all permutations between the nucleons in A and those in a that are order-preserving. The second line in eqn (7.3) is obtained on commuting these permutation operators with V and acting on $\Phi_{B,b}$. This state is still fully antisymmetric in all particles, so that each and every order-preserving permutation leaves $\Phi_{B,b}$ unaffected and gives an equal contribution to the matrix element. Eqn (7.3) is in complete analogy to eqn (6.71).

The next step is to apply the antisymmetrizer once again, and to commute the operators with V:

$$\langle\Phi_{B,b}|V|\Phi_{A,a}\rangle = \binom{A+a}{a}^{+\frac{1}{2}} \binom{B+b}{b}^{-\frac{1}{2}} \langle\Phi_B\Phi_b|V \sum_r{}'(-)^r P_r|\Phi_A\Phi_a\rangle$$

where now

$$\begin{aligned}\sum_r{}'(-)^r P_r &= 1 - \sum_{i=1}^{B}\sum_{j=1}^{b} P_{ij} + \sum_{i<k=1}^{B}\sum_{j<\ell=1}^{b} P_{ij}P_{k\ell} - \cdots \\ &= 1 - \sum_{i=1}^{A}\sum_{j=1}^{b} P_{ij} - \sum_{i=A+1}^{B}\sum_{j=1}^{b} P_{ij} + \sum_{i<k=1}^{A}\sum_{j<\ell=1}^{b} P_{ij}P_{k\ell} + \sum_{i<k=A+1}^{B}\sum_{j<\ell=1}^{b} P_{ij}P_{k\ell} - \cdots\end{aligned} \tag{7.4}$$

Of all the order-preserving permutations, only those which leave the nucleons in A unchanged (assumption of direct reactions), namely the first, third, fifth,... terms in eqn (7.4), give a non-vanishing matrix element, and each of these when acting on Φ_a merely gives back the function Φ_a. So the permutation operators can be replaced by unity and the result multiplied by the number of terms

$$1 + xb + \frac{1}{2}\, x(x-1)\frac{1}{2}\, b(b-1) + \ldots = \sum_s \binom{x}{s}\binom{b}{s} = \binom{a}{x}$$

to give

$$\langle \Phi_{B,b} | V | \Phi_{A,a} \rangle = \binom{A+a}{a}^{\frac{1}{2}} \binom{B+b}{b}^{-\frac{1}{2}} \binom{a}{x} \langle \Phi_B \Phi_b | V | \Phi_A \Phi_a \rangle .$$

Simplifying the binomial coefficients then leads to the result given in eqn (7.2).

The interaction potential is $V = V_{bB} - U_{bB}$ where V_{bB} is a sum of all two-body interactions between the nucleons in the projectile b and those in the nucleus B, and U_{bB} is the one-body optical potential generating $\chi_b^{(-)}$. A standard approximation at this point is to write

$$V_{bB} - U_{bB} = V_{bx} + [V_{bA} - U_{bB}] \approx V_{bx},$$

neglecting the contribution from $V_{bA} - U_{bB}$. We shall not discuss this approximation but a recent appraisal has been given by Smith (1971) and by Tobocman *et al.* (1973).

Returning now to eqn (7.2), let us denote the total angular momenta and their magnetic projections for the nuclei A and B by $J_A M_A$ and $J_B M_B$, and their isospins by $T_A N_A$ and $T_B N_B$, and let us define a spectroscopic amplitude A_{BA} via an expansion of the overlap integral from the form factor:

$$\binom{B}{x}^{\frac{1}{2}} \int \Phi_{M_B}^{J_B}(\xi_B) \Phi_{M_A}^{J_A *}(\xi_A) d\xi_A = \sum_{TJM} \langle J_A M_A J M | J_B M_B \rangle \langle T_A N_A T N | T_B N_B \rangle \times$$

$$\times \sum_{qjQL} A_{BA}(J_B\{|J_A;J)\ |[\phi^{QL}(\underset{\sim}{r}_{xA}) \Phi^{qj}(\xi_x)];JM\rangle . \tag{7.5}$$

The function $\Phi^{qj}(\xi_x)$ represents the intrinsic state of the transferred cluster. It contains orbital functions in the relative coordinates, spin functions and isospin functions, the orbital and spin functions being coupled to angular momentum j, additional label q. The function $\phi^{QL}(\underset{\sim}{r}_{xA})$ describes the relative motion between the cluster and the nucleus A (angular momentum L, additional label Q). The coefficient $A_{BA}(J_B\{|J_A;J)$ is known as the spectroscopic amplitude and is the generalized parentage coefficient describing the breakup of nucleus B into constituents A and x. With an analogous definition for the overlap integral of the projectile functions, an integration over the intrinsic coordinates ξ_x of the cluster can be performed and the form factor is then expressed as

$$F(\underset{\sim}{r}_{aA},\underset{\sim}{r}_{bB}) = \sum_{\substack{qjm\\TJMJ'M'\\QL\Lambda Q'L'\Lambda'}} \langle J_A M_A JM|J_B M_B\rangle\langle j_b m_b J'M'|j_a m_a\rangle \langle L\Lambda jm|JM\rangle\langle L'\Lambda' jm|J'M'\rangle \langle T_A N_A TN|T_B N_B\rangle\langle t_b n_b TN|t_a n_a\rangle A^*_{BA}(J_B\{|J_A;J)\, A_{ab}(j_a\{|j_b;J')\, \phi_\Lambda^{QL^*}(\underset{\sim}{r}_{xA}) V_{bx}\, \phi_{\Lambda'}^{Q'L'}(\underset{\sim}{r}_{bx}) \tag{7.6}$$

Next the angular momentum triangles $\Delta(LjJ)$ and $\Delta(L'jJ')$ are recoupled to give $\Delta(LL'\ell)$ and $\Delta(JJ'\ell)$ using eqn (A.14), and the following definitions introduced (Austern *et al.* 1964):

$$B^{QL,Q'L'}_{JJ'\ell} = \sum_{qj} U(jJL'\ell;LJ')A^*_{BA}\, A_{ab}, \tag{7.7a}$$

$$f^{QL,Q'L'}_{JJ'\ell\lambda}(\underset{\sim}{r}_{aA},\underset{\sim}{r}_{bB}) = i^\ell \frac{\hat{j}_a\hat{\ell}}{\hat{L}\hat{J}'}(-)^{L'+\ell-L} \sum_{\Lambda\Lambda'} (-)^{L'-\Lambda'}\langle L\Lambda L'-\Lambda'|\ell\lambda\rangle \times \times\ \phi_\Lambda^{QL^*}(\underset{\sim}{r}_{xA})V_{bx}\, \phi_{\Lambda'}^{Q'L'}(\underset{\sim}{r}_{bx}), \tag{7.7b}$$

$$i^\ell\, \hat{\ell}\, \beta_{\ell\lambda}(\underset{\sim}{k}_a,\underset{\sim}{k}_b) = \int d\underset{\sim}{r}_{aA} d\underset{\sim}{r}_{bB}\ \chi_b^{*(-)}(\underset{\sim}{k}_b,\underset{\sim}{r}_{bB}) f^{QL,Q'L'}_{JJ'\ell\lambda}(\underset{\sim}{r}_{aA},\underset{\sim}{r}_{bB})\chi_a^{(+)}(\underset{\sim}{k}_a,\underset{\sim}{r}_{aA}). \tag{7.7c}$$

Then on substituting eqn (7.6) back into the expression for the transition amplitude, eqn (7.1), one obtains

$$T_{ba} = \sum_{TJJ'\ell M} C\ c\ \langle J_A M_A JM | J_B M_B \rangle \langle \ell \lambda J'M' | JM \rangle \langle j_a m_a j_b - m_b | J'M' \rangle \times$$

$$\times\ (-)^{j_b - m_b}\ \hat{\ell} \sum_{QLQ'L'} B^{QL,Q'L'}_{JJ'\ell}\ \beta_{\ell\lambda}$$

where C and c are isospin Clebsch-Gordan coefficients: $C = \langle T_A N_A TN | T_B N_B \rangle$ and $c = \langle t_b n_b TN | t_a n_a \rangle$. The differential cross-section for the scattering of unpolarized projectiles from unpolarized target nuclei is given by the square of T_{ba} averaged over initial states and summed over final states:

$$\frac{d\sigma}{d\Omega} = \frac{\mu_a \mu_b}{(2\pi\hbar^2)^2} \frac{k_b}{k_a} \frac{1}{(2J_A+1)(2j_a+1)} \sum_{\substack{M_A M_B \\ m_a m_b}} |T_{ba}|^2$$

$$= \frac{\mu_a \mu_b}{(2\pi\hbar^2)^2} \frac{k_b}{k_a} \frac{(2J_B+1)}{(2J_A+1)(2j_a+1)} \sum_{JJ'\ell\lambda} \Big| \sum_{\substack{QLQ'L' \\ T}} C\ c\ B^{QL,Q'L'}_{JJ'\ell}\ \beta_{\ell\lambda} \Big|^2 \qquad (7.8)$$

where μ_a and μ_b are the reduced masses of the projectiles a and b. Note the separation between the spectroscopic dependence of the cross-section as contained in the amplitude B, and the kinematic dependence contained in $\beta_{\ell\lambda}$.

These latter amplitudes $\beta_{\ell\lambda}$ involve a six-dimensional integral of the form factor f and two distorted waves χ. Three of the angle integrations can be performed algebraically, leaving two radial integrals and one angle integral to be evaluated numerically. If no further approximations are made, then the calculation is said to be a 'full-finite-range DWBA' calculation. Such calculations, however, are very time-consuming on a computer, so it is quite common to introduce approximations reducing the integral in eqn (7.7c) to a single three-dimensional integral. Again, two angle integrations are performed algebraically, leaving just the radial integral to evaluate numerically. For stripping and pick-up reactions initiated by light-ion projectiles, the 'zero-range' approximation has been used (Satchler 1964), in which

$$V_{bx}\ \phi^{Q'L'}_{\Lambda'}(\underset{\sim}{r}_{bx}) \sim D_0\ \delta(\underset{\sim}{r}_{bx})\ \delta_{L',0}$$

$$= D_0\ \delta(\underset{\sim}{r}_{bB} - \frac{A}{B}\underset{\sim}{r}_{aA})\left[\frac{aB}{x(a+A)}\right]^{-3}\ \delta_{L',0}$$

with the strength constant D_0 being related to the asymptotic normalization of the wavefunction for the projectile, a.

For reactions initiated by heavy ions this approximation may be poor, or not even applicable in cases where $L' \neq 0$, and the procedure suggested by Buttle and Goldfarb (1966) was to assume that $\underset{\sim}{r}_{aA}$ and $\underset{\sim}{r}_{bB}$ are parallel, i.e. $\underset{\sim}{r}_{aA} \approx \underset{\sim}{r}_{bA}$, $\underset{\sim}{r}_{bB} \approx \frac{A}{B}\underset{\sim}{r}_{bA}$; then the six-dimensional integral can be replaced by a product of two three-dimensional integrals. However, this procedure, the 'no-recoil' approximation, has not been completely successful (DeVries 1973) for the analysis of heavy-ion transfer reactions. We shall not consider the kinematic amplitudes $\beta_{\ell\lambda}$ in any detail; instead we shall focus our attention for the rest of this chapter on the spectroscopic amplitudes A_{BA} and A_{ab} contained in the factor B, eqn (7.7a).

So far in this treatment, no restriction has been placed on the internal structure of the projectiles and transferred cluster. However, let us consider a deuteron stripping reaction, (d,p), in which one neutron is transferred; then the break-up of the neutron function eqn (7.5) into an intrinsic function $\Phi^{qj}(\xi_x)$ and a function of relative motion $\phi^{QL}(r_{xA})$ can be taken as the division of a single-particle wavefunction into its spin and orbital parts. Thus Φ^{qj} just becomes a spin-1/2 spinor and ϕ^{QL} an orbital function of definite angular momentum L and definite number of nodes, i.e. the label Q can now be taken as a count of the number of nodes. Similarly in the breakup of the deuteron, the function $\phi^{Q'L'}(\underset{\sim}{r}_{bx})$ represents the relative motion of the neutron and proton in the deuteron. If it is assumed that $\phi^{Q'L'}$ is a pure S-state function, L'=0, then $V_{bx}\ \phi^{Q'L'}$ can be written as a scalar function $D(r_{bx})$ and the form factor reduces to

$$f = i^L \left(\frac{3}{2}\right)^{1/2}\ \phi_\Lambda^{QL^*}(r_{xA})\ D(r_{bx}).$$

Similarly putting L'=0 in eqn (7.7a) requires that ℓ=L and j = J'=1/2 and the spectroscopic amplitude B reduces to the simple product $A^*_{BA}\ A_{ab}$. Finally, A_{ab} is trivially evaluated

$A_{ab} = \binom{a}{x}^{1/2} = \sqrt{2}$, such that c A_{ab} is unity for the (d,p) reaction, and the familiar expression for the differential cross-section results:

$$\frac{d\sigma}{d\Omega} = \frac{\mu_a\mu_b}{(2\pi\hbar^2)^2} \cdot \frac{k_b}{k_a} \cdot \frac{(2J_B+1)}{(2J_A+1)(2j_a+1)} C^2 \sum_{\ell J} |A_{BA}|^2 \sum_{\lambda} |\beta_{\ell\lambda}|^2 \quad (7.9)$$

with $J = \ell \pm 1/2$. We have assumed that the interaction potential V and the optical potentials U are spin-independent, so that an incoherent sum over the orbital and total angular momenta, ℓ and J, of the transferred nucleon appears in eqn (7.9).

It is quite frequent that the angular momentum and parity selection rules lead to unique values of ℓ and J; then the spectroscopic amplitude A_{BA} merely becomes a multiplicative factor to the cross-section. Thus if the DWBA theory can be relied upon to estimate accurately the kinematic factors $\beta_{\ell\lambda}$, a comparison of a DWBA calculation with an experimental cross-section leads directly to a measure of the spectroscopic factor $|A_{BA}|^2$. This value may then be compared with the predictions of various nuclear models.

7.2 CALCULATION OF SPECTROSCOPIC AMPLITUDES USING SHELL MODEL WAVEFUNCTIONS

The spectroscopic amplitude A_{BA} was introduced in eqn (7.5). Inverting this expansion gives

$$A_{BA}(J_B\{|J_A;J) = \binom{B}{x}^{\frac{1}{2}} \int [\Phi^{J_A}(\xi_A)[\phi^{QL}(\underset{\sim}{r}_{xA})\Phi^{qj}(\xi_x)]^J]^{J_B *}_{M_B} \Phi^{J_B}_{M_B}(\xi_B)d\xi_B \quad (7.10)$$

where the coordinates satisfy the relation

$$d\xi_B = d\xi_A \, d\xi_x \, d\underset{\sim}{r}_{xA}.$$

The square brackets denote angular momentum coupling as before. The specification of coordinates requires some care, and we shall follow the treatment of Ichimura *et al.* (1973) and where possible adopt their notation. In eqn (7.10), $\Phi^{J_A}(\xi_A)$ represents the internal wavefunction for nucleus A depending on (A-1) internal coordinates. It is more convenient to work with shell model wavefunctions $\Psi^{J_A}(\zeta_A)$ Slater determinants or linear combinations of such), compounded out of single-particle wavefunctions with coordinates expressed relative to the fixed-centre of the shell model potential. We define: σ_i, τ_i, and $\underset{\sim}{r}_i$ as the spin coordinate of the ith nucleon, its isospin coordinate, and its position measured with respect to the centre of the shell model potential respectively; $\underset{\sim}{R}_A = (1/A) \sum_{i=1}^{A} \underset{\sim}{r}_i$ as the position of the centre-of-mass of the nucleus A measured with respect to the centre of the shell model potential; $\zeta_A \equiv \{\underset{\sim}{r}_i, \sigma_i, \tau_i;\ i=1,\ldots,A\}$ as the set of coordinates used for shell model wavefunctions; ξ_A as the set of internal coordinates of the nucleus A. It is assumed that ξ_A is chosen so that

$$d\zeta_A = \pm\, d\xi_A\, d\underset{\sim}{R}_A$$

with analagous definitions for nucleus B and cluster x. Finally $\underset{\sim}{r}_{xA} = \underset{\sim}{R}_x - \underset{\sim}{R}_A$ is defined as the relative coordinate between the centre-of-mass of the nuclei x and A.

The problem of converting from the set of coordinates ξ_A to ζ_A can be solved if each shell model wavefunction is expressed in terms of harmonic oscillator functions. Then the wavefunction is considered to contain no 'spurious' centre-of-mass motion if the following relation holds:

$$\Psi^{J_A}(\zeta_A) = \phi^{00}(\underset{\sim}{R}_A)\Phi^{J_A}(\xi_A). \tag{7.11}$$

With the use of harmonic oscillator wavefunctions, the labels q and Q in eqn (7.10) can now be used to specify the number of oscillator quanta in the intrinsic and relative motion of the cluster respectively. Equivalently the number of radial nodes N could have been specified, where Q = 2N+L. Similarly in eqn (7.11) the function $\phi^{00}(R_A)$ is an oscillator function with no quanta in the centre-of-mass motion.

The spectroscopic amplitude is now expressed as an integral of harmonic oscillator shell-model functions with all radial coordinates referring to the well-centre:

$$A_{BA}(J_B\{|J_A;J) = \left(\frac{B}{B-x}\right)^{\frac{1}{2}Q}\binom{B}{x}^{\frac{1}{2}}\int [\Psi^{J_A}(\zeta_A)[\phi^{QL}(\underset{\sim}{R}_x)\phi^{qj}(\xi_x)]^J]^{J_B^*}_{M_B}\ \Psi^{J_B}_{M_B}(\zeta_B)d\zeta_B. \tag{7.12}$$

the proof of this result has been given by Ichimura *et al.* (1973) and proceeds as follows. Consider the integral I contained in eqn (7.12):

$$I = \int [\Psi^{J_A}(\zeta_A)[\phi^{QL}(\underset{\sim}{R}_x)\phi^{qj}(\xi_x)]^J]^{J_B^*}_{M_B}\ \Psi^{J_B}_{M_B}(\zeta_B)d\zeta_B \tag{7.13}$$

and impose the centre-of-mass condition eqn (7.11) on the wavefunctions for nuclei A and B. Then we obtain

$$I = \int [\Phi^{J_A}(\xi_A)[\phi^{00}(\underset{\sim}{R}_A)\phi^{QL}(\underset{\sim}{R}_x)\phi^{qj}(\xi_x)]^J]^{J_B^*}_{M_B}\ \Phi^{J_B}_{M_B}(\xi_B)\phi^{00}(\underset{\sim}{R}_B)d\xi_B d\underset{\sim}{R}_B.$$

Next the generalized Talmi-Moshinsky transformation (Smirnov 1961) is introduced in which the product $\phi^{00}(\underset{\sim}{R}_A)\phi^{QL}(\underset{\sim}{R}_x)$ is written as a bilinear sum involving functions of the centre-of-mass coordinate $\underset{\sim}{R}_B$ and the relative coordinate $\underset{\sim}{r}_{xA}$:

$$\phi^{00}(\underset{\sim}{R}_A)\phi^{QL}(\underset{\sim}{R}_x) = \sum_{Q'L'q'\ell'} \langle 00,NL:L|n'\ell',N'L':L\rangle[\phi^{q'\ell'}(\underset{\sim}{R}_B)\phi^{Q'L'}(\underset{\sim}{r}_{xA})]^L$$

with Q = 2N+L etc. Inserting this expansion in I and integrating over $d\underset{\sim}{R}_B$, only the term $q'\ell'$ = 00 survives.

$$I = \langle 00,NL:L|00,NL;L\rangle \int [\Phi^{J_A}(\xi_A)[\phi^{QL}(\underset{\sim}{r}_{xA})\phi^{qj}(\xi_x)]^J]^{J_B^*}_{M_B}\ \Phi^{J_B}_{M_B}(\xi_B)d\xi_B.$$

Thus by comparing this value for eqn (7.13) with the definition (eqn (7.10)) and explicitly evaluating the generalized Talmi-Moshinsky bracket

$$\langle 00,NL:L|00,NL:L\rangle = \left(\frac{B-x}{B}\right)^{\frac{1}{2}Q},$$

the result, eqn (7.12), is obtained. This completes the proof of that equation.

The next step is a fractional parentage expansion of Ψ^{J_B}:

$$\binom{B}{x}^{\frac{1}{2}}\Psi^{J_B}(\zeta_B) = \sum_{J_A\Gamma J} S^{\frac{1}{2}}(J_B\{|J_A;\Gamma J)[\Psi^{J_A}(\zeta_A)\Psi^{\Gamma J}(\zeta_x)]^{J_B} \tag{7.14}$$

where the coefficient $S^{\frac{1}{2}}(J_B\{|J_A:\Gamma J)$ is a product of a fractional parentage coefficient and the combinational factor $\binom{B}{x}^{\frac{1}{2}}$. The additional label Γ is introduced to specify the shell model nature of the x transferred nucleons, i.e. identifies the shell model orbitals from which the nucleons originated. Then on inserting eqn (7.14) in eqn (7.12) and integrating over ζ_A, our final result is obtained:

$$A_{BA}(J_B\{|J_A;J) = \left(\frac{B}{B-x}\right)^{\frac{1}{2}Q} \sum_{\Gamma} S^{\frac{1}{2}}(J_B\{|J_A;\Gamma J)G(\Gamma J,qj,QL) \tag{7.15a}$$

where

$$G(\Gamma J,qj,QL) \equiv \int [\phi^{QL}(\underset{\sim}{R}_x)\phi^{qj}(\xi_x)]^{J*}\ \Psi^{\Gamma J}(\zeta_x)d\zeta_x. \tag{7.15b}$$

This result is completely general; no approximation has been made other than the decision to use shell model harmonic oscillator wavefunctions and to use the same oscillator size parameter for all three nuclei, A, B, and x. This latter restriction is not essential, the appropriate modifications have been given by Ichimura *et al.* (1973).

Let us now evaluate the integral G, eqn (7.15b) for the specific cases of one-nucleon and two-nucleon transfer reactions.

One-nucleon transfer reaction. We suppose that in the breakup of the nucleus B into nucleus A and a single particle, the single particle originates from a particular shell model orbital $n_1\ell_1j_1$, say. The function $\Psi^{\Gamma J}(\zeta_x)$ is therefore just a single-particle function, such as eqn (A.1) in the appendix, with the label Γ designating the orbit $\Gamma \equiv n_1\ell_1j_1$ and with total angular momentum necessarily given by $J = j_1$. The overlap of $\Psi^{\Gamma J}$ with an intrinsic and relative function is trivial. The intrinsic function in this case is just the spin part and the relative function the orbital part of the single-particle wavefunction, so the overlap is simply $G = 1$, with $j = 1/2$, $Q = 2n_1+\ell_1$ and $L = \ell_1$. Thus the spectroscopic amplitude is

$$A_{BA}(J_B\{|J_A,J=j_1) = \left(\frac{B}{B-1}\right)^{n_1+\ell_1/2} S^{1/2}(J_B\{|J_A;n_1\ell_1j_1) \tag{7.16}$$

and the differential cross-section, eqn (7.9), is seen to depend on the square of A_{BA}, and therefore essentially on S. The quantity S is called the 'spectroscopic factor' and we discuss its evaluation in more detail in the next section.

There is also a spectroscopic amplitude A_{ab} relating to the overlap of projectile wavefunctions. For light ions, $a \leqslant 4$, this is straightforward to evaluate with harmonic oscillator shell-model functions, on the assumption that only 0s orbitals are occupied. The results for selected reactions are listed in Table 7.1.

TABLE 7.1

Values for $c^2 A_{ab}^2$ *for the single-nucleon transfer reaction (a,b), where* $c = \langle t_b n_b \frac{1}{2} \pm\frac{1}{2} | t_a n_a \rangle$ *and* A_{ab} *the spectroscopic amplitude.*

Reaction	(d,p) or (d,n)	(t,d) or (^{3}He,d)	(α,t) or (α,^{3}He)
$c^2 A_{ab}^2$	1	3/2	2

Two-nucleon transfer reaction. As before, we suppose that the nucleus B breaks up into nucleus A and two particles with the two particles originating from particular shell-model orbitals, say $n_1\ell_1 j_1$ and $n_2\ell_2 j_2$. The function $\Psi^{\Gamma J}(\zeta_x)$ is therefore written as a normalized, antisymmetrized product

$$\Psi^{\Gamma J}(\zeta_x) = N(j_1 j_2 JT) \sum_{m_1 m_2} \langle j_1 m_1 j_2 m_2 | JM \rangle \frac{1}{\sqrt{2}}[\Phi_{j_1 m_1}(\zeta_1)\Phi_{j_2 m_2}(\zeta_2) - \text{exch.}] \tag{7.17}$$

with

$$N(j_1 j_2 JT) = [1 - (-)^{J+T} \delta_{j_1 j_2}]^{-\frac{1}{2}}.$$

The label Γ represents $n_1\ell_1 j_1, n_2\ell_2 j_2$. The isospin functions have been omitted for brevity. In order to overlap this function $\Psi^{\Gamma J}$ with the intrinsic and centre-of-mass functions in eqn (7.15b), eqn (7.17) must first be transformed from j-j to L-S coupling and then a Moshinsky transformation applied to convert from shell model coordinates to relative and centre-of-mass coordinates. Finally after some angular momentum recoupling, the integral in eqn (7.15b) can be performed with the result

$$G(\Gamma J,qj,QL) = \frac{1}{\sqrt{2}}N(j_1 j_2 JT) \sum_{L'} (-)^{\ell-L'+L}\, U(L\ell JS;L'J)\,(1-(-)^{\ell+S+T}) \times \times \langle n\ell NL:L'|n_1\ell_1 n_2\ell_2:L'\rangle \begin{bmatrix} \ell_1 & \ell_2 & L' \\ 1/2 & 1/2 & S \\ j_1 & j_2 & J \end{bmatrix} \tag{7.18}$$

with $Q = 2N+L$ and $q = 2n+\ell$. There is a restriction on the possible values of Q and q, namely $Q+q = q_1+q_2$, coming from the conservation of energy in the Moshinsky transformation.

If an additional assumption is injected at this point, namely that the relative motion of the two-transferred nucleons is predominately s-state, i.e. $\ell=0$, then eqn (7.18) simplifies considerably:

$$G(\Gamma J,qj,QL) = \frac{1}{\sqrt{2}}N(j_1 j_2 JT)\,\delta_{S+T,1}\,\delta_{j,s}\langle n0NL;L|n_1\ell_1 n_2\ell_2:L\rangle \times \times \begin{bmatrix} \ell_1 & \ell_2 & L \\ 1/2 & 1/2 & S \\ j_1 & j_2 & J \end{bmatrix} \tag{7.19}$$

This is essentially the result derived by Glendenning (1965) except that he has not made the assumption that the harmonic oscillator length parameter is the same for projectile and nuclear wavefunctions. His result, therefore, is slightly more general than eqn (7.19), containing an extra radial integral denoted there by Ω_n. The spectroscopic amplitude A_{BA} is now given by eqn (7.15a) with G given either by eqn (7.18) or (7.19).

Similarly there is a spectroscopic amplitude A_{ab} for the projectiles. If we consider only light ions, $a \leqslant 4$, and assume their functions are simply given in terms of 0s harmonic oscillator functions, then A_{ab} is again trivially evaluated and the results are listed in Table 7.2.

TABLE 7.2

Values for $c^2 A_{ab}{}^2$ for the two-nucleon transfer reaction (a,b) where $c = \langle t_b n_b TN | t_a n_a \rangle$ and A_{ab} is the spectroscopic amplitude. $c^2 A_{ab}{}^2 = (2S+1) b_{st}{}^2$ where $b_{ST}{}^2$ is the quantity defined by Glendenning (1965) in connection with two-nucleon transfer reactions.

Reaction	(t,p) or (^{3}He,n)	(^{3}He,p) or (t,n)	(α,d)
$c^2 A_{ab}{}^2$	$\delta_{S,0}\delta_{T,1}$	$\frac{1}{2}(\delta_{S,0}\delta_{T,1}+3\delta_{S,1}\delta_{T,0})$	$3\delta_{S,1}\delta_{T,0}$

7.3 SHELL MODEL EXPRESSIONS FOR THE SPECTROSCOPIC FACTOR

In the last two sections, we have introduced the spectroscopic amplitude, defined in eqn (7.10), and rewritten in eqn (7.15) in terms of shell model coordinates as

$$A_{BA}(J_B\{|J_A;J) = \left(\frac{B}{B-x}\right)^{\frac{1}{2}Q} \sum_{\Gamma} S^{\frac{1}{2}}(J_B\{|J_A;\Gamma J) G(\Gamma J, qj, QL). \quad (7.15)$$

In this section, we want to look a little closer at the factor $S^{\frac{1}{2}}(J_B\{|J_A;\Gamma J)$ and discuss its evaluation in the nuclear shell model. We restrict our discussion to one-nucleon transfer reactions. Then the factor G is unity and the label Γ is unique, representing in this case the particular shell model orbital from which the transferred nucleon originated.

The quantity $S^{\frac{1}{2}}$, defined in eqn (7.14), is essentially the parentage coefficient in the expansion of the wavefunction for nucleus B in terms of that for nucleus A:

$$\binom{B}{1}^{\frac{1}{2}} \Psi^{J_B}(\zeta_B) = \sum_{J_A \Gamma J} S^{\frac{1}{2}}(J_B\{|J_A;\Gamma J)[\Psi^{J_A}(\zeta_A)\Psi^{\Gamma J}(\zeta_1)]^{J_B}. \quad (7.14)$$

The square of this coefficient is called the spectroscopic factor. An inversion of the expansion gives the following expression for the coefficient $S^{\frac{1}{2}}$:

$$S^{\frac{1}{2}}(J_B\{|J_A;\Gamma J) = \begin{pmatrix} B \\ 1 \end{pmatrix}^{\frac{1}{2}} \int [\Psi^{J_A}(\zeta_A)\Psi^{\Gamma J}(\zeta_1)]^* \Psi^{J_B}(\zeta_B)d\zeta_B. \quad (7.20)$$

To continue, the nuclear wavefunctions have to be specified.

Let us suppose that the nucleus B contains n_1+n_2 particles and these particles are distributed over just two shell-model orbitals. Then the wavefunction is expanded in a basis of such functions

$$\Psi^{J_B} = \Sigma_\beta\, b_\beta\, \phi_\beta,$$

exactly as discussed in section (6.9) on configuration mixing. The wavefunction amplitudes b_β are the eigenvectors derived in the standard diagonalization of the Hamiltonian matrix. The basis states were specified in eqn (6.62) as

$$\phi_\beta = |j_1^{n_1} j_2^{n_2}(\underset{\sim}{r}_1,\ldots,\underset{\sim}{r}_{n_1+n_2})X_1J_1,X_2J_2;J_BM_B\rangle$$

with β representing all the quantum numbers X_1J_1,X_2J_2 characterizing the division of $B = n_1+n_2$ particles into the two orbitals. Similarly the wavefunction for nucleus A is written

$$\Psi^{J_A} = \Sigma_\alpha\, a_\alpha\, \phi_\alpha$$

with

$$\phi_\alpha = |j_1^{n_1'} j_2^{n_2'}(\underset{\sim}{r}_1,\ldots,\underset{\sim}{r}_{n_1+n_2-1})X_1'J_1',X_2'J_2';J_AM_A\rangle$$

where now there is one less nucleon in $A = n_1'+n_2' = n_1+n_2-1$. The spectroscopic factor is then quite generally given by

$$S^{\frac{1}{2}}(J_B\{|J_A;\Gamma J) = \sum_{\alpha\beta} a_\alpha b_\beta\, S^{\frac{1}{2}}_{\alpha\beta} \qquad (7.21)$$

where $S^{\frac{1}{2}}_{\alpha\beta}$ is given by an analogous expression to eqn (7.20), with basis functions ϕ replacing total wavefunctions Ψ.

The first step in the evaluation of $S^{\frac{1}{2}}_{\alpha\beta}$ is to replace the fully antisymmetrized basis states ϕ by states χ, in which the n_1 nucleons in the j_1 shell are antisymmetrized amongst themselves and similarly the n_2 nucleons in the j_2 shell, but the two groups are just vector-coupled to form a state of good total angular momentum. The antisymmetrizer projecting the states χ on to ϕ was defined in eqn (6.69) and is given as a summation of order-preserving permutations. Let us suppose that the transferred nucleon originated in the j_2 shell, then

$$S^{\frac{1}{2}}_{\alpha\beta}(j_2) = (n_1+n_2)^{\frac{1}{2}} \langle [\phi_\alpha(j_1^{\,n_1} j_2^{\,n_2-1})\phi(j_2)] | \phi_\beta(j_1^{\,n_1} j_2^{\,n_2}) \rangle$$

$$= (n_1+n_2)^{\frac{1}{2}} \binom{n_1+n_2-1}{n_1}^{-\frac{1}{2}} \langle [\chi_\alpha(j_1^{\,n_1};j_2^{\,n_2-1})\phi(j_2)] | \sum_r{}'(-)^r P_r | \phi_\beta(j_1^{\,n_1} j_2^{\,n_2}) \rangle$$

$$= (n_1+n_2)^{\frac{1}{2}} \binom{n_1+n_2-1}{n_1}^{\frac{1}{2}} \langle [\chi_\alpha(j_1^{\,n_1};j_2^{\,n_2-1})\phi(j_2)] | \phi_\beta(j_1^{\,n_1} j_2^{\,n_2}) \rangle.$$

The last step follows from noting that ϕ_β is fully antisymmetric in all (n_1+n_2) particles, so that each and every order-preserving permutation leaves ϕ_β unaffected and gives an equal contribution to the matrix element. Applying the antisymmetrizer once again gives

$$S^{\frac{1}{2}}_{\alpha\beta}(j_2) = (n_1+n_2)^{\frac{1}{2}} \binom{n_1+n_2-1}{n_1}^{\frac{1}{2}} \binom{n_1+n_2}{n_1}^{-\frac{1}{2}}$$

$$\langle [\chi_\alpha(j_1^{\,n_1};j_2^{\,n_2-1})\phi(j_2)] | \sum_r{}'(-)^r P_r | \chi_\beta(j_1^{\,n_1};j_2^{\,n_2}) \rangle.$$

This time note that out of all the order-preserving permutations involved, only one contributes to the matrix element, the others all vanishing due to the orthogonality of the single-particle wavefunctions making up the determinants in χ. Simplifying the binomial coefficients then leads to the result

$$S^{\frac{1}{2}}_{\alpha\beta}(j_2) = n_2^{\frac{1}{2}} \langle [\chi_\alpha(j_1^{\,n_1};j_2^{\,n_2-1})\phi(j_2)] | \chi_\beta(j_1^{\,n_1};j_2^{\,n_2}) \rangle$$

$$= n_2^{\frac{1}{2}} \langle [j_1^{\,n_1} x_1' J_1', j_2^{\,n_2-1} x_2' J_2']^{J_A} j_2; J_B | j_1^{\,n_1} x_1 J_1, j_2^{\,n_2} x_2 J_2; J_B \rangle. \quad (7.22)$$

The second line is just a change in the notation, illustrating the angular momentum coupling. Finally the right-hand side of the matrix element in eqn (7.22) is rearranged by making first a fractional parentage expansion and second an angular momentum recoupling to bring it to the form of the left-hand side. Then an integration over the n_1+n_2 coordinates gives the required result:

$$S^{\frac{1}{2}}_{\alpha\beta}(j_2) = n_2^{\frac{1}{2}}\langle j_2^{\,n_2}X_2J_2\{|j_2^{\,n_2-1}\;X_2'J_2';j_2\rangle U(J_1J_2'J_Bj_2;J_AJ_2)\delta_{X_1X_1'}\delta_{J_1J_1'}. \tag{7.23}$$

This expression for the spectroscopic amplitude $S^{\frac{1}{2}}_{\alpha\beta}(j_2)$ was first obtained by Macfarlane and French (1960) and further details on the intricacies of the algebraic manipulations and some illustrative examples can be found in this reference. A very similar expression is obtained when the transferred nucleon originates from an inner orbital, say j_1. Then the equation analogous to eqn (7.22) reads

$$S^{\frac{1}{2}}_{\alpha\beta}(j_1) = n_1^{\frac{1}{2}}\langle [j_1^{\,n_1-1}\;X_1'J_1',j_2^{\,n_2}X_2'J_2']^{J_A}\;j_1;J_B|j_1^{\,n_1}X_1J_1,j_2^{\,n_2}\;X_2J_2;J_B\rangle.$$

It remains to make the angular momentum reductions. First a fractional parentage expansion on the group $j_1^{\,n_1}X_1J_1$, secondly an angular momentum recoupling (note the introduction of the Pauli phase $(-)^{n_2}$ as the nucleon j_1 is commuted with the group $j_2^{\,n_2}$), and finally an integration over the n_1+n_2 coordinates to give

$$S^{\frac{1}{2}}_{\alpha\beta}(j_1) = n_1^{\frac{1}{2}}\,(-)^{n_2}(-)^{J_A-J_B+J_1-J_1'}\langle j_1^{\,n_1}X_1J_1\{|j_1^{\,n_1-1}\;X_1'J_1';j_1\rangle\times$$

$$\times\, U(J_2J_1'J_Bj_1;J_AJ_1)\delta_{X_2X_2'}\delta_{J_2J_2'}. \tag{7.24}$$

This should be sufficient illustration of how the amplitudes $S^{\frac{1}{2}}$ are evaluated; more complicated configurations for the basis states can be handled in much the same way.

7.4 SUM RULES FOR SPECTROSCOPIC FACTORS

For a single nucleon stripping reaction A(a,b)B initiated by light ions, such as the (d,p) reaction, the differential cross-section is given by eqn (7.9) with the spectroscopic factor $|A_{BA}|^2$ taken from eqn (7.16):

$$\left(\frac{d\sigma}{d\Omega}\right)_{\text{stripping}} = \frac{\mu_a\mu_b}{(2\pi\hbar^2)^2}\frac{k_b}{k_a}\left(\frac{B}{B-1}\right)^{2n+\ell}\frac{(2J_B+1)}{(2J_A+1)(2j_a+1)}C^2S\sum_\lambda|\beta_{\ell\lambda}|^2$$

$$= \frac{2J_B+1}{2J_A+1}C^2S\,\sigma_{DWBA}. \qquad (7.25)$$

It has been assumed that the transferred nucleon originates from a unique shell model orbital, $n\ell j$. All the details of the kinematics are contained in the DWBA cross-section σ_{DWBA} while the spectroscopy of the nuclear states involved influences only the spectroscopic factor S. This convenient factorization between the kinematic and spectroscopic dependences has made (d,p) stripping reactions very popular in nuclear physics. The typical procedure is to calculate the cross-section σ_{DWBA} using standard DWBA computer codes and then to normalize the result to the experimental data. In this way a spectroscopic factor S is determined, which in turn can be compared with the predictions from various nuclear models.

For the inverse pick-up reaction B(b,a)A, the differential cross-section is essentially that of the stripping reaction, only with a change in the phase space:

$$\left(\frac{d\sigma}{d\Omega}\right)_{\text{pickup}} = \frac{(2J_A+1)(2j_a+1)}{(2J_B+1)(2j_b+1)}\left(\frac{d\sigma}{d\Omega}\right)_{\text{stripping}}$$

$$= \frac{2j_a+1}{2j_b+1}C^2S\,\sigma_{DWBA}. \qquad (7.26)$$

In this section we want to derive sum rules for the spectroscopic factor both for stripping and pick-up reactions and illustrate how they can be useful in analysing experimental data. It is convenient to introduce a slightly different notation. We shall no longer use subscripts A and B to differentiate the initial and final nuclei, but rather the target nucleus, be it stripping or a pick-up reaction, will always be given the quantum numbers J_i and T_i and the final nucleus J_f and T_f. The spectroscopic factor will be written S^+ with a plus superscript for a stripping reaction and S^- for a pick-up reaction.

Let us assume for a start that the wavefunctions for the initial and final nuclei are given by single basis functions ϕ_α and ϕ_β with no configuration mixing:

$$\Psi^{J_i} = \phi_\alpha = |j_1^{n_1}\, X_1 J_1,\ j_2^{n_2}\, X_2 J_2; J_i\rangle$$

$$\Psi^{J_f} = \phi_\beta = |j_1^{n_1}\, X_1' J_1',\ j_2^{n_2\pm 1}\, X_2' J_2'; J_f\rangle.$$

The transferred nucleon in this instance originates from the outermost filled shell, the j_2 shell. Then the spectroscopic factor for the stripping reaction is given by the square of eqn (7.23):

$$S^+ = (n_2+1)\langle j_2^{n_2+1}\, X_2' J_2' \{| j_2^{n_2} X_2 J_2; j_2\rangle^2\ U(J_1 J_2 J_f j_2; J_i J_2')^2\ \delta_{X_1 X_1'}\ \delta_{J_1 J_1'} \qquad (7.27)$$

Note that the inner orbital j_1 plays a passive role; only the total angular momentum J_1 enters eqn (7.27). In fact this equation can be reinterpreted for any number of inner shells, the spectroscopic factor depending only on their total angular momentum, J_1. Furthermore, if all these inner shells are coupled to zero, $J_1=0$, then they have no influence at all, and the spectroscopic factor S^+ reduces to

$$S^+ = (n_2+1)\langle j_2^{n_2+1}\ X_2' J_2' \{| j_2^{n_2}\ X_2 J_2; j_2\rangle^2.$$

The most frequently encountered example of this situation is when the inner shells are completely occupied closed shells. Thus we see that in evaluating spectroscopic factors, one is completely justified in ignoring the closed shells.

Note also that in the particular case of a reaction proceeding from a closed-shell to closed-shell-plus-one nucleus, then $n_2=0$, and the spectroscopic factor is simply

$$S^+ = 1.$$

This result has led to the jargon that whenever the reaction proceeds with a spectroscopic factor of one or close to one, the transition is said to be a 'good single-particle' transition.

The first sum rule to be derived is for the pick-up reaction. We shall sum the spectroscopic factor, S^-, over all final states in the residual nucleus, that is sum over all possible quantum numbers $X_1'J_1'$, $X_2'J_2'$, J_f. The result is

$$\sum_{\substack{X_1'J_1'X_2'J_2'\\J_f}} S^- = n_2 \sum_{X_2'J_2'J_f} \langle j_2^{\,n_2} X_2J_2\{|j_2^{\,n_2-1} X_2'J_2';j_2\rangle^2 \; U(J_1J_2'J_ij_2;J_fJ_2)^2$$

$$= n_2 \sum_{X_2'J_2'} \langle j_2^{\,n_2} X_2J_2\{|j_2^{\,n_2-1} X_2'J_2';j_2\rangle^2$$

$$= n_2, \tag{7.28}$$

where first orthogonality of the U-coefficient, and then orthogonality of the cfp has been used. Thus the sum of spectroscopic factors to all states populated in a single-nucleon pick-up reaction, with the transferred nucleon being identified as arising from a particular shell-model orbital j_2, is equal to n_2, the number of nucleons in the target nucleus occupying that specified orbital.

A similar result can be obtained for the stripping reaction. This time, we sum the spectroscopic factors S^+, modulated by $(2J_f+1)/(2J_i+1)$, over all final states in the residual nucleus to obtain

$$\sum_{\substack{X_1'J_1'X_2'J_2'\\J_f}} \frac{(2J_f+1)}{(2J_i+1)} S^+ = (n_2+1) \sum_{X_2'J_2'J_f} \langle j_2^{\,n_2+1} X_2'J_2'\{|j_2^{\,n_2} X_2J_2;j_2\rangle^2$$

$$\times \frac{2J_f+1}{2J_i+1} U(J_1J_2J_fj_2;J_iJ_2')^2$$

$$= (n_2+1) \sum_{X_2'J_2'} \langle j_2^{\,n_2+1} \; X_2'J_2'\{|j_2^{\,n_2} X_2J_2;j_2\rangle^2 \frac{(2J_2'+1)}{(2J_2+1)}$$

$$= n_{max} - n_2 \tag{7.29}$$

where the hole-particle relation for the cfp has been used. The result states, as might have been expected, that the sum is equal to the number of nucleon holes in the j_2 orbital in the target nucleus. In eqn (7.29), n_{max} is the maximum number of particles in the j_2-orbital allowed by the Pauli exclusion principle.

These two sum rules can be generalized to the case when configuration mixing is present. The nuclear wavefunction is now expressed as a linear combination of basis states

$$\Psi_{J_f}^{(q)} = \sum_\beta b_\beta^{(q)} \phi_\beta$$

with the superscript q differentiating the distinct eigenvalue solutions. The wavefunction amplitudes satisfy orthogonality and closure relations:

$$\begin{aligned} \sum_\beta b_\beta^{(q)} b_\beta^{(q')} &= \delta_{qq'} \\ \sum_q b_\beta^{(q)} b_{\beta'}^{(q)} &= \delta_{\beta\beta'}. \end{aligned} \tag{7.30}$$

Let us now recompute the sum rule for the single-nucleon pick-up reaction, allowing configuration mixing in the final nucleus wavefunctions, but for the moment, keeping just the single basis state ϕ_α for the target nucleus. The sum required is over all the distinct eigenvalue solutions, namely

$$\sum_q S^{-(q)}(j_2) = \sum_q |\sum_\beta b_\beta^{(q)} \; S_{\alpha\beta}^{\frac{1}{2}}(j_2)|^2$$

$$= \sum_{q\beta\beta'} b_\beta^{(q)} b_{\beta'}^{(q)} \; S_{\alpha\beta}^{\frac{1}{2}}(j_2) \; S_{\alpha\beta'}^{\frac{1}{2}}(j_2)$$

$$= \sum_{\beta} S^{-}_{\alpha\beta}(j_2)\ \delta_{\beta\beta'}$$

$$= n_2$$

where the closure relation (7.28) has been used. The result is that the sum rule is unchanged. This again is intuitively what we would have expected. The role of configuration mixing is to redistribute the transition strength over the various states, but the summed strength is conserved. An exactly analogous result holds for stripping reactions.

Next, let us consider configuration mixing in the target wavefunction, and from the preceding discussion it is obvious that there is no loss in generality in using the basis states ϕ_β to describe the final nuclear states. The sum required is over all final states

$$\sum_{\beta} S^{-} = \sum_{\beta} \left|\sum_{\alpha} a_\alpha\ S^{\frac{1}{2}}_{\alpha\beta}(j_2)\right|^2$$

$$= \sum_{\alpha\bar{\alpha}\beta} a_\alpha a_{\bar{\alpha}}\ S^{\frac{1}{2}}_{\alpha\beta}(j_2)\ S^{\frac{1}{2}}_{\bar{\alpha}\beta}(j_2).$$

Using the expression (7.23) for $S^{\frac{1}{2}}(j_2)$ and remembering the sum β is over the configurations $X_1'J_1'X_2'J_2'J_f$ and similarly for α,

$$\sum_{\beta} S^{-}(j_2) = \sum_{\substack{X_1'J_1'X_2'J_2'J_f\\ X_1J_1X_2J_2\\ \bar{X}_1\bar{J}_1\bar{X}_2\bar{J}_2}} a_\alpha a_{\bar{\alpha}} n_2 \langle j_2^{\,n_2} X_2J_2\{|j_2^{\,n_2-1}\ X_2'J_2';j_2\rangle U(J_1J_2'J_ij_2;J_fJ_2)\ \times$$

$$\times \langle j_2^{\,n_2}\bar{X}_2\bar{J}_2\{|j_2^{\,n_2-1}\ X_2'J_2';j_2\rangle U(J_1J_2'J_ij_2;J_f\bar{J}_2)\ \times$$

$$\times\ \delta_{X_1X_1'}\delta_{\bar{X}_1X_1'}\delta_{J_1J_1'}\delta_{\bar{J}_1J_1'}$$

$$= \sum_{X_1J_1X_2J_2} |a_\alpha|^2\ n_2$$

$$= \langle n_2\rangle. \tag{7.31}$$

The result is obtained by again using orthogonality on the U-coefficients and the cfps. Note we are left with a sum over the target wavefunction configurations α of $|a_\alpha|^2$

multiplied by n_2, the number of nucleons in orbital j_2 for configuration α. The label α spans all possible divisions of the n_1+n_2 nucleons into the two orbitals j_1 and j_2, so n_2 does depend on the configuration α. This sum we interpret as the average population of the j_2 orbital, denoted by $\langle n_2 \rangle$.

Again a similar result holds for stripping reactions. These results can be generalized to any degree of sophistication in the configuration mixing, and the sum rules are summarized in the form

$$\sum_f S^-(j) = \langle \text{particles} \rangle_j \qquad \text{pick-up}$$

$$\sum_f \frac{(2J_f+1)}{(2J_i+1)} S^+(j) = \langle \text{holes} \rangle_j \qquad \text{stripping} \quad (7.32)$$

where $\langle \text{particles} \rangle_j$ and $\langle \text{holes} \rangle_j$ represent the average number of particles or holes in the j-orbital in the target nucleus wavefunction.

In deriving these sum rules, we have glossed over one small point. The differential cross-section eqn (7.25) depends on C^2S and not just S, where C is the appropriate isospin Clebsch-Gordan coefficient: $C_- = \langle T_f N_f \frac{1}{2} N_x | T_i N_i \rangle$ for pick-up reactions and $C_+ = \langle T_i N_i \frac{1}{2} N_x | T_f N_f \rangle$ for stripping reactions. (We use a plus/minus notation to differentiate stripping from pick-up.) Thus when summing over all final states, one is summing all possible isospins as well as spins and the presence of C^2 cannot be ignored. So let us repeat the derivation of eqn (7.28), explicitly writing down all the isospin quantum numbers:

$$\sum_{\substack{X_1'J_1'T_1' \\ X_2'J_2'T_2' \\ J_fT_f}} C_-^2 S^- = n_2 \sum_{\substack{X_2'J_2'T_2' \\ J_fT_f}} \langle j_2^{\,n_2} X_2J_2T_2 \{| j_2^{\,n_2-1} X_2'J_2'T_2'; j_2 \rangle^2 \times$$

$$\times\, U(J_1J_2'J_ij_2;J_fJ_2)^2 \; U(T_1T_2'T_i\tfrac{1}{2};T_fT_2)^2 \times$$

$$\times \langle T_fN_f \tfrac{1}{2} N_x | T_iN_i \rangle^2 . \qquad (7.33)$$

As before, the sum over J_f can be performed using the orthogonality of the U-coefficient, but the sum over T_f cannot, in general, be simplified with T_f occurring in both the Clebsch-Gordan and U-coefficients. In certain specialized cases, such as an N=Z target nucleus with $T_i=0$, there is a unique value for T_f, and the sum rule can be evaluated, but for N≠Z targets this is not possible. The way out of the dilemma is to sum $C_-^2S^-$ not only over all final states, but also over N_f and N_x, that is, sum over both proton and neutron reaction channels. Then orthogonality of the Clebsch-Gordan coefficient can be invoked, and the sum rules (7.32) are once again recovered.

However, from the point of view of using the sum rules in the analysis of experimental data, it is most inconvenient to have to consider both neutron and proton pick-up reactions simultaneously. What is required is the knowledge of how the sum rule strength is divided between the two possible isospins of the final nucleus: $T_{f>} = T_i + \frac{1}{2}$ and $T_{f<} = T_i - \frac{1}{2}$. That is, we want to sum $C_-^2S^-$ over all quantum numbers of the final states *except* T_f:

$$\sum_{\substack{X_2'J_2'T_2' \\ J_f}} C_-^2\, S^- = C_-^2 \sum_{X_2'J_2'T_2'} n_2 \langle j_2^{\,n_2}\, X_2J_2T_2\{|j_2^{\,n_2-1}\, X_2'J_2'T_2';j_2\rangle^2 \times$$

$$\times\, U(T_1T_2'T_i\tfrac{1}{2};T_fT_2)^2 .$$

By expressing the cfps as products of spin-orbit and isotopic-spin factors, the sum over $X_2'J_2'$ can be performed (Macfarlane and French 1960, p. 621):

$$\sum_{X_2'J_2'} n_2 \langle j_2^{\,n_2}\, X_2J_2T_2\{|j_2^{\,n_2-1}\, X_2'J_2'T_2';j_2\rangle^2 = \frac{(n_2-2T_2)(T_2+1)}{(2T_2+1)},\quad T_2' = T_2 + \frac{1}{2}\,.$$

$$= \frac{(n_2+2T_2+2)T_2}{(2T_2+1)},\quad T_2' = T_2 - \frac{1}{2}\,.$$

Next the U-coefficient is explicitly evaluated for the four cases of interest $T_2' = T_2 \pm \frac{1}{2}$, $T_f = T_i \pm \frac{1}{2}$,

$$U(T_1\ T_2+\tfrac{1}{2}\ T_i\ \tfrac{1}{2};\ T_i+\tfrac{1}{2}\ T_2)^2 = \frac{T_i+1}{2T_i+1}\left(1 + \frac{T_i}{T_2+1}\mathcal{L}\right)$$

$$U(T_1\ T_2-\tfrac{1}{2}\ T_i\ \tfrac{1}{2};\ T_i+\tfrac{1}{2}\ T_2)^2 = \frac{T_i+1}{2T_i+1}\left(1 - \frac{T_i}{T_2}\mathcal{L}\right)$$

$$U(T_1\ T_2+\tfrac{1}{2}\ T_i\ \tfrac{1}{2};\ T_i-\tfrac{1}{2}\ T_2)^2 = \frac{T_i}{2T_i+1}\left(1 - \frac{T_i+1}{T_2+1}\mathcal{L}\right)$$

$$U(T_1\ T_2-\tfrac{1}{2}\ T_i\ \tfrac{1}{2};\ T_i-\tfrac{1}{2}\ T_2)^2 = \frac{T_i}{2T_i+1}\left(1 + \frac{T_i+1}{T_2}\mathcal{L}\right)$$

where

$$\mathcal{L} = \frac{T_2(T_2+1)+T_i(T_i+1)-T_1(T_1+1)}{2T_i(T_i+1)},$$

resulting in the following expressions for the sum rule:

$$\sum C_-^2 S^-(T_{f>}) = C_-^2\ \frac{T_i+1}{2T_i+1}\ (n_2 - 2T_i\mathcal{L})$$

$$\sum C_-^2 S^-(T_{f<}) = C_-^2\ \frac{T_i}{2T_i+1}\ (n_2 + (2T_i+1)\mathcal{L}). \qquad (7.34)$$

A more elegant proof of these results has been given by French and Macfarlane (1961) using a projection operator technique.

The sum rule, eqn (7.32), has now been split according to the isospin of the final nucleus. It is also possible to differentiate between neutron and proton pick-up reactions by obtaining a value for C^2 which does not depends on T_f:

$$C_-^2(T_{f>}) = \langle T_i+\tfrac{1}{2}\ N_f\ \tfrac{1}{2}\ N_x|T_iN_i\rangle^2 = \frac{(T_i+1)-2N_xN_i}{2(T_i+1)}$$

$$C_-^2(T_{f<}) = \langle T_i-\tfrac{1}{2}\ N_f\ \tfrac{1}{2}\ N_x|T_iN_i\rangle^2 = \frac{T_i+2N_xN_i}{2T_i}\ .$$

Incorporating these expressions in eqn (7.34), and taking the appropriate spectroscopic averages to account for configuration mixing in the target nucleus wavefunction, leads to the final sum rule analogous to eqn (7.32):

$$\sum C_-^2 S^-(T_{f<}) = \frac{T_i+2N_xN_i}{2(2T_i+1)} [\langle \text{particles}\rangle_j + 2(T_i+1)\langle \mathcal{L}\rangle_j]$$

$$\sum C_-^2 S^-(T_{f>}) = \frac{(T_i+1)-2N_xN_i}{2(2T_i+1)} [\langle \text{particles}\rangle_j - 2T_i\langle \mathcal{L}\rangle_j] \quad (7.35)$$

with N_x taking the value $+\frac{1}{2}$ for neutron pick-up and the value $-\frac{1}{2}$ for proton pick-up reactions. These two equations can be added together to give the sum rule we were originally seeking, namely the sum of the spectroscopic strength $C_-^2S^-$ over *all* final states for a particular reaction channel. The result is

$$\sum_f C_-^2 S^- = \frac{1}{2}\langle \text{particles}\rangle_j + 2N_xN_i\langle \mathcal{L}\rangle_j. \quad (7.36)$$

The analogous expressions for the stripping reaction are

$$\sum \frac{(2J_f+1)}{(2J_i+1)} C_+^2 S^+(T_{f<}) = \frac{T_i-2N_xN_i}{2(2T_i+1)} [\langle \text{holes}\rangle_j + 2(T_i+1)\langle \mathcal{L}\rangle_j]$$

$$\sum \frac{(2J_f+1)}{(2J_i+1)} C_+^2 S^+(T_{f>}) = \frac{(T_i+1)+2N_xN_i}{2(2T_i+1)} [\langle \text{holes}\rangle_j - 2T_i\langle \mathcal{L}\rangle_j] \quad (7.37)$$

where as before $N_x = +\frac{1}{2}$ for neutron stripping and $N_x = -\frac{1}{2}$ for proton stripping. Again the two equations can be added to give a sum rule appropriate to all final states:

$$\sum_f \frac{(2J_f+1)}{(2J_i+1)} C_+^2 S^+ = \frac{1}{2}\langle \text{holes}\rangle_j - 2N_xN_i\langle \mathcal{L}\rangle_j. \quad (7.38)$$

As an example on the use of the sum rules (7.35), let us consider the neutron pick-up reaction $^{35}\text{C}\ell(p,d)^{34}\text{C}\ell$. We shall describe the target nucleus in the shell-model approximation by a wavefunction in which the first 28 nucleons occupy closed-shell orbitals and the last 7 nucleons are distributed among the $2s_{1/2}$ and $1d_{3/2}$ orbitals. The leading term in this wavefunction has a configuration

$$|s^{4}_{1/2}\ J_1{=}0\ T_1{=}0,\ d^{3}_{3/2}\ J_2{=}\tfrac{3}{2}\ T_2{=}\tfrac{1}{2};\ J_i{=}\tfrac{3}{2}\ T_i{=}\tfrac{1}{2}\rangle. \qquad (7.37)$$

Let us assume for the moment that this one configuration represents the target nucleus ^{35}Cℓ, then the spectroscopic averages required for eqn (7.35) are $\langle n\rangle_s = 4$, $\langle n\rangle_d = 3$, $\langle \mathcal{L}\rangle_s = 0$, and $\langle \mathcal{L}\rangle_d = 1$, the subscripts referring to the $s_{1/2}$ and $d_{3/2}$ orbitals respectively. The required sum rules are trivially evaluated and are listed in Table 7.3, where they are compared with the experimental data of Vignon *et al.* (1972).

TABLE 7.3

Sum rules for the reaction ^{35}Cℓ(p,d)^{34}Cℓ *evaluated using eqn (7.35) and compared with the experimental data of Vignon et al. (1972)*

Sum	Orbital	Single Configuration[a]	Mixed Configurations[b]	Experiment
$\sum C_-^2 S^-(T_f{=}0)$	$s_{1/2}$	1.0	0.93	0.82
$\sum C_-^2 S^-(T_f{=}0)$	$d_{3/2}$	1.5	1.57	1.48
$\sum C_-^2 S^-(T_f{=}1)$	$s_{1/2}$	1.0	0.89	0.46
$\sum C_-^2 S^-(T_f{=}1)$	$d_{3/2}$	0.5	0.61	0.15

a Eqn (7.37)

b Wavefunctions of Glaudemans *et al.* (1964).

Next let us introduce configuration mixing and use the wavefunction of Glaudemans *et al.* (1964) to describe the target nucleus ^{35}Cℓ. Then the spectroscopic averages are modified slightly: $\langle n\rangle_s = 3.61$, $\langle n\rangle_d = 3.39$, $\langle \mathcal{L}\rangle_s = 0.04$, $\langle \mathcal{L}\rangle_d = 0.96$ and the resulting sum rules are also presented in Table 7.3.

One can see that the experimental data almost exhaust the predicted sum rules for the $T_f{=}0$ states populated in ^{34}Cℓ, but for the $T_f{=}1$ states barely half of the transition strength has been experimentally observed. This is typical of the results one can obtain from this sort of analysis.

One assumption has been made in deriving spectroscopic

factors from the experimental data. This concerns the calculated DWBA cross-section σ_{DWBA}. The distorted wave theory generally gives a reasonable description of the kinematic dependence of the reaction, the angular distribution of reaction products, the energy variations corresponding to the excitation of different final nuclear states, and so on. But the theory has some uncertainty as to the absolute normalization of σ_{DWBA}, due primarily to the use of a zero-range approximation. We shall not go into this subject. It is sufficient to say that another use of the sum rules is to determine the normalization for σ_{DWBA} such that the summed spectroscopic strength equals the model prediction. Then the comparison between theory and experiment is confined to discussing how this spectroscopic strength is distributed among specific final states.

The formulae (7.35) and (7.37) can be written in slightly simpler form on making some assumptions concerning the target nucleus wavefunction. We assume a closed-shell core having a neutron excess $2T_1 = N_c - Z_c$, where N_c and Z_c are the number of neutrons and protons in the core, and coupled to this core is a single valence shell also with a neutron excess $2T_2 = n_v - z_v$ with n_v and z_v the respective neutron and proton population in the valence shell. The total number of neutrons and protons in the target nucleus is then $N = N_c + n_v$ and $Z = Z_c + z_v$ and we shall assume that the total isospin is given by $2T_i = 2N_i = N - Z$, that is T_i is the arithmetic sum rather than the vector sum of T_1 and T_2. Any amount of configuration mixing between states with this isospin structure is allowed.

Within this model, the appropriate spectroscopic averages can be written

$$\langle \mathcal{L} \rangle_j = \frac{\langle n_v \rangle_j - \langle z_v \rangle_j}{N-Z} = \frac{\langle z_v^{-1} \rangle_j - \langle n_v^{-1} \rangle_j}{N-Z}$$

$$\langle \text{particles} \rangle_j = \langle n_v \rangle_j + \langle z_v \rangle_j$$

$$\langle \text{holes} \rangle_j = \langle n_v^{-1} \rangle_j + \langle z_v^{-1} \rangle_j,$$

where n_v^{-1} and z_v^{-1} are shorthand notations for the number of neutron holes and proton holes in the valence shell. Inserting these values for the averages into eqns (7.35) and (7.37), we obtain the sum rules shown in Table 7.4. These results were first obtained by French and Macfarlane (1961).

TABLE 7.4

Isospin splitting of the sum rules for single-nucleon pick-up and stripping reactions on target nuclei with neutron excess $N > Z$.

Sum	Neutron transfer	Proton transfer
$\sum C_-^2 S^-(T_{f<})$	$\langle n_v \rangle_j - \frac{\langle z_v \rangle_j}{N-Z+1}$	0
$\sum C_-^2 S^-(T_{f>})$	$\frac{\langle z_v \rangle_j}{N-Z+1}$	$\langle z_v \rangle_j$
$\sum g\, C_+^2 S^+(T_{f<})$	0	$\langle z_v^{-1} \rangle_j - \frac{\langle n_v^{-1} \rangle_j}{N-Z+1}$
$\sum g\, C_+^2 S^+(T_{f>})$	$\langle n_v^{-1} \rangle_j$	$\frac{\langle n_v^{-1} \rangle_j}{N-Z+1}$

$g = (2J_f+1)/(2J_i+1)$

S^- = spectroscopic factor for pick-up of nucleon from shell-model orbital, j.

S^+ = same, but for stripping reaction.

Finally we want to discuss another appliation of the sum rules. Consider neutron stripping and neutron pick-up reactions on the same target nucleus and sum the transition strength over all final states in both reactions. The result is just the sum of eqns (7.36) and (7.38) and is

$$\sum_f C_-^2 S^- + \sum_f \frac{(2J_f+1)}{(2J_i+1)} C_+^2 S^+ = \frac{1}{2}\langle \text{particles} \rangle_j + \frac{1}{2}\langle \text{holes} \rangle_j$$
$$= 2j+1. \qquad (7.39)$$

The same result holds when summing proton stripping with proton pick-up reactions, but of course is not true for the cross-combination of, say, proton stripping with neutron pick-up. Eqn (7.39) is not in itself very interesting - it merely reflects the number of neutrons that can be placed in a single shell-model orbital. The result nevertheless can be useful in verifying the normalization of spectroscopic factors deduced from experimental data.

More useful, however, is an extension of this idea to find partial sum rules, where the transition strength is summed over all final states of *fixed* spin and isospin, J_f and T_f respectively. Since we are now considering concurrently stripping and pick-up reactions on the same target nucleus, let us denote the quantum numbers of the final states in the stripping reaction by a single prime and those of the final states in the pick-up reaction by a double prime. Consider first the sum

$$\sum_{X_2'J_2'T_2'} \frac{2J_f'+1}{2J_i+1} C_+^2 S^+(j_2)$$

$$= \frac{2J_f'+1}{2J_i+1} C_+^2 \sum_{X_2'J_2'T_2'} (n_2+1)\langle j_2^{\,n_2+1} X_2'J_2'T_2'\{|j_2^{\,n_2} X_2J_2T_2;j_2\rangle^2\, U(J_1J_2J_f'j_2;J_iJ_2')^2 \times$$
$$\times\, U(T_1T_2T_f'\tfrac{1}{2};T_iT_2')^2$$

where the expression (7.27) has been inserted for S^+. Next the $(n_2+1) \rightarrow n_2$ cfp is expanded in terms of a complete set of $n_2 \rightarrow (n_2-1)$ cfps using the recurrence relation, eqn (6.19):

$$= \frac{2J_f'+1}{2J_i+1} C_+^2 \sum_{J_2'T_2'} U(J_1J_2J_f'j_2;J_iJ_2')^2\, U(T_1T_2T_f'\tfrac{1}{2};T_iT_2')^2 -$$

$$- \frac{2J_f'+1}{2J_i+1} C_+^2 \sum_{\substack{J_2'T_2'\\ X_2''J_2''T_2''}} n_2\langle j_2^{\,n_2} X_2J_2T_2\{|j_2^{\,n_2-1} X_2''J_2''T_2'';j_2\rangle^2 \times$$

$$\times\, U(J_2''j_2j_2J_2';J_2J_2)\, U(T_2''\tfrac{1}{2}\tfrac{1}{2}T_2';T_2T_2)(-)^{2J_2-J_2'-J_2''+2T_2-T_2'-T_2''} \times$$

$$\times\, U(J_1J_2J_f'j_2;J_iJ_2')^2\, U(T_1T_2T_f'\tfrac{1}{2};T_iT_2')^2.$$

There are two terms. The first term is trivially evaluated using the orthogonality relation for U-coefficients. In the second term, the sum over J_2' and T_2' produces 9j-coefficients, which are then broken up again into a different sum of U-coefficients. The result is

$$= \frac{2J_f'+1}{2J_i+1} C_+^2 - C_+^2 \sum_{\substack{J_f''T_f'' \\ X_2''J_2''T_2''}} n_2 \langle j_2^{n_2} X_2J_2T_2\{|j_2^{n_2-1} X_2''J_2''T_2''; j_2\rangle^2 \times$$

$$\times\, U(J_1J_2''J_ij_2;J_f''J_2)^2\, U(T_1T_2''T_i\tfrac{1}{2};T_f''T_2)^2\, \frac{\hat{J}_f'}{\hat{J}_f''} \frac{\hat{T}_i\hat{T}_i}{\hat{T}_f'\hat{T}_f''} \times$$

$$\times\, U(j_2J_iJ_ij_2;J_f'J_f'')\, U(\tfrac{1}{2}T_iT_i\tfrac{1}{2};T_f'T_f'').$$

Finally we identify the expression (7.23) for the spectroscopic factor S^- for the pick-up reaction, and thereby obtain the required result:

$$\sum_{X_2'J_2'T_2'} \frac{2J_f'+1}{2J_i+1} C_+^2 S^+ = \frac{2J_f'+1}{2J_i+1} C_+^2 -$$

$$- \sum_{J_f''T_f''} \frac{\hat{J}_f'\hat{T}_i\hat{T}_i}{\hat{J}_f''\hat{T}_f'\hat{T}_f''} U(j_2J_iJ_ij_2;J_f'J_f'')U(\tfrac{1}{2}T_iT_i\tfrac{1}{2};T_f'T_f'') \frac{C_+^2}{C_-^2} \sum_{X_2''J_2''T_2''} C_-^2 S^- \tag{7.40}$$

where $\hat{J} = (2J+1)^{\frac{1}{2}}$.

Let us consider the case of neutron pick-up and neutron stripping on a target nucleus with $T_i=N_i=\frac{1}{2}(N-Z)$. Then T_f', the isospin of the final states in the stripping reaction, can only be $T_i+\frac{1}{2}$, and the isospin dependence of eqn (7.40) is easily evaluated. There are still two possible values for T_f'', the isospin of the final nucleus in the pick-up reaction, but the result is the same in either case. Thus we obtain

$$\sum_{X_2'J_2'T_2'} \frac{2J_f'+1}{2J_i+1} C_+^2 S^+ = \frac{2J_f'+1}{2J_i+1} C_+^2 - \sum_{J_f''T_f''} \frac{\hat{J}_f'}{\hat{J}_f''} U(j_2J_iJ_ij_2;J_f'J_f'') \sum_{X_2''J_2''T_2''} C_-^2 S^- \tag{7.41}$$

This is a fascinating result: the sum of transition strengths in the stripping reaction, leading to states of one particular spin, has been related to a corresponding set of sums for the pick-up reaction. The equation represents in some sense the multipole decomposition of the total sum rule, eqn (7.39), which simply states that the total number of neutron particles and neutron holes in a shell-model orbit j_2 is just $2j_2+1$.

The result, eqn (7.41), is in fact very general, although the derivation given here made use of specific shell-model configurations. No assumption needs to be made about the shell-model structure of the target nucleus, in fact all mention of the target (other than its spin) has disappeared in eqn (7.41). However, the equation is restricted to discussing the spectroscopic factors for the transfer of a neutron from a particular shell model orbital j_2.

The result, eqn (7.41), was first derived by French (1965b) and it has received little attention since. An alternative derivation, which is formally exact, has been recently given by Clement (1973). No assumption is made other than that of the existence of quantum mechanical completeness relations. Clement's result differs slightly from eqn (7.41) in two respects. First, the highly excited states in the stripping reaction are frequently unbound to nucleon emission (the continuum problem) and this presents a difficulty in the theory. Clement introduces a correction term for this problem. Secondly, a centre-of-mass correction is incorporated which is evaluated to first order. This latter correction is closely related to the factor $(B/B-1)^{n_1+\ell_1/2}$ in the spectroscopic amplitude, which we derived in eqn (7.16) using harmonic oscillator functions.

One of the most interesting uses of eqn (7.41) is as a test of the accuracy of present methods of determining spectroscopic factors and possibly as providing additional evidence on which to base spin assignments. Examples of this type of analysis have been given by Hodgson and Millener (1972) and by Clement and Perez (1973). Stripping and pick-up experiments on the same odd-mass target are needed.

Consider the target nucleus ^{45}Sc for which rather complete data on the (d,t) pick-up reaction (Ohnuma and Sourkes

1971) and the (d,p) stripping reaction (Rapaport *et al.* 1966) are available. This is the example discussed by Clement and Perez. Listed in Table 7.5 are the spectroscopic factors $C_-^2S^-$ for the strong $\ell=3$ ($j=7/2$ predominantly) transitions seen in the (d,t) reaction. Using these values we compute the

TABLE 7.5

Prediction (column 7) for the partial sums of spectroscopic factors $C_+^2S^+$ *for the stripping reaction* ^{45}Sc(d,p)^{46}Sc *obtained using eqn (7.41) and the experimental* ^{45}Sc(d,t)^{44}Sc *pick-up data (columns 1, 2, and 3), compared with the experimental data (columns 4, 5 and 6).*

Pick-up data[a]			Stripping data[b]			Prediction
Spin	E_x(keV)	$C_-^2S^-$	Spin	E_x(keV)	$C_+^2S^+$	$C_+^2S^+$
0^+	2784	0.22	0^+	-	-	-0.05
1^+	667	0.32	1^+	-	-	0.05
2^+	0	0.35	2^+	444	0.31	0.37
3^+	763 }	0.46	3^+	227	0.62	0.37
3^+	1186 }		4^+	0 }	0.80	0.75
4^+	350	0.35	4^+	833 }		
5^+	1052 }		5^+	722	0.61	0.68
5^+	1532 }	0.70	6^+	51	1.33	1.23
(5^+)	2912 }		7^+	975	0.50	0.44
6^+	271	0.48				
7^+	974	1.29				

a Ohnuma and Sourkes (1971)

b Rapaport *et al.* (1966)

corresponding stripping transition strengths $C_+^2S^+$ using the formula (7.41), and compare with the experimental data. (We have arbitrarily assigned 5^+ to the excitation at 2.912 MeV in ^{44}Sc; the spins on all the other states are more or less established.) The comparison is really quite remarkable and lends weight to the spin assignments in ^{46}Sc. For further discussion of this type of analysis, the reader is referred to Clement and Perez (1973).

APPENDIX A

A.1 SINGLE PARTICLE WAVEFUNCTIONS

In this book, we have worked exclusively in a j-j coupled representation and defined the single-particle wavefunction as

$$\psi_i(\underset{\sim}{x}) \equiv |i\rangle \equiv |j_i m_i\rangle$$

$$= R_{n_i \ell_i j_i}(r) \sum_{m_{\ell_i} m_{s_i}} \langle \ell_i m_{\ell_i} \tfrac{1}{2} m_{s_i} | j_i m_i \rangle [i^{\ell_i} Y_{\ell_i m_i}(\hat{\underset{\sim}{r}})] \chi_{m_{s_i}} \chi_{m_{t_i}} \qquad \text{(A.1)}$$

This is the assumed form for the solution of the Schrödinger equation in spherical coordinates for a central potential: $R(r)$ is a radial function, $Y(\hat{\underset{\sim}{r}})$ a spherical harmonic, $\chi_{m_{s_i}}$ a spin-1/2 function in spin space, and $\chi_{m_{t_i}}$ a spin-1/2 function in isospin space. The Clebsch-Gordan coefficient vector couples the angle and spin functions to form a state of good total angular momentum. The label i stands for all the quantum numbers necessary to describe the single-particle state, in this case $i \equiv n_i \ell_i j_i m_i m_{t_i}$. Here n_i, ℓ_i, and j_i are the principal, orbital, and total angular momentum quantum numbers, m_{ℓ_i}, m_{s_i}, and m_i are the magnetic projection quantum numbers of orbital, spin, and total angular momentum respectively, and m_{t_i} is the magnetic projection of isospin defined such that $m_{t_i} = +\tfrac{1}{2}$ corresponds to a neutron state and $m_{t_i} = -\tfrac{1}{2}$ to a proton state.

A.2 PROTON-NEUTRON VERSUS ISOSPIN FORMALISM

It has been assumed in eqn (A.1) that protons and neutrons have identical single-particle functions and the isospin label distinguishes between them. In some applications it may be preferable to treat the protons and neutrons separately, a composite wavefunction then being a vector coupling of a proton function with a neutron function. Most of the results to be derived in this book can be obtained in either formalism by essentially the same steps. We prefer the isospin formalism since one is then dealing with just one type of particle and questions of antisymmetry are somewhat easier to handle. However, many of the formulae become rather cumbersome as there are twice as many labels on all the wavefunctions. For example, the vector coupling of two single-particle functions to form a composite state of angular momentum J and isospin T is written

$$|JM;TN\rangle = \sum_{\substack{m_1 m_2 \\ m_{t_1} m_{t_2}}} \langle j_1 m_1 j_2 m_2 | JM \rangle \langle \tfrac{1}{2} m_{t_1} \tfrac{1}{2} m_{t_2} | TN \rangle |j_1 m_1; \tfrac{1}{2} m_{t_1}\rangle |j_2 m_2; \tfrac{1}{2} m_{t_2}\rangle \quad \text{(A.2)}$$

We have used lower case letters to denote the quantum numbers of single-particle states and upper case letters for composite states. Since the orbital-spin space (frequently abbreviated to just the spin space) is distinct from the isospin space, wavefunctions always occur as simple direct products of spin space functions and isospin space functions. Similarly, operators have the same direct product form. Thus some economy in the notation can be achieved by only writing down the operators, wavefunctions, and labels appropriate to the spin space, the presence of the isospin space being understood. For example, we identify

$$|JM\rangle \equiv |JM;TN\rangle$$

$$(-)^J \equiv (-)^{J+T}$$

$$\hat{J} = (2J+1)^{\frac{1}{2}} \equiv [(2J+1)(2T+1)]^{\frac{1}{2}}$$

$$\langle j_1 m_1 j_2 m_2 | JM\rangle \equiv \langle j_1 m_1 j_2 m_2 | JM\rangle \langle \tfrac{1}{2} m_{t_1} \tfrac{1}{2} m_{t_2} | TN\rangle \quad \text{(A.3)}$$

$$U(j_1 j_2 j_3 j_4; JJ') \equiv U(j_1 j_2 j_3 j_4; JJ') U(\tfrac{1}{2}\tfrac{1}{2}\tfrac{1}{2}\tfrac{1}{2}; TT')$$

$$\langle j_1 \| O^J \| j_2 \rangle \equiv \langle j_1 \| O^J \| j_2 \rangle \langle \tfrac{1}{2} \| O^T \| \tfrac{1}{2} \rangle \text{ etc.}$$

This economy of notation was first suggested by French (1966a); however, he preferred to use Greek letters to identify angular momentum quantum numbers, with the interpretation of the Greek letter being either J and T or just J depending upon whether the isospin formalism is in use or not. We shall not use Greek letters, but shall use J with the interpretations of eqn (A.3). Where there is any breakdown in the notation, or any possibility of ambiguity, the full expression will be given.

A.3 TIME REVERSAL

Returning to eqn (A.1), which defines the single-particle wavefunction, note the presence of the phase i^{ℓ_i} with the spherical harmonic $Y_{\ell_i m_i}(\hat{\underset{\sim}{r}})$. This is included so that the wavefunction has convenient time-reversal properties (Brink and Satchler 1968). The time-reversal operator θ can be expressed as a product of a unitary operator U and an operator K which takes the complex conjugate of all complex numbers.

$$\theta = UK. \tag{A.4}$$

This is an example of what is called an antiunitary operator. For spin-$\frac{1}{2}$ functions, a convenient choice for U is $i\sigma_y$ where σ_y is one of the Pauli spin operators, then

$$\theta\chi_{m_s} = i\sigma_y\,\chi_{m_s} = (-)^{\frac{1}{2}+m_s}\,\chi_{-m_s}. \tag{A.5}$$

Furthermore, operating with θ on the orbital function, we have

$$\theta\,|i^\ell\,Y_{\ell m}\rangle = K|i^\ell\,Y_{\ell m}\rangle = (-i)^\ell\,|Y_{\ell m}\rangle^*$$

$$= (-)^{\ell+m}\,|i^\ell\,Y_{\ell -m}\rangle. \tag{A.6}$$

In eqn (A.1) the angle function and spin function are vector coupled to form a state of good total angular momentum, and the time-reversal operator acting on this state gives

$$\theta|jm\rangle = (-)^{j+m}|j\ -m\rangle, \tag{A.7}$$

providing the vector coupling coefficient follows the usual Condon and Shortley conventions, namely

$$\langle \ell\ m_\ell\ \tfrac{1}{2}\ m_s|jm\rangle = (-)^{\ell+\frac{1}{2}-j}\langle \ell\ -m_\ell\ \tfrac{1}{2}-m_s|j-m\rangle.$$

Eqn (A.7) is the required time-reversal property for our single-particle functions, yet notice the crucial role the phase i^ℓ played in eqn (A.6) in order to obtain this result.

The time-reversal operator θ does not influence the isospin spinor χ_{m_t}. In order that the properties of the isospin functions should closely follow the properties of the spin space functions, as required for our abbreviated notation, a more general time-reversal operator has to be introduced. This operator, Θ, defined as

$$\Theta = \exp(i\pi T_y)\theta = R^{-1}\theta \tag{A.8}$$

is a product of time reversal and a rotation through the angle $-\pi$ about the y-axis in isospin space. Operating with Θ on the single-particle function, eqn (A.1), gives

$$\Theta|jmm_t\rangle = (-)^{j+m+\frac{1}{2}+m_t}\,|j\ -m\ -m_t\rangle \tag{A.9}$$

where for once we write the isospin quantum numbers explicitly. We denote the time-reversed state with a tilde, thus eqn (A.9) repeated in the abbreviated notation becomes

$$|\widetilde{jm}\rangle = \Theta|jm\rangle = (-)^{j+m}|j-m\rangle.$$

Operating twice with Θ,

$$\Theta^2|jm\rangle = (-)^{j+m}\,\Theta|j-m\rangle = (-)^{2j}|jm\rangle, \tag{A.10}$$

gives the result $\Theta^2 = +1$ in the isospin formalism, but in the proton-neutron formalism $\Theta^2 = \theta^2 = -1$, since j is always a half-integer.

A.4 RECOUPLING COEFFICIENTS

It is convenient to introduce a bracket notation to represent angular momentum coupling. For example, eqn (A.2) can be abbreviated to read

$$|JM\rangle = |[j_1 j_2];JM\rangle \tag{A.2}$$

where the pair of square brackets denotes the presence of a Clebsch-Gordan coefficient and a sum over the appropriate magnetic quantum numbers. Interchanging the order of coupling just gives the standard (Condon-Shortley) sign relation between Clebsch-Gordan coefficients, viz.

$$|[j_1 j_2];JM\rangle = (-)^{j_1+j_2-J} |[j_2 j_1];JM\rangle. \tag{A.11}$$

The coupling of three angular momenta is written

$$|[[j_1 j_2]^{J_1} j_3];JM\rangle = \sum_{m_1 m_2 M_1 m_3} \langle j_1 m_1 j_2 m_2|J_1 M_1\rangle\langle J_1 M_1 j_3 m_3|JM\rangle \times$$
$$\times |j_1 m_1\rangle|j_2 m_2\rangle|j_3 m_3\rangle.$$

Interchanging the order of coupling in this case introduces Jahn's recoupling coefficient

$$|[[j_1 j_2]^{J_1} j_3];JM\rangle = \sum_{J_2} U(j_1 j_2 J j_3;J_1 J_2)|[j_1[j_2 j_3]^{J_2}];JM\rangle, \tag{A.12}$$

which is related to the W-coefficient (Brink and Satchler 1968) and the 6j-symbol (de Shalit and Talmi 1963) by

$$U(j_1 j_2 J j_3;J_1 J_2) = \hat{J}_1 \hat{J}_2 \; W(j_1 j_2 J j_3;J_1 J_2)$$

$$= (-)^{j_1+j_2+J+j_3} \hat{J}_1 \hat{J}_2 \begin{Bmatrix} j_1 & j_2 & J_1 \\ j_3 & J & J_2 \end{Bmatrix} \tag{A.13}$$

where $\hat{J} = (2J+1)^{1/2}$. Another useful recoupling of three angular momenta is

$$|[[j_1j_2]^{J_1}j_3];JM\rangle = \sum_{J_3} (-)^{J_1+J_3-j_1-J}\, U(j_1j_2j_3J;J_1J_3)\,|[[j_1j_3]^{J_3}j_2];JM\rangle. \tag{A.14}$$

The U-coefficient is defined as a sum of four Clebsch-Gordan coefficients:

$$U(j_1j_2Jj_3;J_1J_2) = \sum_{\substack{m_1m_2m_3\\ M_1M_2}} \langle j_1m_1j_2m_2|J_1M_1\rangle\langle j_2m_2j_3m_3|J_2M_2\rangle \times$$
$$\times \langle J_1M_1j_3m_3|JM\rangle\langle j_1m_1J_2M_2|JM\rangle,$$

or as more frequently encountered,

$$\langle j_1m_1J_2M_2|JM\rangle\, U(j_1j_2Jj_3;J_1J_2)$$
$$= \sum_{m_2m_3M_1} \langle j_1m_1j_2m_2|J_1M_1\rangle\langle j_2m_2j_3m_3|J_2M_2\rangle\langle J_1M_1j_3m_3|JM\rangle. \tag{A.15}$$

A number of properties of the U-coefficient are listed below, the proofs of which are to be found in standard books on angular momentum.

Symmetry

$$U(abcd;ef) = U(badc;ef) = U(cdab;ef) = U(dcba;ef)$$
$$= \frac{\hat{e}\hat{f}}{\hat{a}\hat{d}}(-)^{e+f-a-d}\, U(ebcf;ad). \tag{A.16a}$$

Orthogonality

$$\sum_f U(abcd;ef)\, U(abcd;gf) = \delta_{e,g}. \tag{A.16b}$$

Sum rules

$$\sum_b \frac{\hat{b}}{\hat{c}}\, U(adda;bc) = 1. \tag{A.16c}$$

$$\sum_d (-)^{b-a-d}\frac{\hat{b}}{\hat{a}\hat{d}}\, U(adad;bc) = \delta_{c,0}. \tag{A.16d}$$

$$\sum_f (-)^{e+f+g}\, U(abcd;ef)\, U(adcb;gf) = (-)^{a+b+c+d}\, U(abdc;eg). \tag{A.16e}$$

$$\sum_k U(dbhj;fk)U(dehg;ck)U(abgk;ej) = U(abcd;ef)U(afgh;cj). \tag{A.16f}$$

Special values

$$U(0bcd;ef) = \delta_{b,e}\delta_{c,f}. \tag{A.16g}$$

$$U(abcd;0f) = \frac{\hat{f}}{\hat{a}\hat{c}} (-)^{a+c-f}\, \delta_{a,b}\delta_{c,d} \tag{A.16h}$$

In a similar way the coupling of four angular momenta can be discussed, in particular interchanging the order of coupling introduces a normalized 9j-coefficient

$$|[[j_1j_2]^{J_1}\, [j_3j_4]^{J_2}];JM\rangle$$

$$= \sum_{J_3J_4} \begin{bmatrix} j_1 & j_2 & J_1 \\ j_3 & j_4 & J_2 \\ J_3 & J_4 & J \end{bmatrix} |[[j_1j_3]^{J_3}\, [j_2j_4]^{J_4}];JM\rangle, \tag{A.17}$$

where it is assumed that $|j_2m_2\rangle$ and $|j_3m_3\rangle$ commute. The normalized 9j-coefficient is related to the standard 9j-coefficient of de Shalit and Talmi (1963) by

$$\begin{bmatrix} j_1 & j_2 & J_1 \\ j_3 & j_4 & J_2 \\ J_3 & J_4 & J \end{bmatrix} = \hat{J}_1\hat{J}_2\hat{J}_3\hat{J}_4 \begin{Bmatrix} j_1 & j_2 & J_1 \\ j_3 & j_4 & J_2 \\ J_3 & J_4 & J \end{Bmatrix} \tag{A.18}$$

Another useful recoupling formula is

$$|[[[j_1j_2]^{J_1} j_3]^{J_2} j_4];JM\rangle$$

$$= \sum_{J_3J_4} (-)^{J_1+J_4-J_2-J_3}\, \frac{\hat{J}_1\hat{J}_3}{\hat{j}_2\hat{j}_4} \begin{bmatrix} j_3 & J_1 & J_2 \\ J_3 & j_1 & j_4 \\ J_4 & j_2 & J \end{bmatrix} |[[[j_1j_4]^{J_3} j_3]^{J_4} j_2];JM\rangle. \tag{A.19}$$

The normalized 9j-coefficient is defined as the sum over six Clebsch-Gordan coefficients:

$$\begin{bmatrix} j_1 & j_2 & J_1 \\ j_3 & j_4 & J_2 \\ J_3 & J_4 & J \end{bmatrix} = \frac{2j_1+1}{2J+1} \sum \langle j_1m_1j_2m_2|J_1M_1\rangle\langle j_2m_2j_4m_4|J_4M_4\rangle\langle J_1M_1J_2M_2|JM\rangle$$

$$\langle j_1m_1j_3m_3|J_3M_3\rangle\langle j_3m_3j_4m_4|J_2M_2\rangle\langle J_3M_3J_4M_4|JM\rangle, \quad \text{(A.20}$$

summed over all magnetic projections *except* m_1, or equivalently as a sum over three U-coefficients:

$$\begin{bmatrix} j_1 & j_2 & J_1 \\ j_3 & j_4 & J_2 \\ J_3 & J_4 & J \end{bmatrix} = \sum_x (-)^{x+j_4-J_2-J_4} U(j_1j_3JJ_4;J_3x)U(j_2j_4xj_3;J_4J_2)U(J_2Jj_2j_1;J_1x \quad \text{(A.21}$$

Some properties of the normalized 9j-coefficient follow.

Symmetry

$$\begin{bmatrix} a & b & c \\ d & e & f \\ g & h & i \end{bmatrix} = S\,\frac{\hat{c}\hat{f}\hat{g}\hat{h}}{\hat{\gamma}\hat{\phi}\hat{\lambda}\hat{\mu}} \begin{bmatrix} \alpha & \beta & \gamma \\ \delta & \varepsilon & \phi \\ \lambda & \mu & \nu \end{bmatrix} \quad \text{(A.22}$$

where the elements $\alpha, \beta, \ldots$ are the set $a, b, \ldots$ reordered by any number of permutations of the rows or columns. If the total permutation is even, $S = +1$, if it is odd $S = (-)^\sigma$ where σ is the sum of all nine elements.

Orthogonality

$$\sum_{cf} \begin{bmatrix} a & b & c \\ d & e & f \\ g & h & i \end{bmatrix} \begin{bmatrix} a & b & c \\ d & e & f \\ j & k & i \end{bmatrix} = \delta_{g,j}\,\delta_{h,k}. \quad \text{(A.22}$$

um rules

$$\sum_{cf} (-)^{e+f-b-k} \begin{bmatrix} a & b & c \\ d & e & f \\ g & h & i \end{bmatrix} \begin{bmatrix} a & b & c \\ e & d & f \\ j & k & i \end{bmatrix} = (-)^{e+b+h} \begin{bmatrix} a & e & j \\ d & b & k \\ g & h & i \end{bmatrix}. \tag{A.22c}$$

$$\begin{bmatrix} a & b & c \\ d & e & f \\ x & h & i \end{bmatrix} U(hida;xg) = (-)^{h+f-g-e}\, U(abif;cg)U(degb;fh). \tag{A.22d}$$

$$\sum_{y} \begin{bmatrix} a & b & c \\ d & x & y \\ g & h & i \end{bmatrix} U(cfim;ky)U(deym;fx)U(behm;jx)$$

$$= \begin{bmatrix} a & b & c \\ d & e & f \\ g & j & k \end{bmatrix} U(gjim;kh). \tag{A.22e}$$

pecial values

$$\begin{bmatrix} a & a & 0 \\ b & c & d \\ e & f & d \end{bmatrix} = \frac{\hat{e}}{\hat{a}\hat{b}} U(cdae;bf). \tag{A.22f}$$

$$\begin{bmatrix} a & b & a+b \\ c & d & c+d \\ a+c & b+d & a+b+c+d \end{bmatrix} = 1. \tag{A.22g}$$

There are a number of special relationships connecting he various recoupling coefficients. We shall not attempt to resent a complete list, but rather we will pick out a few esults of relevance to this book.

Consider two single-particle wavefunctions $\psi_1(\underset{\sim}{x})$, $\psi_2(\underset{\sim}{x})$ f type given in eqn (A.1) with $\underset{\sim}{\ell}_1 + \frac{1}{2} = \underset{\sim}{j}_1$ and $\underset{\sim}{\ell}_2 + \frac{1}{2} = \underset{\sim}{j}_2$, ien we shall frequently meet the following Clebsch-Gordan oefficients:

$$a(j_1j_2J) = (-)^{j_1-\frac{1}{2}} \frac{\hat{j}_1\hat{j}_2}{\hat{J}} \langle j_1\tfrac{1}{2}j_2-\tfrac{1}{2}|J0\rangle$$

$$b(j_1j_2J) = (-)^{\ell_1+j_1-j_2+1} \frac{\hat{j}_1\hat{j}_2}{\hat{J}} \langle j_1\tfrac{1}{2}j_2\tfrac{1}{2}|J1\rangle \tag{A.23}$$

with $\hat{J} = (2J+1)^{\frac{1}{2}}$. The coefficients $a(j_1j_2J)$ and $b(j_1j_2J)$ satisfy a symmetry requirement

$$a(j_1j_2J) = (-)^{j_1-j_2} a(j_2j_1J)$$
$$b(j_1j_2J) = (-)^{j_1-j_2+x+1} b(j_2j_1J) \tag{A.24}$$

where $x = \ell_1+\ell_2+J$. In evaluating the expectation value of one-body operators between states $\psi_2(\underset{\sim}{x})$ and $\psi_1(\underset{\sim}{x})$, the following combinations of recoupling coefficients arise (Brink and Satchler 1968):

$$\langle \ell_10\ell_20|J0\rangle\, U(j_2J\tfrac{1}{2}\ell_1;j_1\ell_2) = (-)^{\ell_1} \frac{\hat{J}}{\hat{\ell}_1\hat{j}_2}\, \frac{1}{2}(1+(-)^x)a(j_1j_2J). \tag{A.25a}$$

$$\hat{\ell}_1\hat{\ell}_2\langle \ell_10\ell_20|J0\rangle \begin{bmatrix} \ell_1 & \ell_2 & J \\ \frac{1}{2} & \frac{1}{2} & 0 \\ j_1 & j_2 & J \end{bmatrix} = (-)^{\ell_1} \frac{1}{\sqrt{2}}\, \frac{1}{2}(1+(-)^x)\hat{J}\ a(j_1j_2J). \tag{A.25b}$$

$$\hat{\ell}_1\hat{\ell}_2\langle \ell_10\ell_20|J0\rangle \begin{bmatrix} \ell_1 & \ell_2 & J \\ \frac{1}{2} & \frac{1}{2} & 1 \\ j_1 & j_2 & J \end{bmatrix} = (-)^{\ell_2} \frac{1}{\sqrt{2}}\, \frac{1}{2}(1+(-)^x)\hat{J}\ b(j_1j_2J). \tag{A.25c}$$

$$\hat{\ell}_1\hat{\ell}_2\langle \ell_10\ell_20|J-10\rangle \begin{bmatrix} \ell_1 & \ell_2 & J-1 \\ \frac{1}{2} & \frac{1}{2} & 1 \\ j_1 & j_2 & J \end{bmatrix} = \frac{1}{\sqrt{2}}\, \frac{1}{2}(1-(-)^x)\ \left[\frac{2J-1}{2J+1}\right]^{\frac{1}{2}} \times$$
$$\times \{(-)^{\ell_1} J^{\frac{1}{2}}\, a(j_1j_2J)-(-)^{\ell_2}(J+1)^{\frac{1}{2}}\, b(j_1j_2J)\}. \tag{A.25d}$$

$$\hat{\ell}_1\hat{\ell}_2\langle \ell_10\ell_20|J+10\rangle \begin{bmatrix} \ell_1 & \ell_2 & J+1 \\ \frac{1}{2} & \frac{1}{2} & 1 \\ j_1 & j_2 & J \end{bmatrix} =-\frac{1}{\sqrt{2}}\, \frac{1}{2}(1-(-)^x)\ \left[\frac{2J+3}{2J+1}\right]^{\frac{1}{2}} \times$$
$$\times \{(-)^{\ell_1}(J+1)^{\frac{1}{2}}\, a(j_1j_2J)+(-)^{\ell_2} J^{\frac{1}{2}}\, b(j_1j_2J)\}. \tag{A.25e}$$

Finally in this section, we give a few sum rules for the coefficients $a(j_1j_2J)$ and $b(j_1j_2J)$:

$$\sum_J (-)^{j_1+j_2+j_3+j_4} \frac{\hat{J}}{\hat{J}'} U(j_1j_2j_4j_3;JJ')\ a(j_1j_2J)\ a(j_3j_4J) = (-)^{\ell_1+\ell_3}\ b(j_1j_4J')\ b(j_3j_2J'). \tag{A.26a}$$

$$\sum_J (-)^J (-)^{j_1+j_2+j_3+j_4} \frac{\hat{J}}{\hat{J}'} U(j_1j_2j_4j_3;JJ')\ a(j_1j_2J)\ a(j_3j_4J) = (-)^{J'+1}\ a(j_1j_4J')\ a(j_3j_2J'). \tag{A.26b}$$

$$\sum_J (-)^{j_1+j_2+j_3+j_4} \frac{\hat{J}}{\hat{J}'} U(j_1j_2j_4j_3;JJ')\ b(j_1j_2J)\ b(j_3j_4J) = (-)^{\ell_1+\ell_3}\ a(j_1j_4J')\ a(j_3j_2J'). \tag{A.26c}$$

A.5 MATRIX ELEMENTS OF TENSOR OPERATORS

Matrix elements of a spherical tensor operator $F_q^{(k)}$ of rank k and magnetic projection q have a simple geometrical dependence on the magnetic quantum numbers as given by the Wigner-Eckhart theorem:

$$\langle X_2J_2M_2|F_q^{(k)}|X_1J_1M_1\rangle = (-)^{2k}\langle J_1M_1kq|J_2M_2\rangle\langle X_2J_2\|F^{(k)}\|X_1J_1\rangle. \tag{A.27}$$

We use the definitions of Brink and Satchler (1968). For most applications k is an integer and the phase factor can be dropped. Here X_1 and X_2 are additional labels (not angular momentum quantum numbers) necessary to complete the specification of the basis states. The result (A.27) shows that all the directional properties are contained in the Clebsch-Gordan coefficients and the dynamics of the system appear in the reduced matrix element $\langle X_2J_2\|F^{(k)}\|X_1J_1\rangle$.

As an example consider the matrix element of the angular momentum operator J whose spherical components J_q, $q = 0, \pm 1$ form a tensor of rank 1.

$$J_{\pm 1} = \mp \frac{1}{\sqrt{2}} (J_x \pm i\, J_y) \qquad J_0 = J_z .$$

The reduced matrix element of $\underset{\sim}{J}$ is found by considering the J_z component

$$\langle X_2 J_2 M_2 | J_z | X_1 J_1 M_1 \rangle = M\, \delta_{X_1 X_2}\, \delta_{J_1 J_2}\, \delta_{M_1 M_2}$$

$$= \langle J_1 M_1 10 | J_2 M_2 \rangle \langle X_2 J_2 \| \underset{\sim}{J} \| X_1 J_1 \rangle$$

and explicitly evaluating the Clebsch-Gordan coefficient giving the result

$$\langle X_2 J_2 \| \underset{\sim}{J} \| X_1 J_1 \rangle = \delta_{X_1 X_2}\, \delta_{J_1 J_2}\, [J_1 (J_1 + 1)]^{1/2} . \tag{A.28a}$$

Some other useful reduced matrix elements are

$$\langle X_2 J_2 \| 1 \| X_1 J_1 \rangle = \delta_{X_1 X_2}\, \delta_{J_1 J_2} , \tag{A.28b}$$

$$\langle \ell_2 \| \underset{\sim}{L} f(r) \| \ell_1 \rangle = \delta_{\ell_1 \ell_2}\, [\ell_1 (\ell_1 + 1)]^{\frac{1}{2}} \langle R_2 | f(r) | R_1 , \tag{A.28c}$$

$$\langle \ell_2 \| f(r) i^L Y_L \| \ell_1 \rangle = i^{\ell_1 + L - \ell_2} \langle \ell_1 0 L 0 | \ell_2 0 \rangle \frac{\hat{\ell}_1 \hat{L}}{\hat{\ell}_2} (4\pi)^{-\frac{1}{2}} \langle R_2 | f(r) | R_1 \rangle , \tag{A.28d}$$

where $|\ell m\rangle$ is the orbital part of a wavefunction: $|\ell m\rangle = R(r)\, i^{\ell} Y_{\ell m}(\hat{\underset{\sim}{r}})$, and $\langle R_2 | f(r) | R_1 \rangle$ is the radial integral $\int_0^\infty r^2 R_2(r) f(r) R_1(r) dr$. The momentum operator $\underset{\sim}{p} = -i\underset{\sim}{\nabla}$ is also a spherical tensor of rank 1:

$$\langle \ell_2 \| \underset{\sim}{p} \| \ell_1 \rangle = i^{\ell_1 - \ell_2 - 1} \frac{\hat{\ell}_1}{\hat{\ell}_2} (\ell_1 010 | \ell_2 0) \{ \delta_{\ell_2, \ell_1 + 1} \langle R_2 | \left(\frac{\partial}{\partial r} - \frac{\ell_1}{r} \right) | R_1 \rangle +$$

$$+ \delta_{\ell_2, \ell_1 - 1} \langle R_2 | \left(\frac{\partial}{\partial r} + \frac{\ell_1 + 1}{r} \right) | R_1 \rangle \} . \tag{A.28e}$$

More often $\underset{\sim}{p}$, operating on an arbitrary function $f(r)$, occurs in the symmetric combination

$$\langle \ell_2 \| \underset{\sim}{p} f(r) \| \ell_1 \rangle + \langle \ell_2 \| f(r) \underset{\sim}{p} \| \ell_1 \rangle$$

$$= i^{\ell_1 - \ell_2 - 1} \frac{\hat{\ell}_1}{\hat{\ell}_2} (\ell_1 010 | \ell_2 0) \{ \delta_{\ell_2, \ell_1 + 1} [\langle R_2 | f \left(\frac{\partial}{\partial r} - \frac{\ell_1}{r} \right) | R_1 \rangle -$$

$$- \langle R_1 | f \left(\frac{\partial}{\partial r} + \frac{\ell_2 + 1}{r} \right) | R_2 \rangle] + \delta_{\ell_2, \ell_1 - 1} [\langle R_2 | f \left(\frac{\partial}{\partial r} + \frac{\ell_1 + 1}{r} \right) | R_1 \rangle -$$

$$- \langle R_1 | f \left(\frac{\partial}{\partial r} - \frac{\ell_2}{r} \right) | R_2 \rangle] \} \tag{A.28f}$$

where the radial integrals have been manipulated by integrating by parts and the surface term dropped on the assumption that $rR(r) \to 0$ as $r \to \infty$.

For the spin space the basic reduced matrix element is

$$\langle \tfrac{1}{2} \| \underset{\sim}{\sigma} \| \tfrac{1}{2} \rangle = \sqrt{3} \tag{A.28g}$$

where $\underset{\sim}{\sigma} = 2\underset{\sim}{s}$ and $|\tfrac{1}{2} m_s\rangle$ is the spinor $\chi^{\frac{1}{2}}_{m_s}$. Similarly in isospin space $\langle \tfrac{1}{2} \| \underset{\sim}{\tau} \| \tfrac{1}{2} \rangle$ equals $\sqrt{3}$.

The Hermitian conjugate or adjoint $F^{(k)\dagger}_q$ of an operator $F^{(k)}_q$ is defined by expressing its matrix elements in terms of matrix elements of $F^{(k)}_q$ as

$$\langle X_2 J_2 M_2 | F^{(k)\dagger}_q | X_1 J_1 M_1 \rangle = \langle X_1 J_1 M_1 | F^{(k)}_q | X_2 J_2 M_2 \rangle^* \tag{A.29a}$$

or in terms of reduced matrix elements

$$\hat{J}_2 \langle X_2 J_2 \| F^{(k)} \| X_1 J_1 \rangle = (-)^{J_1 - J_2 + p} \hat{J}_1 \langle X_1 J_1 \| F^{(k)} \| X_2 J_2 \rangle^* . \tag{A.29b}$$

The phase factor $(-)^p$ depends on the Hermitian property of the operator

$$F^{(k)\dagger}_q = (-)^{p+q} F^{(k)}_{-q} . \tag{A.30}$$

If p-k is an even integer, the operator is said to be Hermitian and if p-k is an odd integer, the operator is anti-Hermitian. With this definition, operators 1, $i^L Y_{LM}$, $\underset{\sim}{\nabla}$ are Hermitian and $\underset{\sim}{J}$, $\underset{\sim}{L}$, $\underset{\sim}{\sigma}$, $\underset{\sim}{\tau}$, and $\underset{\sim}{p}$ are all anti-Hermitian.

The reduced matrix element $\langle X_1J_1\|F^{(k)}\|X_2J_2\rangle^*$ is real if the basis states $|XJM\rangle$ transform under time reversal as $\theta|XJM\rangle = |\widetilde{XJM}\rangle = (-)^{J+M}|XJ{-M}\rangle$ and if the operator $F_q^{(k)}$ transforms as $\theta F_q^{(k)}\theta^\dagger = (-)^{k_T+q} F_{-q}^{(k)}$ with k_T-k being an even integer. The proof makes use of the property of anti-linear operators (Messiah 1964) that

$$\langle \widetilde{X_1J_1M_1}|\tilde{F}_q^{(k)}|\widetilde{X_2J_2M_2}\rangle = \langle X_1J_1M_1|F_q^{(k)}|X_2J_2M_2\rangle^*$$

where $\tilde{F}_q^{(k)} = \theta F_q^{(k)}\theta^\dagger$. Thus we have

$$\begin{aligned}\langle X_1J_1\|F^{(k)}\|X_2J_2\rangle^* &= \sum_{M_2q}\langle \widetilde{X_1J_1M_1}|\tilde{F}_q^{(k)}|\widetilde{X_2J_2M_2}\rangle\langle J_2M_2kq|J_1M_1\rangle \\ &= \sum_{M_2q}(-)^{-J_1-M_1+k_T+q+J_2+M_2}\langle X_1J_1{-M_1}|F_{-q}^{(k)}|X_2J_2{-M_2}\rangle\langle J_2M_2kq|J_1M_1\rangle \\ &= (-)^{-J_1+k_T+J_2}\langle X_1J_1\|F^{(k)}\|X_2J_2\rangle(-)^{J_2+k-J_1} \\ &= \langle X_1J_1\|F^{(k)}\|X_2J_2\rangle \quad \text{if } k-k_T = \text{even integer.}\end{aligned} \tag{A.31}$$

All the operators considered in this section have $k = k_T$ and so their reduced matrix elements are real. For example, $\theta\underset{\sim}{r}\theta^\dagger = \underset{\sim}{r}$ and $\theta\underset{\sim}{p}\theta^\dagger = -\underset{\sim}{p}$ by definition of time-reversal, so that $\theta\underset{\sim}{L}\theta^\dagger = \theta(\underset{\sim}{r}\wedge\underset{\sim}{p})\theta^\dagger = -\underset{\sim}{L}$. Similarly $\theta\underset{\sim}{\sigma}\theta^\dagger = -\underset{\sim}{\sigma}$, $\theta\underset{\sim}{J}\theta^\dagger = -\underset{\sim}{J}$ and $\theta[i^L Y_{LM}]\theta^\dagger = (-i)^L Y_{LM}^* = (-)^{L+M}[i^L Y_{L-M}]$.

From the set of spherical tensors considered so far, more complicated tensors can be constructed in the same way as more complicated angular momentum states are built up, by vector coupling

$$F_q^{(k)}(k_1k_2) = \sum_{q_1q_2}\langle k_1q_1k_2q_2|kq\rangle\, T_{q_1}^{(k_1)}\, U_{q_2}^{(k_2)}\,. \tag{A.32}$$

These composite tensors also satisfy the Wigner-Eckhart theorem and have reduced matrix elements which can be expressed in terms of the reduced matrix elements of the constituents:

$$\langle X_2J_2 \| F_q^{(k)}(k_1k_2) \| X_1J_1 \rangle = \sum_{X'J'} (-)^{k-k_1-k_2} U(J_2J_1k_1k_2;kJ') \times$$
$$\times \langle X_2J_2 \| T^{(k_1)} \| X'J' \rangle \langle X'J' \| U^{(k_2)} \| X_1J_1 \rangle . \tag{A.33}$$

For example, a very important tensor is the vector spherical harmonic $\underset{\sim}{Y}_{JLM}$ defined as

$$i^L \underset{\sim}{Y}_{JLM} = \sum_{\lambda\rho} \langle L\lambda 1\rho | JM \rangle \, i^L Y_{L\lambda} \, e_\rho \tag{A.34}$$

where e_ρ is a spherical unit tensor of rank one satisfying $e^*_\rho = (-)^\rho e_{-\rho}$ and $e^*_{\rho'} \cdot e_\rho = (-)^{\rho'} e_{-\rho'} \cdot e_\rho = \delta_{\rho\rho'}$. Thus the scalar product of any tensor of rank one with the vector spherical harmonic forms a tensor of type (A.32). For example

$$i^L \underset{\sim}{Y}_{JLM} \cdot \underset{\sim}{L} = \sum_{\lambda\rho} \langle L\lambda 1\rho | JM \rangle \, i^L Y_{L\lambda} L_\rho$$

and

$$\langle \ell_2 \| i^L \underset{\sim}{Y}_{JL} \cdot \underset{\sim}{L} \| \ell_1 \rangle = \sum_{\ell'} (-)^{J-L-1} U(\ell_2\ell_1L1;J\ell') \langle \ell_2 \| i^L Y_L \| \ell' \rangle \langle \ell' \| \underset{\sim}{L} \| \ell_1 \rangle$$
$$= (-)^{J-L-1} [\ell_1(\ell_1+1)]^{\frac{1}{2}} U(\ell_2\ell_1L1;J\ell_1) \langle \ell_2 \| i^L Y_L \| \ell_1 \rangle . \tag{A.35}$$

But beware of a trap: $Y_{L\lambda}$ and L_ρ do not commute, so that the matrix element of $\underset{\sim}{L} \cdot i^L \underset{\sim}{Y}_{JLM}$ is *not* given by eqn (A.35) but rather by

$$\underset{\sim}{L} \cdot i^L \underset{\sim}{Y}_{JLM} = (\underset{\sim}{L} \cdot i^L \underset{\sim}{Y}_{JLM}) + i^L \underset{\sim}{Y}_{JLM} \cdot \underset{\sim}{L}$$
$$= i^L [L(L+1)]^{\frac{1}{2}} \delta_{JL} Y_{LM} + i^L \underset{\sim}{Y}_{JLM} \cdot \underset{\sim}{L}.$$

A second example, which is met in connection with electromagnetic transitions, is

$$\langle \ell_2 \| f(r) i^L \underset{\sim}{Y}_{LL} \cdot \underset{\sim}{p} \| \ell_1 \rangle$$
$$= -i^{\ell_1+L-1-\ell_2} \hat{L} \frac{\hat{\ell}_1}{\hat{\ell}_2} (4\pi)^{-\frac{1}{2}} \sum_{\ell} U(\ell_2\ell_1L1;L\ell) \langle \ell 0 L 0 | \ell_2 0 \rangle \langle \ell_1 0 1 0 | \ell 0 \rangle \times$$

$$\times \{\delta_{\ell,\ell_1+1}\langle R_2|f\left(\frac{\partial}{\partial r}-\frac{\ell_1}{r}\right)|R_1\rangle + \delta_{\ell,\ell_1-1}\langle R_2|f\left(\frac{\partial}{\partial r}+\frac{\ell_1+1}{r}\right)|R_1\rangle\}$$

$$= (-)^{\ell_2}\, i^{\ell_1+L-1-\ell_2}\,\hat{\ell}_1[\ell_1(\ell_1+1)]^{\frac{1}{2}}\,(4\pi)^{-\frac{1}{2}}\langle \ell_2 0 \ell_1 1|L1\rangle\langle R_2|\frac{f}{r}|R_1\rangle \qquad (A.36)$$

where the U-coefficient is explicitly evaluated to obtain the final form.

Another situation frequently met is that the composite spherical tensor, $F_q^{(k)}$ of eqn (A.32), is constructed from two tensors, with each tensor operating on a separate subsystem. For example, the tensors may be operating on distinct particles or even on the same particle but on different aspects of that particle, such as the orbital and spin properties. The spin-orbit operator $\underset{\sim}{L}\cdot\underset{\sim}{\sigma}$ is a typical such tensor. Let us evaluate the reduced matrix elements of $F_q^{(k)}(k_1,k_2)$ between states which are similarly composed of two parts vector-coupled to good total angular momentum:

$$\langle [J_1J_2]J\| F^{(k)}(k_1,k_2)\| [J_1'J_2']J'\rangle$$

$$= \begin{bmatrix} J_1' & k_1 & J_1 \\ J_2' & k_2 & J_2 \\ J' & k & J \end{bmatrix} \langle J_1\| T^{(k_1)}\| J_1'\rangle\,\langle J_2\| U^{(k_2)}\| J_2'\rangle . \qquad (A.37)$$

This very important result shows how the matrix element of a composite tensor can be expressed in terms of the matrices of the component systems. The result, referred to as the *factorization theorem* has been used many times in this book (see, for example, the evaluation of many-particle matrix elements discussed in section 6.5).

Two limiting forms of eqn (A.37) are obtained on first setting $k_2 = 0$ and then $k_1 = 0$:

$$\langle [J_1J_2]J\| T^{(k_1)}\| [J_1'J_2']J'\rangle$$

$$= (-)^{J_1'+J-J_1-J'}\, U(Jk_1J_2'J_1';J'J_1)\langle J_1\| T^{(k_1)}\| J_1'\rangle\,\delta_{J_2J_2'} \qquad (A.38a)$$

$$\langle [J_1J_2]J\| U^{(k_2)}\| [J_1'J_2']J'\rangle = U(J_2'k_2J_1'J;J_2J')\langle J_2\| U^{(k_2)}\| J_2'\rangle\,\delta_{J_1J_1'} . \qquad (A.38b)$$

These results are used in the situation when the tensor operates on only one part of a composite system. Another special case is k = 0; the tensor $F^{(k)}$ is then proportional to the scalar product of two tensors, viz.

$$F^{(0)}(k,k) = (-)^k (2k+1)^{-\frac{1}{2}} \underset{\sim}{T}^{(k)} \cdot \underset{\sim}{U}^{(k)}$$

and the reduced matrix element is

$$\langle [J_1 J_2] J \| T^{(k)} \cdot U^{(k)} \| [J_1' J_2'] J' \rangle$$

$$= (-)^k \left[\frac{2J_2+1}{2J_2'+1}\right]^{\frac{1}{2}} U(J_2 J k J_1' ; J_1 J_2') \langle J_1 \| T^{(k)} \| J_1' \rangle \langle J_2 \| U^{(k)} \| J_2' \rangle \; \delta_{JJ'} . \quad \text{(A.38c)}$$

Consider the spin-orbit operator $\underset{\sim}{L} \cdot \underset{\sim}{\sigma}$ evaluated for single-particle states. We obtain from eqn (A.38c)

$$\langle [\ell \tfrac{1}{2}] j \| \underset{\sim}{L} \cdot \underset{\sim}{\sigma} \| [\ell \tfrac{1}{2}] j \rangle$$

$$= -U(\tfrac{1}{2} j 1 \ell ; \ell \tfrac{1}{2}) \langle \ell \| \underset{\sim}{L} \| \ell \rangle \langle \tfrac{1}{2} \| \sigma \| \tfrac{1}{2} \rangle$$

$$= j(j+1) - \ell(\ell+1) - \frac{3}{4}$$

where eqns (A.28c) and (A.28g) have been used for the reduced matrix elements and the explicit value for the U-coefficient inserted. Of course, the matrix element of the spin-orbit operator could have been obtained more simply by noting that

$$\underset{\sim}{L} \cdot \underset{\sim}{\sigma} = 2\underset{\sim}{L} \cdot \underset{\sim}{S} = \underset{\sim}{J}^2 - \underset{\sim}{L}^2 - \underset{\sim}{S}^2$$

and writing the expectation value of J^2 as $j(j+1)$ and so on.

Two of the more common matrix elements to be evaluated involve the operators $i^L Y_L$ and $i^L \underset{\sim}{Y}_{JL} \cdot \underset{\sim}{\sigma}$. In the first case, we obtain using eqn (A.38a)

$$\langle [\ell_2 \tfrac{1}{2}] j_2 \| f(r) \, i^L Y_L \| [\ell_1 \tfrac{1}{2}] j_1 \rangle$$

$$= (-)^{\ell_1 + j_2 - \ell_2 - j_1} U(j_2 L \tfrac{1}{2} \ell_1 ; j_1 \ell_2) \langle \ell_2 \| f(r) i^L Y_L \| \ell_1 \rangle$$

$$= i^{\ell_1+L-\ell_2}(-)^{j_2-j_1+L}\frac{\hat{L}}{\hat{j}_2}\frac{1}{2}(1+(-)^x)a(j_1j_2L)(4\pi)^{-\frac{1}{2}}\langle R_2|f(r)|R_1\rangle \qquad (A.39)$$

where $x = \ell_1+\ell_2+L$, and the coefficient $a(j_1j_2L)$, essentially a Clebsch-Gordan coefficient, was defined in eqn (A.23). The special relation (A.25a) was used to reach the final result (A.39).

In the second case, involving the spherical harmonic coupled to the spin operator, we obtain using eqn (A.37)

$$\langle[\ell_2\tfrac{1}{2}]j_2\| f(r)\; i^L \underset{\sim}{Y}_{JL}\cdot\underset{\sim}{\sigma}\| [\ell_1\tfrac{1}{2}]j_1\rangle$$

$$= \begin{bmatrix} \ell_1 & L & \ell_2 \\ \frac{1}{2} & 1 & \frac{1}{2} \\ j_1 & J & j_2 \end{bmatrix} \langle \ell_2\| f(r) i^L Y_L\| \ell_1\rangle\; \langle \tfrac{1}{2}\| \underset{\sim}{\sigma}\| \tfrac{1}{2}\rangle$$

$$= i^{\ell_1+L-\ell_2}(-)^{j_2-j_1+J}\frac{\hat{J}}{\hat{j}_2}(4\pi)^{-\frac{1}{2}}\langle R_2|f(r)|R_1\rangle A \qquad (A.40)$$

where

$$A = -\frac{1}{2}(1-(-)^x)\hat{J}^{-1}\{J^{\frac{1}{2}}a(j_1j_2J) + (-)^J(J+1)^{\frac{1}{2}}b(j_1j_2J)\}, \quad L=J-1$$

$$= \frac{1}{2}(1+(-)^x)\;(-)^{J+1}b(j_1j_2J) \qquad , \quad L=J$$

$$= \frac{1}{2}(1-(-)^x)\hat{J}^{-1}\{(J+1)^{\frac{1}{2}}a(j_1j_2J)+(-)^{J+1}J^{\frac{1}{2}}b(j_1j_2J)\}\; , \quad L=J+1$$

with $x = \ell_1+\ell_2+L$. Again the special relations (A.25c), (A.25d), and (A.25e) were used to reach the final result. Expressing the single-particle reduced matrix elements in terms of the coefficients $a(j_1j_2J)$ and $b(j_1j_2J)$ has an important application for the schematic model description of particle-hole states as discussed in Chapter 4.

A.6 TRANSITION RATES FOR GAMMA DECAY

The transition probability per unit time for the emission of a photon with wave number $\underset{\sim}{k}$ with circular polarization $\underset{\sim}{\varepsilon}_q$ ($q = \pm 1$ for right- or left-handed polarization) associated with a transition from an initial state $|J_1M_1\rangle$ to a final state $|J_2M_2\rangle$ is given from first-order perturbation theory by the expression

$$\omega(J_1M_1 \rightarrow J_2M_2 + (\underset{\sim}{k},\underset{\sim}{\varepsilon}_q)) = \frac{k}{2\pi\hbar} \,|\langle J_2M_2|H_e(\underset{\sim}{k},\underset{\sim}{\varepsilon}_q)|J_1M_1\rangle|^2 \quad (A.41a)$$

$$= \frac{k}{2\pi\hbar} \,|\langle J_1M_1|H_a(\underset{\sim}{k},\underset{\sim}{\varepsilon}_q)|J_2M_2\rangle|^2. \quad (A.41b)$$

Here $H_e(\underset{\sim}{k},\underset{\sim}{\varepsilon}_q)$ is the interaction Hamiltonian for the spontaneous emission of a photon and is related to the interaction Hamiltonian for photon absorption:

$$H_e(\underset{\sim}{k},\underset{\sim}{\varepsilon}_q) = [H_a(\underset{\sim}{k},\underset{\sim}{\varepsilon}_q)]^\dagger.$$

The equivalence of eqns (A.41a) and (A.41b) is just a statement of the principle of detailed balance. If $|J_1M_1\rangle$ and $|J_2M_2\rangle$ are discrete nuclear states, then the transition probability for absorption is simply related to the transition probability for emission. We shall adopt the convention, introduced by Rose and Brink (1967), of using the form (A.41b) with the initial state for the emission process written on the left of the matrix element and the final state on the right.

The next step is to make a multipole decomposition of the interaction Hamiltonian (Rose and Brink 1967):

$$H_a(\underset{\sim}{k},\underset{\sim}{\varepsilon}_q) = -\sum_{LM\pi} q^\pi \, T^{\langle\pi\rangle}_{LM}\mathscr{D}^L_{Mq}(R) \quad (A.42)$$

where the summation runs over all possible multipoles L,M,π which can contribute to the transition in question. The $T^{\langle\pi\rangle}_{LM}$ are one-body operators, spherical tensors of rank L, and π differentiates between electric ($\pi=0$) and magnetic ($\pi=1$) radiation. The rotation matrix $\mathscr{D}^L_{Mq}(R)$ transforms from a laboratory frame of reference to one where the z-axis is aligned parallel to $\underset{\sim}{k}$, the photon direction.

If the orientation of the spin J_2 of the final nucleus is not observed, then we must sum over all magnetic substates, M_2. Inserting eqn (A.42) back in eqn (A.41b) and using the Wigner-Eckhart theorem we obtain

$$\sum_{M_2} \omega(J_1M_1 \to J_2M_2+(\underset{\sim}{k},\underset{\sim}{\varepsilon}_q)) =$$

$$= \frac{k}{2\pi\hbar} \sum_{L\pi L'\pi' K} \frac{\hat{J}_1}{\hat{L}} (-)^{J_1-M_1} \langle J_1M_1J_1-M_1|K0\rangle \times$$

$$\times (-)^{L'-q} \langle LqL'-q|K0\rangle q^{\pi+\pi'} \langle J_1\| T_L^{\langle\pi\rangle} \| J_2\rangle \overline{\langle J_1\| T_{L'}^{\langle\pi'\rangle} \| J_2\rangle} \times$$

$$\times U(J_1J_2KL';LJ_1)\mathscr{D}^K_{00}(R). \tag{A.43}$$

If the initial nuclear state is randomly oriented, then we must average over the magnetic substates M_1. Finally to obtain the total transition probability per unit time (i.e. the reciprocal lifetime) we integrate over all angles:

$$\int \mathscr{D}^K_{00}(R)\,dR = 4\pi\ \delta_{K,0}$$

and sum over the two directions of polarization to obtain

$$\begin{aligned}\frac{1}{\tau} &= \frac{1}{2J_1+1} \sum_{M_1qM_2} \int dR\ \omega(J_1M_1 \to J_2M_2 + (\underset{\sim}{k},\underset{\sim}{\varepsilon}_q)) \\ &= \frac{4k}{\hbar} \sum_{L\pi} |\langle J_1\| T_L^{\langle\pi\rangle} \| J_2\rangle|^2/(2L+1).\end{aligned} \tag{A.44}$$

The gamma width Γ_γ is just $\Gamma_\gamma = \hbar/\tau$.

Next we discuss the form of the interaction multipole operators $T_{LM}^{\langle\pi\rangle}$. They are expressible in terms of the solutions to the vector wave equation

$$\nabla^2 \underset{\sim}{A}_{LM}^{\langle\pi\rangle} + k^2 \underset{\sim}{A}_{LM}^{\langle\pi\rangle} = 0$$

by the formula (Rose and Brink 1967):

$$T_{LM}^{\langle\pi\rangle} = -\frac{1}{\sqrt{2}}\beta\{g_\ell(\underset{\sim}{p}\cdot\underset{\sim}{A}_{LM}^{\langle\pi\rangle} + \underset{\sim}{A}_{LM}^{\langle\pi\rangle}\cdot\underset{\sim}{p}) + g_s\ \underset{\sim}{S}\cdot\underset{\sim}{\nabla}_\wedge\underset{\sim}{A}_{LM}^{\langle\pi\rangle}\} \tag{A.45}$$

where g_ℓ and g_s are the orbital and spin g-factors, and β the nuclear magneton, $\beta = e\hbar/2mc = 0.1262\ \mathrm{MeV}^{1/2}\ \mathrm{fm}^{3/2}$, m being the proton mass. The g-factors have the values $g_\ell = 1$ or 0 and $g_s = 5.586$ or -3.826 for a free proton or a free neutron respectively.

The vector fields are defined as

$$\underset{\sim}{A}_{LM}^{\langle el\rangle} = \frac{1}{k}[L(L+1)]^{-\frac{1}{2}}\ \underset{\sim}{\nabla}_\wedge \underset{\sim}{L}\ \phi_{LM}$$

$$= i^{L+1}(4\pi)^{\frac{1}{2}}\ (L+1)^{\frac{1}{2}} j_{L-1}(kr)\underset{\sim}{Y}_{LL-1M} - L^{\frac{1}{2}} j_{L+1}(kr)\underset{\sim}{Y}_{LL+1M}\} \tag{A.46a}$$

$$\underset{\sim}{A}_{LM}^{\langle mag\rangle} = [L(L+1)]^{-\frac{1}{2}}\ \underset{\sim}{L}\ \phi_{LM}$$

$$= i^{L}\ (2L+1)^{\frac{1}{2}}\ (4\pi)^{\frac{1}{2}}\ j_L(kr)\ \underset{\sim}{Y}_{LLM} \tag{A.46b}$$

where

$$\phi_{LM} = i^L(2L+1)^{\frac{1}{2}}\ (4\pi)^{\frac{1}{2}}\ j_L(kr)Y_{LM}$$

is a solution to the scalar wave equation $\nabla^2\phi_{LM}+k^2\phi_{LM} = 0$. These fields $\underset{\sim}{A}_{LM}^{\langle el\rangle}$ and $\underset{\sim}{A}_{LM}^{\langle mag\rangle}$ are known as the electric and magnetic components of the transverse field; they are solenoidal ($\underset{\sim}{\nabla}\cdot\underset{\sim}{A}_{LM}^{\langle\pi\rangle} = 0$) and have parities $(-)^{L+1}$ and $(-)^L$ respectively. A third independent solution of the vector wave equation is the longitudinal field

$$\underset{\sim}{A}_{LM} = \frac{1}{ik}\ \underset{\sim}{\nabla}\ \phi_{LM}$$

$$= -i^{L+1}(4\pi)^{\frac{1}{2}}\{L^{\frac{1}{2}} j_{L-1}(kr)\underset{\sim}{Y}_{LL-1M} + (L+1)^{\frac{1}{2}} j_{L+1}\underset{\sim}{Y}_{LL+1M}\} \tag{A.46c}$$

which is irrotational ($\underset{\sim}{\nabla}_\wedge\underset{\sim}{A}_{LM} = 0$) and has parity $(-)^{L+1}$.

This latter solution has one very important property, namely the symmetric combination $\underset{\sim}{A}_{LM}$ with $\underset{\sim}{p}$ required in eqn (A.45) can be simplified as follows:

$$\underset{\sim}{p}\cdot\underset{\sim}{A}_{LM} + \underset{\sim}{A}_{LM}\cdot\underset{\sim}{p} = -\frac{1}{k}\ \{\vec{\underset{\sim}{\nabla}}\cdot(\underset{\sim}{\nabla}\phi) + (\underset{\sim}{\nabla}\phi)\cdot\vec{\underset{\sim}{\nabla}}\}$$

$$= -\frac{1}{k}\{\overleftarrow{\nabla}^2\phi - \phi\overrightarrow{\nabla}^2\}$$

$$= \frac{2m}{\hbar^2 k}[H,\phi] \tag{A.47}$$

where in the last step, the replacement of the kinetic energy operator by the total Hamiltonian H of the system is possible if the nuclear potential V is local and commutes with the scalar field ϕ. Then the evaluation of a matrix element between an initial state of spin J_1 and a final state of spin J_2 becomes

$$\langle J_1M_1|\underset{\sim}{p}\cdot\underset{\sim}{A}_{LM} + \underset{\sim}{A}_{LM}\cdot\underset{\sim}{p}|J_2M_2\rangle$$
$$= \frac{2m}{\hbar^2 k}(E_1-E_2)\langle J_1M_1|\phi_{LM}|J_2M_2\rangle$$
$$= \frac{e}{\beta}\langle J_1M_1|\phi_{LM}|J_2M_2\rangle \tag{A.48}$$

where (E_1-E_2) is the difference in energy between the initial and final nuclear states and is the energy released to the photon, i.e. $E_1-E_2 = \hbar ck$.

Let us return now to the transverse vector fields $\underset{\sim}{A}_{LM}^{\langle el\rangle}$ and $\underset{\sim}{A}_{LM}^{\langle mag\rangle}$ appropriate for photon emission, and let us introduce the long-wavelength approximation. Then the spherical Bessel function is replaced, in the limit that $kr \ll 1$, by

$$j_L(kr) \to \frac{(kr)^L}{(2L+1)!!}.$$

This is justified for photon emission in nuclei, where a typical photon energy may be 10 MeV and a typical nuclear radius 7 fm or less; then kr is of order 0.36 and the ratio $j_{L+1}(kr)/j_{L-1}(kr)$ is of order 1 per cent. In this approximation, the second term in eqn (A.46a) for a transverse electric field is dropped, and $\underset{\sim}{A}_{LM}^{\langle el\rangle}$ can now be expressed in terms of the longitudinal field $\underset{\sim}{A}_{LM}$, viz.

$$\underset{\sim}{A}_{LM}^{\langle el\rangle} \xrightarrow[kr\ll 1]{} -\left(\frac{L+1}{L}\right)^{1/2}\underset{\sim}{A}_{LM}. \tag{A.49}$$

Thus the matrix elements for the orbital part of $\underset{\sim}{A}_{LM}^{\langle el\rangle}$ can be related to those of $\underset{\sim}{A}_{LM}$, which in turn are related to those of the scalar field ϕ via eqn (A.48). This replacement is known as Siegert's theorem (Sachs 1953). The result is clearly useful since it eliminates the technical complications of the derivative operator when evaluating electromagnetic transition rates in nuclear model calculations (Eisenberg and Greiner 1970). However, in using the theorem with nuclear models, one needs to exercise some caution since inconsistencies can occur. For example, when configuration mixing calculations are being performed with the shell model, the unperturbed wavefunctions refer to a different excitation energy from that which is produced when the residual interaction is turned on, so that $E_1 - E_2$ may not equal $\hbar ck$.

Another difficulty to watch out for is that the Hermitian property, eqn (A.30), for the operator ϕ_{LM} is

$$\phi^{\dagger}_{LM} = (-)^{L+M} \phi_{L-M}$$

whereas for $T^{\langle el \rangle}_{LM}$ the Hermitian property is

$$T^{\langle \pi \rangle \dagger}_{LM} = (-)^{L+M+1} T^{\langle \pi \rangle}_{L-M}.$$

The latter is the correct form. This point is relevant when, for example, one is comparing a matrix element for a given particle configuration with that in the conjugate hole configuration (see eqn (6.61)). This example is discussed in some detail in Rose and Brink (1967).

We are now ready to write down the form of the interaction multipole operators $T^{\langle \pi \rangle}_{LM}$ after using Siegert's theorem and invoking the long-wavelength approximation. The results are

$$T^{\langle el \rangle}_{LM} = \frac{k^L}{(2L-1)!!} \left(\frac{L+1}{2L}\right)^{\frac{1}{2}} \{Q_{LM} + Q'_{LM}\} \tag{A.50a}$$

$$Q_{LM} = e\, g_{\ell} \left(\frac{4\pi}{2L+1}\right)^{\frac{1}{2}} r^L\, i^L\, Y_{LM} \tag{A.50b}$$

$$Q'_{LM} = -\beta\, g_s k \left(\frac{L}{L+1}\right)^{\frac{1}{2}} \left(\frac{4\pi}{2L+1}\right)^{\frac{1}{2}} r^L\, i^L \underset{\sim}{Y}_{LLM} \cdot \underset{\sim}{S} \tag{A.50c}$$

and

$$T_{LM}^{\langle mag\rangle} = \frac{k^L}{(2L-1)!!}\left(\frac{L+1}{2L}\right)^{\frac{1}{2}} \{M_{LM} + M'_{LM}\} \tag{A.50d}$$

$$M_{LM} = -2\, g_\ell \beta \left(\frac{L}{L+1}\right)^{\frac{1}{2}} \left(\frac{4\pi}{2L+1}\right)^{\frac{1}{2}} r^L\, i^L\, \underset{\sim}{Y}_{LLM}\cdot \underset{\sim}{p} \tag{A.50e}$$

$$M'_{LM} = g_s\, \beta\, L^{\frac{1}{2}}\, (4\pi)^{\frac{1}{2}}\, r^{L-1}\, i^{L-1}\, \underset{\sim}{Y}_{LL-1M}\cdot \underset{\sim}{S}\,. \tag{A.50f}$$

In deriving the form of the spin operators Q'_{LM} and M'_{LM}, we used the result that $\underset{\sim}{\nabla}\wedge\underset{\sim}{A}_{LM}^{\langle el\rangle} = k\,\underset{\sim}{A}_{LM}^{\langle mag\rangle}$ and $\underset{\sim}{\nabla}\wedge\underset{\sim}{A}_{LM}^{\langle mag\rangle} = k\,\underset{\sim}{A}_{LM}^{\langle el\rangle}$.

The last step is to evaluate the single-particle reduced matrix elements for Q_{LM}, Q'_{LM}, M_{LM}, and M'_{LM} using the results (A.36), (A.39), and (A.40):

$$\langle j_2\|Q_L\|j_1\rangle = e\, g_\ell\, i^{\ell_1+L-\ell_2}\, (-)^{j_2-j_1+L}\, \hat{j}_2^{-1}\, \tfrac{1}{2}(1+(-)^x)\, a(j_1j_2L)\, \langle R_2|r^L|R_1\rangle \tag{A.51a}$$

$$\langle j_2\|Q'_L\|j_1\rangle = \tfrac{1}{2}\beta g_s k\, i^{\ell_1+L-\ell_2}(-)^{j_2-j_1}\, \hat{j}_2^{-1} \left(\frac{L}{L+1}\right)^{\frac{1}{2}} \tfrac{1}{2}(1+(-)^x)\, b(j_1j_2L)\, \langle R_2|r^L|R_1\rangle \tag{A.51b}$$

$$\langle j_2\|M_L\|j_1\rangle = 2\, g_\ell \beta\, i^{\ell_1+L-1-\ell_2}\, (-)^{j_2-j_1+\ell_2}\, U(j_2 L \tfrac{1}{2}\ell_1; j_1 \ell_2)\, \langle \ell_2 0 \ell_1 1|L1\rangle \times$$
$$\times \left[\frac{L(2\ell_1+1)\ell_1(\ell_1+1)}{(L+1)(2L+1)}\right]^{\frac{1}{2}} \tfrac{1}{2}(1-(-)^x)\, \langle R_2|r^{L-1}|R_1\rangle \tag{A.51c}$$

$$\langle j_2\|M'_L\|j_1\rangle = -\frac{1}{2}\,\beta g_s\, i^{\ell_1+L-1-\ell_2}\, (-)^{j_2-j_1+L}\, \hat{j}_2^{-1}\, L^{\frac{1}{2}}\, \tfrac{1}{2}(1-(-)^x) \times$$
$$\times \{L^{\frac{1}{2}}\, a(j_1j_2L) + (-)^L\, (L+1)^{\frac{1}{2}}\, b(j_1j_2L)\}\, \langle R_2|r^{L-1}|R_1\rangle \tag{A.51d}$$

where $x = \ell_1+\ell_2+L$, $\hat{j} = (2j+1)^{\frac{1}{2}}$ and where the Clebsch-Gordan coefficients $a(j_1j_2L)$ and $b(j_1j_2L)$ were defined in eqn (A.23).

Notice that each of the multipole one-body operators Q, Q', M, and M' takes the form of a g-factor multiplied by a spherical tensor, where the g-factor has the value, say, g_p or g_n when the particle is a proton or a neutron, and where the spherical tensor depends only on the orbital and spin co-ordinates of the particle. The multipole operators therefore have the structure

$$M_{LM} = \sum_{\text{protons}} g_p U_{LM} + \sum_{\text{neutrons}} g_n U_{LM}.$$

It is sometimes more convenient to recast M_{LM} into an isospin formalism by introducing the isospin operator τ_z, such that

$$M_{LM} = \sum_{i=1}^{A} \tfrac{1}{2}(1-\tau_z) g_p U_{LM} + \sum_{i=1}^{A} \tfrac{1}{2}(1+\tau_z) g_n U_{LM}$$

$$= \tfrac{1}{2}(g_p+g_n) \sum_{i=1}^{A} U_{LM} - \tfrac{1}{2}(g_p-g_n) \sum_{i=1}^{A} U_{LM}\, \tau_z.$$

The first term is a scalar and the second term a vector in isospin space. Our convention is that the expectation value of τ_z is +1 for a neutron state and -1 for a proton state. Consider now a transition between states of spin and isospin J_1,T_1 and J_2,T_2 in a nucleus of mass A (neutron number N, proton number Z, and $T_Z = \frac{1}{2}(N-Z)$). The reduced matrix element (reduced in spin space only) must be evaluated:

$$\langle J_1 T_1 T_Z \| T_L^{\langle\pi\rangle} \| J_2 T_2 T_Z\rangle$$

$$= \tfrac{1}{2}(g_p+g_n) \langle J_1 T_1 ||| U_L^{\langle\pi\rangle} ||| J_2 T_2\rangle\, \delta_{T_1,T_2} -$$

$$- \tfrac{1}{2}(g_p-g_n) \langle T_2 T_Z 1 0 | T_1 T_Z\rangle \langle J_1 T_1 ||| U_L^{\langle\pi\rangle} \underset{\sim}{\tau} ||| J_2 T_2\rangle,$$

the triple bar indicating that the matrix element is reduced in both spin and isospin space.

Finally we write down the gamma width $\Gamma_\gamma(L\pi)$ for a particular multipole $L\pi$ from eqn (A.44). For electric transitions, we have

$$\Gamma_\gamma(L;el) = k^{2L+1} \frac{2(L+1)}{L[(2L+1)!!]^2} (2L+1) |\langle J_1 \| Q_L \| J_2\rangle|^2 \qquad (A.52)$$

where the spin term Q'_{LM} has been dropped, since its contribution to the width is negligible. A convenient measure of whether a particular transition is fast or slow compared to a norm of the same energy and type is afforded by the Weisskopf unit (Weisskopf 1951, Wilkinson 1960). This unit is based on an extreme one-particle model, in which a single proton of orbital angular momentum L, and total spin $j = L+\frac{1}{2}$ makes a transition to a final state of zero orbital angular momentum. The reduced matrix element is trivially evaluated from eqn (A.50b)

$$(2L+1)\,|\langle j\|Q_L\|\tfrac{1}{2}\rangle|^2 = e^2\langle r^L\rangle^2$$

and the radial integral, $\langle r^L\rangle$, estimated by taking the radial wavefunctions as simple rectangles of radial extension, R, which is identified with the radius of the nucleus. Then the Weisskopf unit becomes

$$\Gamma_{W\gamma}(L;el) = e^2 k^{2L+1}\,\frac{2(L+1)}{L[(2L+1)!!]^2}\left(\frac{3}{L+3}\right)^2 R^{2L} \qquad (A.53)$$

with $R = r_0 A^{1/3}$ and $r_0 = 1.2$ fm.

Another popular measure of transition strength is given by the reduced transition probability, B(EL), introduced by Bohr and Mottelson (1953), and defined as

$$B(EL;J_1\to J_2) = \frac{2L+1}{4\pi}|\langle J_1\|Q_L\|J_2\rangle|^2. \qquad (A.54)$$

This quantity is independent of energy factors.

Considering now magnetic transitions, the gamma width is given by the expression

$$\Gamma_\gamma(L;mag) = k^{2L+1}\,\frac{2(L+1)}{L[(2L+1)!!]^2}(2L+1)\,|\langle J_1\|M_L+M'_L\|J_2\rangle|^2. \qquad (A.55)$$

The Weisskopf unit is derived by comparing the operators $M_L + M'_L$ with Q_L for a single proton transition to obtain a correction factor

$$(g_\ell + \tfrac{1}{2}g_s)^2\left(\frac{2\beta}{e}\right)^2\frac{1}{R^2} \approx 10(\hbar/mcR)^2$$

which is then applied to the electric estimate (eqn (A.53)):

$$\Gamma_{\gamma W}(L;\text{mag}) = 10\left(\frac{e\hbar}{mcR}\right)^2 k^{2L+1} \frac{2(L+1)}{L[(2L+1)!!]^2} \left(\frac{3}{L+3}\right)^2 R^{2L} \qquad (A.56)$$

Finally the reduced transition probability of Bohr and Mottelson B(ML) is given by

$$B(ML;J_1\to J_2) = \frac{2L+1}{4\pi} \; |\langle J_1\| M_L + M'_L \| J_2\rangle|^2 \qquad (A.57)$$

A.7 STATIC MOMENTS

The static electric and magnetic multipole moments of a system with total angular momentum J are given by the diagonal M=J matrix elements of the spherical tensors Q_L, M_L, and M'_L defined in eqn (A.50). For example, the magnetic dipole moment is

$$\mu \equiv \beta\langle JJ|g_\ell \underset{\sim}{L} + g_s \underset{\sim}{S}|JJ\rangle$$

$$= \left(\frac{J}{J+1}\right)^{1/2} \langle J\| M_1 + M'_1 \| J\rangle . \qquad (A.58)$$

For a single particle in shell model orbital j, we obtain from eqns (A.51c) and (A.51d) the result

$$\mu = \beta j[g_\ell \pm (g_s - g_\ell)/(2\ell+1)] \quad \text{as} \quad j = \ell \pm \tfrac{1}{2}, \qquad (A.59)$$

which is known as the Schmidt value for the magnetic moment of a particle.

The electric quadrupole moment is conventionally defined as

$$Q = e\langle JJ|g_\ell(3z^2 - r^2)|JJ\rangle$$

$$= -2\left[\frac{J(2J-1)}{(J+1)(2J+3)}\right]^{1/2} \langle J\| Q_2 \| J\rangle , \qquad (A.60)$$

the minus sign coming from the i^L factor in the definition (A.50b) of the operator Q_{LM}. Again for a single particle in shell model orbital j, we have

$$Q = -eg_\ell \frac{(2j-1)}{(2j+2)} \langle r^2 \rangle \tag{A.61}$$

where $\langle r^2 \rangle$ is a radial integral.

The other static moments of interest are the magnetic octupole moment Ω and electric hexadecapole moment $Q^{(4)}$:

$$\Omega = \left[\frac{J(J-1)(2J-1)}{(J+1)(J+2)(2J+3)}\right]^{1/2} \langle J \| M_3 + M_3' \| J \rangle \tag{A.62}$$

$$Q^{(4)} = \left[\frac{J(J-1)(2J-1)(2J-3)}{(J+1)(J+2)(2J+3)(2J+5)}\right]^{1/2} \langle J \| Q_4 \| J \rangle . \tag{A.63}$$

A.8 TWO-BODY MATRIX ELEMENTS

As part of a nuclear structure calculation, it is frequently required to evaluate two-body matrix elements of the residual interaction, $\langle j_1 j_2;JT|V|j_3 j_4;JT\rangle$, between normalized, antisymmetrized two-particle states. For simple, modelistic interactions, V, this may be possible algebraically and we give some examples in the next few sections.

The interaction V is a scalar in the overall product of orbital, spin, and isospin spaces. However, if V is charge independent then the only isospin operators that can be incorporated in it have to be scalars commuting with the total isospin of the two-particle state, T. The orbital and spin operators themselves do not have to be scalars, but they do have to be combined into scalar products. Thus a general form for the interaction V is

$$V = \sum_r [V^{(r)}(\underset{\sim}{r}_1, \underset{\sim}{r}_2) \cdot S^{(r)}(\underset{\sim}{S}_1, \underset{\sim}{S}_2)] T^{(0)}(\underset{\sim}{\tau}_1, \underset{\sim}{\tau}_2) \tag{A.64}$$

where $V^{(r)}(\underset{\sim}{r}_1, \underset{\sim}{r}_2)$ is a spherical tensor of rank r constructed from the position coordinates $\underset{\sim}{r}_1$ and $\underset{\sim}{r}_2$ of the two particles, $S^{(r)}(\underset{\sim}{S}_1, \underset{\sim}{S}_2)$ is a similar tensor in spin space constructed from coordinates $\underset{\sim}{S}_1$ and $\underset{\sim}{S}_2$, and $T^{(0)}(\underset{\sim}{\tau}_1, \underset{\sim}{\tau}_2)$ is a scalar tensor in isospin space. For central forces, the tensor $V^{(r)}(\underset{\sim}{r}_1, \underset{\sim}{r}_2)$ is constructed using the relative coordinate $\underset{\sim}{r} = \underset{\sim}{r}_1 - \underset{\sim}{r}_2$ only.

The first step in evaluating the two-body matrix element

of V is to rewrite the matrix elements in terms of unnormalized, unsymmetrized two-particle states and transform from j-j to L-S coupling:

$$\langle j_1 j_2; JT|V|j_3 j_4; JT\rangle_A$$

$$= [(1+\delta_{12})(1+\delta_{34})]^{-1/2} \sum_{\substack{LS\\L'S'}} \begin{bmatrix} \ell_1 & \ell_2 & L \\ \frac{1}{2} & \frac{1}{2} & S \\ j_1 & j_2 & J \end{bmatrix} \begin{bmatrix} \ell_3 & \ell_4 & L' \\ \frac{1}{2} & \frac{1}{2} & S' \\ j_3 & j_4 & J \end{bmatrix} \times$$

$$\times \{\langle \ell_1 \ell_2; LSJT|V|\ell_3 \ell_4; L'S'JT\rangle + (-)^{1+\ell_3+\ell_4+L'+S'+T} \langle \ell_1 \ell_2; LSJT|V|\ell_4 \ell_3; L'S'JT\rangle\} \tag{A.65}$$

Here δ_{12} is unity if orbit 1 is identical to orbit 2, i.e. $n_1\ell_1 j_1 \equiv n_2\ell_2 j_2$, and zero otherwise. The first term in the braces is called the direct matrix element and the second term the exchange matrix element. Inserting eqn (A.64) into the direct matrix element and using the factorization theorem, eqn (A.37), gives

$$\langle \ell_1 \ell_2; LSJT|V|\ell_3 \ell_4; L'S'JT\rangle$$

$$= \sum_r (-)^r (2r+1)^{\frac{1}{2}} \begin{bmatrix} L' & r & L \\ S' & r & S \\ J & 0 & J \end{bmatrix} \langle \ell_1 \ell_2; L\|V^{(r)}\|\ell_3 \ell_4; L'\rangle \times$$

$$\times \langle \tfrac{1}{2}\tfrac{1}{2}; S\|S^{(r)}\|\tfrac{1}{2}\tfrac{1}{2}; S'\rangle \langle \tfrac{1}{2}\tfrac{1}{2}; T\|T^{(0)}\|\tfrac{1}{2}\tfrac{1}{2}; T\rangle \tag{A.66}$$

with a similar expression for the exchange matrix element. We consider first the simpler case of scalar, charge-independent forces.

A.9 SCALAR CHARGE-INDEPENDENT FORCES

For scalar forces, r = 0, the matrix element is diagonal in L and S, implying that the spin-dependent part of the interaction V commutes with the total spin S of the two-particle state. The only independent two-particle spin-operators which have this property are 1 and $\underset{\sim}{\sigma}_1 \cdot \underset{\sim}{\sigma}_2$. Similarly there are only two independent isospin scalar operators, 1 and $\underset{\sim}{\tau}_1 \cdot \underset{\sim}{\tau}_2$. Thus the most general spin-isospin dependence of a scalar charge-independent force has at most four terms, for example

$$a_0 + a_\sigma \, \underset{\sim}{\sigma}_1 \cdot \underset{\sim}{\sigma}_2 + a_\tau \, \underset{\sim}{\tau}_1 \cdot \underset{\sim}{\tau}_2 + a_{\sigma\tau} (\underset{\sim}{\sigma}_1 \cdot \underset{\sim}{\sigma}_2)(\underset{\sim}{\tau}_1 \cdot \underset{\sim}{\tau}_2). \tag{A.67a}$$

Other linear combinations of these operators can of course be chosen, two popular versions being

$$A_{00} P_{00} + A_{01} P_{01} + A_{10} P_{10} + A_{11} P_{11} \tag{A.67b}$$

$$W + BP_\sigma - MP_\sigma P_\tau - HP_\tau . \tag{A.67c}$$

Here P_{ST} is a projection operator whose matrix elements in the two-particle states are diagonal, i.e.

$$\langle \tfrac{1}{2}\tfrac{1}{2}S';\tfrac{1}{2}\tfrac{1}{2}T' | P_{ST} | \tfrac{1}{2}\tfrac{1}{2}S';\tfrac{1}{2}\tfrac{1}{2}T' \rangle = \delta_{SS'} \delta_{TT'} .$$

Explicitly these operators are

$$P_{00} = \frac{1}{16} (1 - \underset{\sim}{\sigma}_1 \cdot \underset{\sim}{\sigma}_2)(1 - \underset{\sim}{\tau}_1 \cdot \underset{\sim}{\tau}_2)$$

$$P_{01} = \frac{1}{16} (1 - \underset{\sim}{\sigma}_1 \cdot \underset{\sim}{\sigma}_2)(3 + \underset{\sim}{\tau}_1 \cdot \underset{\sim}{\tau}_2)$$

$$P_{10} = \frac{1}{16} (3 + \underset{\sim}{\sigma}_1 \cdot \underset{\sim}{\sigma}_2)(1 - \underset{\sim}{\tau}_1 \cdot \underset{\sim}{\tau}_2)$$

$$P_{11} = \frac{1}{16} (3 + \underset{\sim}{\sigma}_1 \cdot \underset{\sim}{\sigma}_2)(3 + \underset{\sim}{\tau}_1 \cdot \underset{\sim}{\tau}_2)$$

and they project out the singlet-odd, singlet-even, triplet-even, and triplet-odd parts of the force respectively. The word singlet or triplet designates whether the spin state is antisymmetric (S=0) or symmetric (S=1). The second word odd or even refers to the parity of the L-value and designates whether the orbital space is antisymmetric or symmetric. Remember the two-particle state must be antisymmetric in the overall product of orbital, spin, and isospin spaces. Thus, specifying the symmetry in any two of these spaces fixes the symmetry in the third space.

In the third version, eqn (A.67c), of the spin-isospin part of the interaction, W, B, M, and H represent the Wigner (pure central), Bartlett (spin exchange), Majorana (space exchange), and Heisenberg (isospin exchange) parts of the force. The operators P_σ and P_τ are the spin-exchange and charge-exchange operators defined as

$$P_\sigma = \frac{1}{2}(1 + \underset{\sim}{\sigma}_1 \cdot \underset{\sim}{\sigma}_2)$$

$$P_\tau = \frac{1}{2}(1 + \underset{\sim}{\tau}_1 \cdot \underset{\sim}{\tau}_2)$$

such that P_σ has eigenvalues +1 and -1 for states symmetric (S=1) and antisymmetric (S=0) in the two spin coordinates. Explicitly

$$\langle \tfrac{1}{2}\tfrac{1}{2};S|P_\sigma|\tfrac{1}{2}\tfrac{1}{2};S\rangle = (-)^{S+1}$$

with a similar expression for the matrix element of P_τ. Using the version (A.67c) for the spin-isospin part of the force in eqn (A.66) and inserting in eqn (A.65), the two-body matrix element for scalar charge-independent interactions becomes

$$\langle j_1 j_2;JT|V|j_3 j_4;JT\rangle_A$$

$$= [(1+\delta_{12})(1+\delta_{34})]^{-\frac{1}{2}} \sum_{LS} \begin{bmatrix} \ell_1 & \ell_2 & L \\ \frac{1}{2} & \frac{1}{2} & S \\ j_1 & j_2 & J \end{bmatrix} \begin{bmatrix} \ell_3 & \ell_4 & L \\ \frac{1}{2} & \frac{1}{2} & S \\ j_3 & j_4 & J \end{bmatrix} \times$$

$$\times\ (W + B\,(-)^{1+S} + M\,(-)^{1+S+T} + H\,(-)^{T}) \times$$

$$\times\ \{\langle \ell_1\ell_2;L|V^{(0)}(\underset{\sim}{r})|\ell_3\ell_4;L\rangle + (-)^{1+\ell_3+\ell_4+L+S+T}\langle \ell_1\ell_2;L|V^{(0)}(\underset{\sim}{r})|\ell_4\ell_3;L\rangle\}. \tag{A.68}$$

Similar expressions for the other two versions (A.67a) and (A.67b) of the exchange force can be obtained using the transformations

$$\begin{pmatrix} W \\ B \\ M \\ H \end{pmatrix} = \begin{pmatrix} 1 & -1 & -1 & 1 \\ 0 & 2 & 0 & -2 \\ 0 & 0 & 0 & -4 \\ 0 & 0 & -2 & 2 \end{pmatrix} \begin{pmatrix} a_0 \\ a_\sigma \\ a_\tau \\ a_{\sigma\tau} \end{pmatrix} = \frac{1}{4} \begin{pmatrix} 1 & 1 & 1 & 1 \\ -1 & 1 & 1 & -1 \\ -1 & 1 & -1 & 1 \\ 1 & 1 & -1 & -1 \end{pmatrix} \begin{pmatrix} A_{00} \\ A_{01} \\ A_{10} \\ A_{11} \end{pmatrix}.$$

The usual normalization assumption is W+B+M+H = 1, which leads to normalization conditions of $a_0+a_\sigma-3a_\tau-3a_{\sigma\tau} = 1$ and $A_{01} = 1$ for the other variants.

The next step is to evaluate the orbital matrix elements contained in eqn (A.68). For this the Slater method has been developed. One starts by expressing $V^{(0)}(\underset{\sim}{r})$ in a series of Legendre polynomials,

$$V^{(0)}(\underset{\sim}{r}) = \sum_k v_k(r_1,r_2)P_k(\cos\omega), \tag{A.69}$$

where ω is the angle between $\underset{\sim}{r}_1$ and $\underset{\sim}{r}_2$. The Legendre polynomials may be expanded by the addition theorem

$$P_k(\cos\omega) = \frac{4\pi}{2k+1} Y_k(\hat{\underset{\sim}{r}}_1)\cdot Y_k(\hat{\underset{\sim}{r}}_2)$$

so that the direct matrix element can be written

$$\langle \ell_1\ell_2;L|V^{(0)}(r)|\ell_3\ell_4;L\rangle = \sum_k f_k\ F^k \tag{A.70}$$

where f_k is an integral over the angular variables,

$$f_k = \frac{4\pi}{2k+1}\langle \ell_1\ell_2;L|Y_k(\hat{\underset{\sim}{r}}_1)\cdot Y_k(\hat{\underset{\sim}{r}}_2)|\ell_3\ell_4;L\rangle\ i^{\ell_3+\ell_4-\ell_1-\ell_2}$$

$$= (-)^k \frac{4\pi}{\hat{k}} \begin{bmatrix} \ell_3 & k & \ell_1 \\ \ell_4 & k & \ell_2 \\ L & 0 & L \end{bmatrix} \langle \ell_1\|Y_k\|\ell_3\rangle\langle \ell_2\|Y_k\|\ell_4\rangle\ i^{\ell_3+\ell_4-\ell_1-\ell_2}$$

$$= (-)^{k+L} \frac{\hat{\ell}_1\hat{\ell}_2\hat{\ell}_3\hat{\ell}_4}{\hat{L}\hat{k}\hat{k}\hat{k}} \langle \ell_1 0\ell_3 0|k0\rangle\langle \ell_2 0\ell_4 0|k0\rangle\, U(\ell_1\ell_2\ell_3\ell_4;Lk)\, i^{\ell_3+\ell_4-\ell_1-\ell_2} \tag{A.71}$$

with $\hat{k} = (2k+1)^{1/2}$, and F^k is a radial integral,

$$F^k = \int r_1^{\,2}dr_1 \int r_2^{\,2}dr_2\ v_k(r_1,r_2)R_{\ell_1}(r_1)R_{\ell_2}(r_2)R_{\ell_3}(r_1)R_{\ell_4}(r_2). \tag{A.72}$$

Here $R_\ell(r)$ is the single-particle radial wavefunction. This function is defined in eqn (A.1); note, however, the use of the phase i^ℓ with the spherical harmonic, $Y_\ell(\hat{\underset{\sim}{r}})$. This has been incorporated so that the single-particle wavefunction has convenient time-reveral properties, and is the origin of the factor $i^{\ell_3+\ell_4-\ell_1-\ell_2}$ in the definition of f_k (eqn (A.71)).

Note that the angle integral f_k has the following symmetries

$$f_k(\ell_1\ell_2\ell_3\ell_4;L) = f_k(\ell_3\ell_4\ell_1\ell_2;L) = f_k(\ell_2\ell_1\ell_4\ell_3;L),$$

and the sum over k is limited to $k \leqslant \ell_1+\ell_3$ and $k \leqslant \ell_2+\ell_4$ with the condition that $\ell_1+\ell_3+k$ and $\ell_2+\ell_4+k$ be even. This implies that the sum over k is over either even or odd multipoles. The exchange matrix element can be handled in the same way:

$$\langle \ell_1\ell_2;L|V^{(0)}(\underset{\sim}{r})|\ell_4\ell_3;L\rangle = \sum_k g_k\ G^k$$

where g_k and G_k are obtained from f_k and F^k on interchanging ℓ_3 and ℓ_4. Thus the contribution of the orbital matrix elements to eqn (A.68) can be cast in the form

$$\sum_k f_kF^k + (-)^{1+\ell_3+\ell_4+L+S+T} \sum_k g_kG^k \tag{A.73}$$

where it remains to specify how the Slater integrals are obtained. When harmonic oscillator wavefunctions are used, the Slater integrals can often be obtained analytically. For the case of Gaussian interactions, Ford and Konopinski (1959) have expressed these integrals in terms of a few finite double sums. More recently Wegner (1970) gave an improved algorithm involving only single sums and applicable to Gaussian, Yukawa, Coulomb, and zero-range interactions. We shall not give any details here; however, the case of zero-range interactions is particularly simple so we shall treat this case explicitly.

A.10 ZERO-RANGE INTERACTIONS

For a delta-function interaction $V^{(0)}(\underset{\sim}{r}) = V_0\delta(\underset{\sim}{r}_1-\underset{\sim}{r}_2)$ the radial coefficients in eqn (A.69), $v_k(r_1,r_2)$, are simply

$$v_k(r_1,r_2) = V_0\,\frac{2k+1}{4\pi r_1^2}\,\delta(r_1-r_2)$$

and the Slater integrals F^k and G^k have a trivial dependence on the multipolarity k, viz.

$$F_k = G_k = (2k+1)\,\frac{V_0}{4\pi}\int R_{\ell_1}(r_1)R_{\ell_2}(r_1)R_{\ell_3}(r_1)R_{\ell_4}(r_1)dr_1$$

$$\equiv (2k+1)I(\ell_1\ell_2\ell_3\ell_4).$$

Referring now to eqn (A.70), the sum over k can be performed and the direct matrix element reduces to

$$\langle\ell_1\ell_2;L|V_0\;\delta(\underset{\sim}{r}_1-\underset{\sim}{r}_2)|\ell_3\ell_4;L\rangle$$
$$= i^{\ell_3+\ell_4-\ell_1-\ell_2}\,\frac{\hat{\ell}_1\hat{\ell}_2\hat{\ell}_3\hat{\ell}_4}{\hat{L}\,\hat{L}}\langle\ell_1 0\ell_2 0|L0\rangle\,\langle\ell_3 0\ell_4 0|L0\rangle\,I(\ell_1\ell_2\ell_3\ell_4). \tag{A.74}$$

Notice that this expression is invariant with respect to interchanging ℓ_3 and ℓ_4, so that the exchange matrix element exactl equals the direct matrix element. Inserting eqn (A.74) back into (A.68), the antisymmetric two-body matrix element is now

$$\langle j_1j_2;JT|V_0\delta(\underset{\sim}{r}_1-\underset{\sim}{r}_2)|j_3j_4;JT\rangle_A$$
$$= [(1+\delta_{12})(1+\delta_{34})]^{-\frac{1}{2}}\,i^{\ell_3+\ell_4-\ell_1-\ell_2}[1+(-)^{1+S+T}]\;[(W+M)\;+\;(-)^T(B+H)]\;\times$$
$$\times\sum_L\frac{\hat{\ell}_1\hat{\ell}_2\hat{\ell}_3\hat{\ell}_4}{\hat{L}\,\hat{L}}\langle\ell_1 0\ell_2 0|L0\rangle\,\langle\ell_3 0\ell_4 0|L0\rangle\begin{bmatrix}\ell_1 & \ell_2 & L\\ \frac{1}{2} & \frac{1}{2} & S\\ j_1 & j_2 & J\end{bmatrix}\begin{bmatrix}\ell_3 & \ell_4 & L\\ \frac{1}{2} & \frac{1}{2} & S\\ j_3 & j_4 & J\end{bmatrix}I(\ell_1\ell_2\ell_3\ell_4) \tag{A.75}$$

where it is noticed that the expression vanishes unless S+T = odd, that is, only singlet-even and triplet-even states have non-zero matrix elements for delta-function forces. It is possible to simplify eqn (A.75) still further by writing out each term in the sum over L, L=J, J±1, explicitly, and then replacing the product of a Clebsch-Gordan and recoupling coefficient by a single Clebsch-Gordan coefficient using eqns (A.25). The final expression for the antisymmetric two-body matrix element, written out separately for T=0 and T=1 states, then becomes

$$\langle j_1 j_2; J\ T{=}1 | V_0 \delta(\underset{\sim}{r}_1 - \underset{\sim}{r}_2) | j_3 j_3; J\ T{=}1\rangle_A$$

$$= [(1+\delta_{12})(1+\delta_{34})]^{-1/2}\ i^{\ell_3+\ell_4-\ell_1-\ell_2} [(W+M)-(B+H)] \times$$

$$\times (-)^{\ell_1+\ell_3} \tfrac{1}{2}(1+(-)^x)\ a(j_1 j_2 J)\ a(j_3 j_4 J)\ I(\ell_1 \ell_2 \ell_3 \ell_4) \quad \text{(A.76a)}$$

$$\langle j_1 j_2; J\ T{=}0 | V_0 \delta(\underset{\sim}{r}_1 - \underset{\sim}{r}_2) | j_3 j_4; J\ T{=}0\rangle_A$$

$$= [(1+\delta_{12})(1+\delta_{34})]^{-1/2}\ i^{\ell_3+\ell_4-\ell_1-\ell_2} [(W+M)\ +\ (B+H)]\ (-)^{\ell_1+\ell_3} \times$$

$$\times \{b(j_1 j_2 J) b(j_3 j_4 J) + \tfrac{1}{2}(1-(-)^x) a(j_1 j_2 J) a(j_3 j_4 J)\} I(\ell_1 \ell_2 \ell_3 \ell_4)$$

$$\text{(A.76b)}$$

where $x = \ell_1+\ell_2+J$ and the coefficients $a(j_1 j_2 J)$ and $b(j_1 j_2 J)$ were defined in eqn (A.23).

A.11 NON-SCALAR CHARGE-INDEPENDENT FORCES

Returning to the more general expressions (A.65) and (A.66), the terms with r=1 and 2 are referred to as the vector and tensor forces respectively. The only vector term linear in the spins and momenta satisfying certain symmetry requirements is

$$V_{LS}(r)\ \underset{\sim}{L} \cdot \underset{\sim}{S} \quad \text{(A.77)}$$

where $\underset{\sim}{S} = \frac{1}{2}(\underset{\sim}{\sigma}_1 + \underset{\sim}{\sigma}_2)$ is the total spin and $\underset{\sim}{L} = \frac{1}{2}(\underset{\sim}{r}_1 - \underset{\sim}{r}_2) \wedge (\underset{\sim}{p}_1 - \underset{\sim}{p}_2)$ is the relative orbital angular momentum of the pair. This is called the vector spin-orbit force and is only non-zero in triplet states (S=1).

There are a number of tensor terms which can be constructed; however, the only one which does not involve momentum operators, i.e. derivative operators, has the form $3\, V_T(r) r^{-2} \underset{\sim}{R}^{(2)} \cdot \underset{\sim}{S}^{(2)}$ where $\underset{\sim}{R}^{(2)} = \{(\underset{\sim}{r}_1 - \underset{\sim}{r}_2) \times (\underset{\sim}{r}_1 - \underset{\sim}{r}_2)\}^{(2)}$ and $\underset{\sim}{S}^{(2)} = \{\underset{\sim}{\sigma}_1 \times \underset{\sim}{\sigma}_2\}^{(2)}$ are second-rank tensors obtained by vector coupling two first-rank tensors. This tensor force can also be written as

$$V_T(r) S_{12} = 3\, V_T(r) r^{-2} \underset{\sim}{R}^{(2)} \cdot \underset{\sim}{S}^{(2)}$$

$$= V_T(r)\{3(\underset{\sim}{\sigma}_1 \cdot \hat{\underset{\sim}{r}})(\underset{\sim}{\sigma}_2 \cdot \hat{\underset{\sim}{r}}) - (\underset{\sim}{\sigma}_1 \cdot \underset{\sim}{\sigma}_2)\} \qquad (A.78)$$

where $\hat{\underset{\sim}{r}}$ is a vector of unit length, $\underset{\sim}{r} = \underset{\sim}{r}_1 - \underset{\sim}{r}_2$. This term is again non-zero only in triplet states. Such a tensor force can be derived from a static, one-pion exchange model for nuclear forces, yet the deduced strength for this force is too weak to explain the low-energy scattering data. It is known, however, that the one-pion exchange potential is reliable only at long range and that multiple-pion exchange processes are important at shorter range. These latter processes can generate the rank-one spin-orbit forces. Furthermore a relatvistic correction to the one-pion exchange potential gives rise to a quadratic spin-orbit term of the form

$$V_{LL}(r) L_{12} = V_{LL}(r)\{(\underset{\sim}{\sigma}_1 \cdot \underset{\sim}{\sigma}_2)\underset{\sim}{L}^2 - \tfrac{1}{2}[(\underset{\sim}{\sigma}_1 \cdot \underset{\sim}{L})(\underset{\sim}{\sigma}_2 \cdot \underset{\sim}{L}) + (\underset{\sim}{\sigma}_2 \cdot \underset{\sim}{L})(\underset{\sim}{\sigma}_1 \cdot \underset{\sim}{L})]\}. \qquad (A.79)$$

Phenomenological potentials have been constructed from such scalar, spin-orbit, tensor, and quadratic spin-orbit terms. Typical of these is the Hamada-Johnston (Hamada and Johnston 1962) potential which contains the one-pion exchange potential plus a parameterized mixture of shorter-range potentials of these types. Altogether, 28 parameters were used to fit all available two-body data below 315 MeV.

To evaluate the matrix elements of the vector and tensor forces (eqns (A.77), (A.78), and (A.79)) poses some difficulties in a single-particle representation due to the presence of the relative orbital angular momentum operator, $\underset{\sim}{L}$, in the spin-orbit forces. The solution, therefore, is to transform the two-particle wavefunction from a single-particle representation to a relative and centre-of-mass coordinate system, which enables an integration over the centre-of-mass coordinate to be performed immediately. The resulting matrix element now depends only on the relative coordinate, and the operator $\underset{\sim}{L}$ can be handled in a trivial manner. The only problem is to find the coefficients of the transformation from the one coordinate system to the other. In the case of harmonic oscillator wavefunctions, the transformation has been well studied and the coefficients are available either in tabular form (Brody and Moshinsky 1960), or in terms of recursion relations (Baranger and Davies 1966), or in a general closed form (Talman 1970). For other single-particle wavefunctions, they can always be expanded in terms of harmonic oscillator functions and the transformation coefficients found in that way. However, for the case of Saxon-Woods functions, a faster algorithm has been given by Bayman and Kallio (1967).

We will consider only harmonic oscillator wavefunctions for which the transformation is written

$$|n_1\ell_1,n_2\ell_2;L\rangle = \sum_{n\ell NL} \langle n\ell,\mathcal{NL};L|n_1\ell_1,n_2\ell_2;L\rangle\, |n\ell,\mathcal{NL};L\rangle \qquad (A.80)$$

where the sum is restricted by an angular momentum coupling requirement

$$\underset{\sim}{\ell}_1 + \underset{\sim}{\ell}_2 = \underset{\sim}{L} = \underset{\sim}{\ell} + \underset{\sim}{\mathcal{L}}$$

and by a conservation of energy requirement

$$2n_1+\ell_1 + 2n_2 + \ell_2 = 2n + \ell + 2\mathcal{N} + \mathcal{L}.$$

The coefficients, or transformation brackets as they have been called, have the orthogonality property

$$\sum_{n\ell\mathcal{NL}} \langle n\ell,\mathcal{NL},L|n_1\ell_1,n_2\ell_2;L\rangle\, \langle n\ell,\mathcal{NL},L|n_1'\ell_1',n_2'\ell_2';L\rangle = \delta_{n_1n_1'}\delta_{\ell_1\ell_1'}\delta_{n_2n_2'}\delta_{\ell_2\ell_2'}$$

and the following symmetry relations

$$\langle n\ell, \mathcal{NL}; L|n_1\ell_1, n_2\ell_2; L\rangle = (-)^{\mathcal{L}-L} \langle n\ell, \mathcal{NL}; L|n_2\ell_2, n_1\ell_1; L\rangle$$
$$= (-)^{\ell_1 - L} \langle \mathcal{NL}, n\ell; L|n_1\ell_1, n_2\ell_2; L\rangle$$
$$= (-)^{\ell_2 + \mathcal{L}} \langle n_1\ell_1, n_2\ell_2; L|n\ell, \mathcal{NL}; L\rangle . \quad \text{(A.81)}$$

Returning to eqn (A.65) for the antisymmetric two-body matrix elements of a central force V, and inserting eqn (A.80), recoupling the relative orbital angular momentum ℓ to the spin S to form $\mathcal{J}$, and integrating over the centre-of-mass coordinate gives

$$\langle j_1 j_2; JT|V|j_3 j_4; JT\rangle_A$$

$$= [(1+\delta_{12})(1+\delta_{34})]^{-1/2} \sum_{LL'S} \begin{bmatrix} \ell_1 & \ell_2 & L \\ \frac{1}{2} & \frac{1}{2} & S \\ j_1 & j_2 & J \end{bmatrix} \begin{bmatrix} \ell_3 & \ell_4 & L' \\ \frac{1}{2} & \frac{1}{2} & S \\ j_3 & j_4 & J \end{bmatrix} i^{\ell_3+\ell_4-\ell_1-\ell_2} \times$$

$$\times \sum_{n\ell n'\ell'\mathcal{NL}} (-)^{\ell-\ell'-L+L'} \langle n\ell, \mathcal{NL}; L|n_1\ell_1, n_2\ell_2; L\rangle \langle n'\ell', \mathcal{NL}, L'|n_3\ell_3, n_4\ell_4; L'\rangle \times$$

$$\times [1-(-)^{\ell+S+T}] \sum_{\mathcal{J}} U(\mathcal{L}\ell JS; L\mathcal{J}) U(\mathcal{L}\ell' JS; L'\mathcal{J}) \langle n\ell, S; \mathcal{J}|V|n'\ell', S; \mathcal{J}\rangle . \quad \text{(A.82)}$$

The exchange matrix element has been handled using the symmetry relations (A.81), and assuming that V conserves parity. It is found that S must be diagonal and ℓ+S+T = odd. It remains to evaluate the matrix element $\langle n\ell, S; \mathcal{J}|V|n'\ell', S; \mathcal{J}\rangle$ for spin-orbit, tensor, and quadratic spin-orbit forces. The spin orbit force is simple:

$$\langle n\ell, S, \mathcal{J}|V_{LS}(r)\underset{\sim}{L}\cdot\underset{\sim}{S}|n'\ell', S, \mathcal{J}\rangle$$

$$= \frac{1}{2} \langle \ell S\ \mathcal{J}|\underset{\sim}{J}^2 - \underset{\sim}{L}^2 - \underset{\sim}{S}^2|\ell' S\ \mathcal{J}\rangle\ I_{LS}(n\ell n'\ell')$$

$$= \frac{1}{2} [\mathcal{J}(\mathcal{J}+1) - \ell(\ell+1) - 2] \delta_{\ell,\ell'} \delta_{S,1}\ I_{LS}(n\ell n'\ell'). \quad \text{(A.83)}$$

The tensor force becomes

$$\langle n\ell, S, \mathcal{J}|\, 3V_T(r)r^{-2}\ \underset{\sim}{R}^{(2)}\cdot\underset{\sim}{S}^{(2)}\, |n'\ell', S', \mathcal{J}\rangle$$

$$= 3\ U(S\mathcal{J}2\ell';\ell S')\,\langle \ell \| (\hat{\underset{\sim}{r}}\times\hat{\underset{\sim}{r}})^{(2)} \| \ell'\rangle\,\langle S\|(\underset{\sim}{\sigma}_1\times\underset{\sim}{\sigma}_2)^{(2)}\|S'\rangle\ \delta_{SS'}\ I_T(n\ell n'\ell') \quad (A.84)$$

where

$$\langle \ell \| (\hat{\underset{\sim}{r}}\times\hat{\underset{\sim}{r}})^{(2)} \| \ell'\rangle = \sum_{\ell''} U(11\ell\ell';2\ell'')\,\langle \ell\|\hat{\underset{\sim}{r}}\|\ell''\rangle\,\langle \ell''\|\hat{\underset{\sim}{r}}\|\ell'\rangle$$

$$= (-)^{\ell}\left[\frac{2(2\ell'+1)}{15}\right]^{1/2}\ \langle \ell 0 \ell' 0|20\rangle$$

$$\langle \tfrac12\tfrac12 S\|(\underset{\sim}{\sigma}_1\times\underset{\sim}{\sigma}_2)^{(2)}\|\tfrac12\tfrac12 S\rangle = \begin{bmatrix} \tfrac12 & 1 & \tfrac12 \\ \tfrac12 & 1 & \tfrac12 \\ S & 2 & S \end{bmatrix}\langle \tfrac12\|\underset{\sim}{\sigma}_1\|\tfrac12\rangle\langle \tfrac12\|\underset{\sim}{\sigma}_2\|\tfrac12\rangle$$

$$= \left(\frac{20}{3}\right)^{\frac12}\ \delta_{S,1}$$

Finally the quadratic spin-orbit force can be recast in the form $L_{12} = \underset{\sim}{L}^2[\delta_{\ell\mathcal{J}} + (\underset{\sim}{\sigma}_1\cdot\underset{\sim}{\sigma}_2)] - (\underset{\sim}{L}\cdot\underset{\sim}{S})^2$ which is then trivially evaluated:

$$\langle n\ell S\mathcal{J}|V_{LL}(r)L_{12}|n'\ell'S'\mathcal{J}\rangle$$

$$= [\delta_{\ell\ell'}\delta_{SS'}(\delta_{\ell,\mathcal{J}}-3\delta_{S,0}+\delta_{S,1})\ell(\ell+1) - \langle \ell S\mathcal{J}|\underset{\sim}{L}\cdot\underset{\sim}{S}|\ell S\mathcal{J}\rangle^2]I_{LL}(n\ell n'\ell') \quad (A.85)$$

In eqns (A.83), (A.84), and (A.85), the $I(n\ell n'\ell')$ are the appropriate radial integrals,

$$I_x(n\ell n'\ell') = \int_0^\infty R_{n\ell}(r)\ V_x(r)\ R_{n'\ell'}(r)r^2dr.$$

A.12 MULTIPOLE—MULTIPOLE FORCES

So far we have only considered potentials constructed from scalar products of spherical tensors in orbital and spin space. This is a natural construction to use in L-S coupling, but a little unwieldy for j-j coupling. When one is only interested in modelistic forms of V, then for j-j coupling it might be preferable to construct potentials of the type

$$V = \left[\sum_k U^{(k)}(\underset{\sim}{r}_1,\underset{\sim}{S}_1)\cdot U^{(k)}(\underset{\sim}{r}_2,\underset{\sim}{S}_2)\right] T^{(0)}(\tau_1,\tau_2) \tag{A.86}$$

where $U^{(k)}(\underset{\sim}{r}_1,\underset{\sim}{S}_1)$ is a spherical tensor of rank k formed from the position and spin coordinates of particle 1 and $U^{(k)}(\underset{\sim}{r}_2,\underset{\sim}{S}_2)$ is an equivalent tensor in the coordinates of particle 2. That the two tensors should have the same form comes from the symmetry requirement $V(1,2) = V(2,1)$. Note that the isospin dependence is as before, namely some linear combination of the scalars 1 and $\underset{\sim}{\tau}_1\cdot\underset{\sim}{\tau}_2$. Potentials of the type (A.86) are called multipole-multipole forces. One example, which has been extremely popular in nuclear physics, is

$$V = \sum_k r_1^{\,k}\, r_2^{\,k}\, P_k(\cos\omega)$$

$$= \sum_k r_1^{\,k}\, r_2^{\,k}\, \frac{4\pi}{2k+1}\, Y_k(\hat{\underset{\sim}{r}}_1)\cdot Y_k(\hat{\underset{\sim}{r}}_2), \tag{A.87}$$

the k=2 member being the quadrupole-quadrupole force. In this case $U^{(k)} = r_1^{\,k}\,(4\pi/2k+1)^{\frac{1}{2}}\,Y_k$ and is just the one-body operator describing electric γ-transitions of multipole k. Alternatively one could choose $U^{(k)}$ to be the one-body operator describing magnetic transitions. By constructing potentials in this way, and hoping that they have some approximate correspondence with the nuclear force operable in such a model calculation, one can trace the collective properties of certain states. For example, the decay of particle-hole states, discussed in chapter 4, exhibits enhancements and hindrances to γ-transitions which can be related to the behaviour of a certain multipole in the nuclear force.

To evaluate two-body matrix elements of the potential eqn (A.86), one starts by writing the matrix element in an unnormalized, unsymmetrized form

$$\langle j_1 j_2;JT|V|j_3 j_4;JT\rangle_A =$$

$$= [(1+\delta_{12})(1+\delta_{34})]^{-1/2}\{\langle j_1 j_2;JT|V|j_3 j_4;JT\rangle + $$
$$+ (-)^{j_3+j_4+J+T}\langle j_1 j_2;JT|V|j_4 j_3;JT\rangle\} \qquad (A.88)$$

and then uses the factorization theorem, eqn (A.37), to write the direct matrix element as

$$\langle j_1 j_2;JT|V|j_3 j_4;JT\rangle = \sum_k (-)^k (2k+1)^{1/2} \begin{bmatrix} j_3 & k & j_1 \\ j_4 & k & j_2 \\ J & 0 & J \end{bmatrix} \langle j_1\|U^{(k)}\|j_3\rangle \times$$
$$\times \langle j_2\|U^{(k)}\|j_4\rangle \langle \tfrac{1}{2}\tfrac{1}{2};T\|T^{(0)}\|\tfrac{1}{2}\tfrac{1}{2};T\rangle . \qquad (A.89)$$

A similar expression for the exchange matrix element is obtained on interchanging j_3 and j_4 in eqn (A.89). We choose to write the isospin dependence in the form $T^{(0)} = a_0+a_\tau \underset{\sim}{\tau}_1\cdot\underset{\sim}{\tau}_2$, then the final expression for the two-body matrix element is

$$\langle j_1 j_2;JT|V|j_3 j_3;JT\rangle_A$$
$$= [(1+\delta_{12})(1+\delta_{34})]^{-1/2} \sum_k (-)^{j_1+j_4} \frac{\hat{j}_1\hat{j}_2}{\hat{J}\,\hat{k}} [a_0-a_\tau(1+2(-)^T)] \times$$
$$\times \{(-)^J\, U(j_3 j_1 j_4 j_2;kJ)\, \langle j_1\|U^{(k)}\|j_3\rangle \langle j_2\|U^{(k)}\|j_4\rangle +$$
$$+ (-)^{T+1}\, U(j_3 j_2 j_4 j_1;kJ) \langle j_1\|U^{(k)}\|j_4\rangle \langle j_2\|U^{(k)}\|j_3\rangle\}. \qquad (A.90)$$

The reduced matrix elements of $U^{(k)}$ are just standard one-body matrix elements and are easily evaluated.

Notice that although the direct and the exchange terms are each individually separable in the coordinates of particles 1 and 2, the combined result of direct plus exchange terms is not. Occasionally, however, the total matrix element is separable when for some reason either the direct or the exchange term itself vanishes. One example of this occurs for negative parity two-particle states and a multipole-multipole force constructed from electric one-body operators. The parity selection rule in the direct matrix element is that $\ell_1+\ell_3+k$ = even, while in the exchange matrix element it is $\ell_1+\ell_4+k$ =

even. Since for negative parity states ℓ_3 has the opposite parity to ℓ_4, one of the direct or the exchange matrix elements will vanish. Thus the direct matrix element contains only even multipoles of k (assuming $(-)^{\ell_1} = (-)^{\ell_3}$), and the exchange matrix element contains only odd multipoles. For positive-parity states both direct and exchange matrix elements vanish for odd multipoles, and both are present for even multipoles. If the multipole-multipole force is constructed out of magnetic one-body operators, an analogous set of selection rules is obtained, identical to those above except that the words even and odd are interchanged.

A.13 PARTICLE-HOLE MATRIX ELEMENTS

In chapter 4, the relationship between a particle-hole and a particle-particle matrix element was derived (eqn (4.13)). We repeat the formula here, inserting all the isospin labels:

$$\langle j_3^{-1} j_2; J'T'|V|j_1^{-1} j_4; J'T'\rangle_A$$
$$= -\sum_{JT} (-)^{j_1+j_2+j_3+j_4} \frac{\hat{J}\hat{T}}{\hat{J}'\hat{T}'} U(j_3 j_2 j_4 j_1; J'J) U(\tfrac{1}{2}\tfrac{1}{2}\tfrac{1}{2}\tfrac{1}{2}; T'T) \times$$
$$\times \langle j_1 j_2; JT|V|j_3 j_4; JT\rangle_A . \qquad \text{(A.91)}$$

where the two-body matrix element on the right-hand side is antisymmetrized but *not* normalized. Thus once a set of particle-particle matrix elements has been obtained, it is quite trivial to find an equivalent set of particle-hole matrix elements. In certain cases, for example, multipole-multipole forces and zero-range forces, the sum over J can be performed algebraically. We will consider just these two examples.

For the multipole-multipole force, eqn (A.90), the appropriate J and T sums are straightforward:

$$\sum_J (-)^J U(j_3 j_1 j_4 j_2; kJ) U(j_3 j_2 j_4 j_1; J'J) = (-)^{j_1+j_2+j_3+j_4+k+J'} U(j_3 j_1 j_2 j_4; kJ')$$

$$\sum_J U(j_3 j_2 j_4 j_1; kJ) U(j_3 j_2 j_4 j_1; J'J) = \delta_{k,J'}$$

$$\sum_T (-)^T \frac{\hat{T}}{\hat{T}'} U(\tfrac{1}{2}\tfrac{1}{2}\tfrac{1}{2}\tfrac{1}{2};T'T) = -2\delta_{T',0}$$

$$\sum_T \frac{\hat{T}}{\hat{T}'} U(\tfrac{1}{2}\tfrac{1}{2}\tfrac{1}{2}\tfrac{1}{2};T'T) = 1,$$

and the particle-hole matrix element becomes

$$\langle j_3^{-1} j_2;J'T'|V|j_1^{-1} j_4;J'T'\rangle_A = -\sum_k \hat{j}_1 \hat{j}_2/\hat{J}'\hat{k} \times$$
$$\times \{[a_0+a_\tau(1+2(-)^{T'})](-)^{j_1+j_4+k+J'} U(j_3 j_1 j_2 j_4;kJ')\langle j_1\|U^{(k)}\|j_3\rangle\langle j_2\|U^{(k)}\|j_4\rangle +$$
$$+ [a_0+a_\tau+(-)^{T'}(a_0-a_\tau)]\delta_{k,J'}(-)^{j_2+j_3}\langle j_1\|U^{(k)}\|j_4\rangle\langle j_2\|U^{(k)}\|j_3\rangle\}. \quad (A.92)$$

Notice that the contribution from the exchange part of the multipole-multipole matrix element is particularly simple and separates into a product of two one-body matrix elements connecting the particle-hole states with the closed shell. This can be seen from the following relations

$$\langle j_3^{-1} j_2;k\|U^{(k)}\|0\rangle = (-)^{j_2+j_3-k}\frac{\hat{j}_2}{\hat{k}}\langle j_2\|U^{(k)}\|j_3\rangle$$

$$\langle j_1^{-1} j_4;k\|U^{(k)}\|0\rangle = (-)^{k+p+2j_1}\frac{\hat{j}_1}{\hat{k}}\langle j_1\|U^{(k)}\|j_4\rangle$$

where in the last expression we have used the Hermitian property of reduced matrix elements:

$$\hat{j}_4\langle j_4\|U^{(k)}\|j_1\rangle = (-)^{j_1-j_4+p}\hat{j}_1\langle j_1\|U^{(k)}\|j_4\rangle,$$

the phase $(-)^p$ relating to the Hermitian property of the tensor operator, viz. $U_m^{(k)} = (-)^{p-m}U_{-m}^{(k)\dagger}$. For electric operators, p=k and for magnetic operators, p = k+1. Thus the particle-hole matrix element arising from the *exchange* part of the multipole-multipole force *only* is

$$\langle j_3^{-1} j_2;J'T'|V|j_1^{-1} j_4;J'T'\rangle_A = (-)^p[a_0+a_\tau+(-)^{T'}(a_0-a_\tau)]\delta_{k,J'}$$
$$\times\langle j_3^{-1} j_2;k\|U^{(k)}\|0\rangle\langle j_1^{-1} j_4;k\|U^{(k)}\|0\rangle \quad (A.93)$$

which is exactly the schematic model factorization discussed in chapter 4.

For zero-range forces the algebra follows in an analogous manner. The J-dependence of the particle-particle matrix elements (A.76) is contained in factors of the type: $a(j_1j_2J)a(j_3j_4J)$ and $b(j_1j_2J)b(j_3j_4J)$ where the functions a and b, defined in eqn (A.23), are essentially Clebsch-Gordan coefficients. The sum over J of these factors with the U-coefficient in (A.91) is just a standard recoupling (see eqn (A.26)) and the result is expressed in terms of either $a(j_1j_4J')a(j_3j_2J')$ or $b(j_1j_3J')b(j_3j_2J')$. We will not persevere with the details here, but merely quote the final result for the particle-hole matrix element as

$$\langle j_3^{-1}j_2;J'T'|V|j_1^{-1}j_4;J'T'\rangle_A$$

$$= \lambda_{T'}\, a(j_1j_4J')a(j_3j_2J') + \mu_{T'}\, b(j_1j_4J')b(j_3j_2J'). \tag{A.94}$$

Here the coefficients λ and μ depend on the strength and exchange mixture of the zero-range interaction and the parity and isospin of the particle-hole state. Specifically

$$\lambda_{T'=0} = \{\tfrac{1}{2}(W+M)[1+2(-)^y] + \tfrac{1}{2}(B+H)[1-(-)^y]\} \times$$
$$\times\, I(\ell_1\ell_2\ell_3\ell_4)i^{\ell_3+\ell_4-\ell_1-\ell_2}$$

$$\lambda_{T'=1} = \{-\tfrac{1}{2}(W+M) - \tfrac{1}{2}[1+(-)^y](B+H)\}I(\ell_1\ell_2\ell_3\ell_4)i^{\ell_3+\ell_4-\ell_1-\ell_2}$$

$$\mu_{T'=0} = \{-\tfrac{1}{2}(W+M) + (B+H)\}\, I(\ell_1\ell_2\ell_3\ell_4)i^{\ell_3+\ell_4-\ell_1-\ell_2}$$

$$\mu_{T'=1} = -\tfrac{1}{2}(W+M)I(\ell_1\ell_2\ell_3\ell_4)i^{\ell_3+\ell_4-\ell_1-\ell_2}$$

where $y = \ell_1+\ell_4+J'$ and $I(\ell_1\ell_2\ell_3\ell_4)$ is the radial integral

$$I(\ell_1\ell_2\ell_3\ell_4) = \frac{V_0}{4\pi}\int R_{\ell_1}(r)R_{\ell_2}(r)R_{\ell_3}(r)R_{\ell_4}(r)dr.$$

Here V_0 is the strength of the zero-range interaction (negative for attractive forces). This result, eqn (A.94), again exhibits the schematic model factorization, but one further assumption is required, namely that the radial integrals themselves are factorizable in the same way: $I(\ell_1\ell_2\ell_3\ell_4) = I(\ell_1\ell_4)I(\ell_2\ell_3)$. This rarely can be justified, but one plausible approximation is to replace $I(\ell_1\ell_2\ell_3\ell_4)$ by some constant average value independent of the orbitals ℓ_1, ℓ_2, ℓ_3, and ℓ_4.

The functions a and b in eqn (A.94) can be related to the reduced matrix elements of a particle-hole state decaying to the closed shell via some one-body operator. Thus eqn (A.94) is entirely analogous to the result obtained with the multipole-multipole force. The consequences for the properties of particle-hole states are discussed in chapter 4.

APPENDIX B

B.1 CLOSED-SHELL DIAGRAMS

Listed below is a complete set of rules for interpreting a Feynman-Goldstone diagram. These rules are essentially those listed by Brandow (1967) from Appendix B of his review article, but with a few additional remarks and some clarifying examples. We distinguish between 'closed-shell diagrams' and 'open-shell diagrams'. 'Closed-shell' diagrams have no external lines, i.e. all lines are joined up with no free ends. All three diagrams in Fig. 1.5 are of this type. On the other hand, the diagrams in Fig. 1.6 all have one free particle line in the initial state and one free particle line in the final state, and these are examples of open-shall diagrams.

We list first the rules for closed-shell diagrams and then discuss the modifications required for open-shell diagrams.

(1) The overall sign factor of the diagram is $(-)^{\ell+h}$ where ℓ is the number of closed loops and h the number of downgoing or hole line segments. For example, two very common components are illustrated in Fig. B.1 (a) and (b). These both contribute a plus sign to the overall sign factor since they both comprise one closed loop and one hole line, whereas diagram (c) of Fig. B.1 has an overall minus sign coming from its two closed loops and three hole line segments. A closed loop is any continuous closed path of particle or hole line segments passing through any number of vertices; however, no part of this path may be made up from dashed, i.e. interaction, line segments. Thus diagram (d) of Fig. B.1 has one closed loop, comprising lines imjknp taken in that order.

(2) Include a factor of $(\frac{1}{2})$ for each 'equivalent pair' of lines. Two lines form an equivalent pair if they (a) both begin at the same interaction, (b) both end at the same interaction, and (c) both go in the same direction. For example in

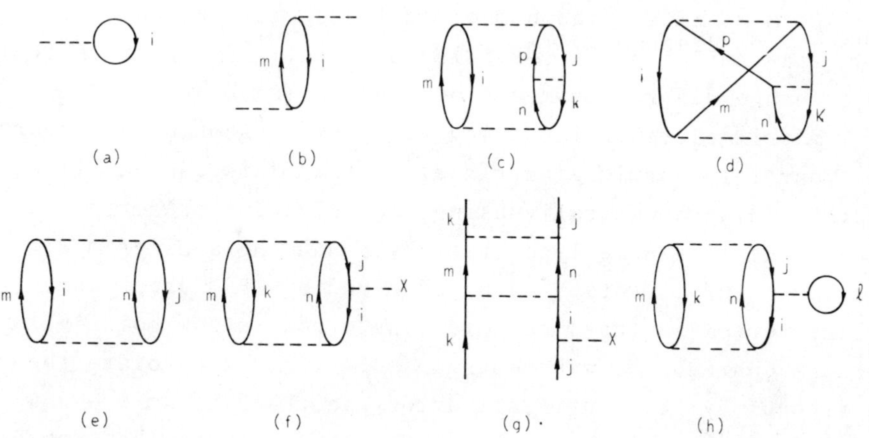

FIG. B.1. Illustrative closed-shell diagrams. Diagrams (a) and (b) are examples of closed loops. Diagram (c) has two closed loops and three hole lines. Diagram (d) has one closed loop. Diagram (e) has two sets of equivalent pairs of lines. To help in the determination of the correct two-body matrix elements, hole lines in diagram (f) have been cut and rotated through 180° to produce diagram (g), as discussed in the text.

diagram (e) of Fig. B.1 the appropriate factor is $\frac{1}{4}$ since particle lines m and n and hole lines i and j are equivalent.

(3) For each intermediate state include an energy denominator given by the sum of all hole line energies minus the sum of all particle line energies. If the diagram has n interaction lines there will be (n-1) energy denominators. For example, the energy denominator for diagram (e) in Fig. B.1 is $\varepsilon_i + \varepsilon_j - \varepsilon_m - \varepsilon_n$.

(4) Each dashed line segment with a vertex at one end and a cross at the other end represents the matrix element of a one-body operator. Each dashed line segment with a vertex at both ends represents an antisymmetrized matrix element of a two-body operator, i.e. a 'direct-minus-exchange' matrix element. Only one diagram is drawn from the set of all those obtained from each other when 'direct' interactions are replaced by 'exchange' interactions or vice versa. All other members of this set are ignored. It is usually convenient to choose a diagram with the maximum number of closed loops; such

a diagram minimizes the number of lines crossing each other. Thus diagram (d) of Fig. B.1 is not usually drawn since it is an exchange diagram formed from diagram (c).

Matrix elements involving hole states should be converted into particle-particle matrix elements. This can easily be achieved diagrammatically using the following prescription. Make a cut in a hole line, thereby producing a diagram with two loose ends. Rotate each loose end by 180° about an appropriate vertex so that the hole line arrow is now pointing upwards. Continue to make cuts in hole lines and rotate the loose ends by 180° until all arrows in the diagram are now pointing upwards. The original Goldstone diagram has now been converted into a Rajaraman diagram; this procedure was first introduced by Rajaraman (1963*b*) in connection with three-body clusters in nuclear matter. As an example, diagram (f) is converted into diagram (g) of Fig. B.1 by making cuts in hole lines k and j. It is now a simple matter to read off the appropriate matrix elements from the Rajaraman diagram. For a one-body operator the matrix element is written

$$\langle \text{out}|F|\text{in}\rangle$$

and for a two-body operator, the prescription is

$$\langle \text{left out, right out}|G|\text{left in, right in}\rangle_A$$

where 'in' and 'out' refer to lines entering and leaving a vertex. Remember all arrows are now pointing up the page. Thus the matrix elements entering diagram (f) are obtained from diagram (g) of Fig. B.1 as

$$\langle kj|G|mn\rangle_A \langle mn|G|ki\rangle_A \langle i|F|j\rangle .$$

No extra sign factor is introduced by this procedure. Note that in the two-body matrix elements, the two-particle wavefunctions have not been coupled to states of good total angula momentum.

(5) Sum over all intermediate states by summing in the

original diagram each upgoing line independently over all particle states and each downgoing line independently over all hole states. Ignore the Pauli exclusion principle in performing this summation.

Thus as our final example, diagram (h) of Fig. B.1, when evaluated according to these rules, gives the result

$$-\frac{1}{2}\sum_{\substack{ijk\ell\\mn}}\frac{\langle kj|G|mn\rangle_A\ \langle mn|G|ki\rangle_A\ \langle i\ell|G|j\ell\rangle_A}{(\varepsilon_j+\varepsilon_k-\varepsilon_m-\varepsilon_n)(\varepsilon_i+\varepsilon_k-\varepsilon_m-\varepsilon_n)}\,.$$

The overall sign is minus, rule (1), since the number of closed loops is three and the number of hole lines is four. There is a factor of $\frac{1}{2}$ as particle lines m and n are equivalent. When a pair of equivalent lines are being summed, it is possible to rewrite the sum as

$$\frac{1}{2}\sum_{mn} = \sum_{m\leqslant n} 1/(1+\delta_{mn})$$

so that no overall factor appears in front of the sum.

B.2 OPEN-SHELL DIAGRAMS

We now discuss the modifications required to the above listed rules for evaluating open-shell diagrams. The main problem is to obtain the correct overall sign. Each external line present at the top of the diagram corresponds to a creation operator (either a particle creation operator $a_m^\dagger$ or a hole creation operator $b_j^\dagger$), and each external line present at the bottom of the diagram corresponds to an annihilation operator. The order in which the operators appear can be obtained from the diagram by reading clockwise round the diagram starting on the upper left-hand corner. For example, the external lines in diagram (a) of Fig. B.2 correspond to a list of operators $a_m^\dagger\, b_i^\dagger\, b_j\, a_n$. These operators are always in normal order, i.e. creation operators to the left of annihilation operators; however, the precise order of the creation operators among themselves and the annihilation operators among themselves is

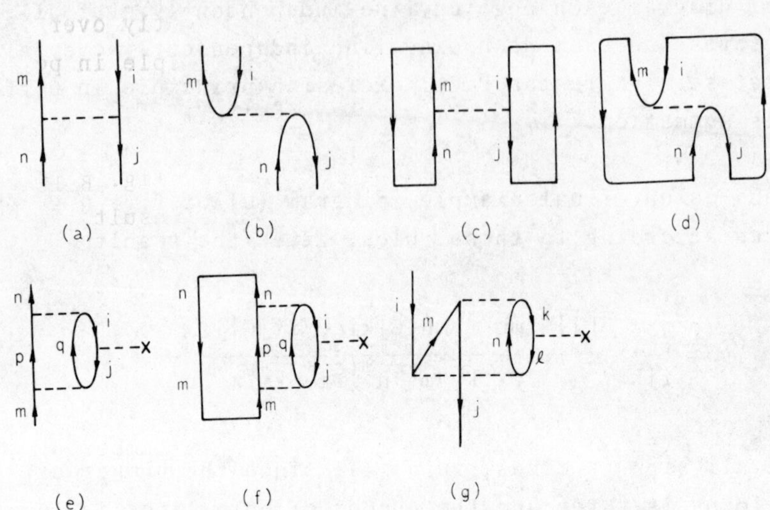

FIG. B.2. Illustrative open-shell diagrams. To help in the interpretation, open-shell diagrams are 'closed' as discussed in the text. Diagrams (c), (d), and (f) are the closed versions of diagrams (a), (b), and (e) respectively.

of course arbitrary. But if one is solving a particular physical problem using perturbation theory and one is summing a whole sequence of diagrams, one must ensure that the external lines are in the same order for each diagram in the sequence. If a particular diagram does not have the external lines in the required order, this can be corrected by multiplying the value of the diagram by $(-)^p$ where p is the number of interchanges required to bring the external lines into the desired order. Remember we always like to draw the diagram with the maximum number of closed loops. In this connection we define a particle-hole state as

$$|j^{-1}m\rangle = a_m^\dagger\ b_j^\dagger|C\rangle$$

where the particle creation operator has been put to the left of the hole operator. (This is purely an arbitrary choice; we could equally well choose the reverse order.) In terms of diagrams, this choice requires that external particle lines be drawn on the left of external hole lines.

The next step is to construct an equivalent 'closed form' to the original open-shell diagram. This is done in the following stages. First, check whether there are the same number of external particle lines at the top as there are at the bottom of the diagram, and the same number of external hole lines at the top as at the bottom. If this is not the case, then one or more particle-hole pairs have been created or annihilated, and the first stage in closing the diagram is to join up the loose ends from these additional particle-hole pairs. The second stage is to join the particle lines at the top to those at the bottom of the diagram, and similarly to join the hole lines at the top with those at the bottom. The standard order must be observed in forming these connections, i.e. the 'first' line from the top must be joined to the 'first' line from the bottom, etc., even though their actual labels may be different. These new diagrams are called the 'closed forms' of the original open-shell diagrams. For example the 'closed forms' of diagrams (a) and (b) in Fig. B.2 are diagrams (c) and (d) respectively. The rules for evaluating open-shell diagrams are now as follows:

(1') The overall sign factor is $(-)^{\ell+h}$ where ℓ is the number of closed loops and h the number of hole line segments in the 'closed form' of the diagram. An additional factor $(-)^p$ is included if the external lines in the original diagram are not in the required order. Thus the sign factor for diagram (c) in Fig. B.2 is minus, $\ell=2$ $h=3$, whereas for diagram (d) it is plus, $\ell=1$ $h=3$. Diagram (c) is the exchange version of diagram (d). Notice in obtaining the 'closed form' of the diagram that whenever two external particle lines are joined, a hole line is created and h is increased by one. However, when two external hole lines are joined an upgoing line is created which does not contribute to h.

(2') Include a factor of $(\frac{1}{2})$ for each 'equivalent pair' of lines in the original open-shell diagram.

(3') For each intermediate state include an energy denominator given by the unperturbed energy of the initial state minus the unperturbed energy of the intermediate state. A simple rule of thumb is as follows.

Consider the 'closed form' of the diagram, in particular consider the closure lines drawn at the second stage of its construction. These closure lines each carry two labels, one from the top and one from the bottom of the original diagram. Discard the labels which came from the top of the open diagram, keeping only the bottom labels. Then the energy denominators are given by the usual Goldstone prescription, viz. the sum of all hole line energies minus the sum of all particle line energies including the closure lines. This is the rule for diagrams in which every dashed, i.e. interaction, line represents the perturbation in question. When all diagrams of a particular order in the perturbation expansion are summed, the result is usually Hermitian, i.e. invariant with respect to switching the labels on the external lines at the top of a diagram with those at the bottom. In this case the prescription used for labelling the closure line becomes unimportant: on the other hand if the summed result is not Hermitian, then the result depends on the labelling of the closure line. In this situation, one frequently Hermitizes the result, i.e. averages the result obtained with both possible labellings of the closure lines.

In the case when one is evaluating a transition matrix element, i.e. at the nth order in the perturbation expansion the diagrams being summed all have n interaction lines representing the perturbation and one interaction line representing the transition, then the closure lines are labelled in the following way. Make a mark on each closure line at the level of the interaction line representing the transition. Label that part of each closure line above the mark with the label coming from the top of the diagram, and that part below the mark with the label coming from the bottom of the diagram. Then the energy denominators are given by the same rule, viz. the sum of all hole-line energies minus the sum of all particle-line energies. This prescription originates from the discussion at the end of section 1.12. As an example consider diagram (e) in Fig. B.2. This diagram is closed in (f) and the two energy denominators occurring are $(\varepsilon_n+\varepsilon_i-\varepsilon_p-\varepsilon_q)$ and

$(\varepsilon_m+\varepsilon_j-\varepsilon_p-\varepsilon_q)$.

(4') To obtain the correct particle matrix element corresponding to each dashed line, convert the open form of the diagram into a Rajaraman diagram and then read off the appropriate matrix element as discussed in rule (4) above. However, for each external hole line i, the time-reversed state $\tilde{i}$ is written into the matrix element. Remember $|\tilde{i}\rangle = S_i|-i\rangle = b_i^\dagger|C\rangle$. We could have required that the time-reversed state be written into the matrix elements for all hole lines in the diagram; however, hole lines in intermediate states are summed in evaluating the diagram, and it makes no difference whether one sums over all the time-reversed states $|\tilde{i}\rangle$ or over all $|i\rangle$, since both sums span the same set of states.

(5') As in rule (5), sum over all intermediate states by summing in the open form of the diagram each upgoing line independently over all particle states and each downgoing line independently over all hole states. Do not sum over the external lines. As a last example, we evaluate diagram (g) of Fig. B.2 to give the result

$$+\frac{1}{2}\sum_{mnk\ell}\frac{\langle\tilde{j}k|G|mn\rangle_A\ \ \langle\ell|F|k\rangle\ \langle mn|G|\tilde{i}\ell\rangle_A}{(\varepsilon_j+\varepsilon_k-\varepsilon_m-\varepsilon_n)(\varepsilon_i+\varepsilon_\ell-\varepsilon_m-\varepsilon_n)}$$

We make one final comment. It is frequently stated that the Pauli principle may be disregarded in intermediate states. For example consider diagram (g) of Fig. B.2. The Pauli principle dictates that in the sum over the particle labels m and n, the term with m=n should be excluded since this term corresponds to two particles occupying the same shell-model state. (Remember that the label m stands for all the quantum numbers of the state.) On the other hand, it is very inconvenient to have to write restricted sums in the formulae. Fortunately, if one always uses antisymmetrized two-body matrix elements, then the contribution from the Pauli violating terms is zero, since the contribution from the direct part of the matrix element exactly cancels the exchange part. Hence the maxim that the

Pauli principle may be disregarded in intermediate states can be safely adopted. This question is discussed further in section (3.2), in connection with the Goldstone expansion.

B.3 FOLDED DIAGRAMS

Folded diagrams were discussed in chapter 5 in connection with effective interactions. The rules for their evaluation are very similar to those just discussed. In particular the same procedure for 'closing' an open-shell diagram by joining the external lines at the top of the diagram with corresponding lines at the bottom is used to close folded diagrams. Furthermore, we shall talk about the 'folded' and 'unfolded' version of the diagram. This terminology was introduced in section (5.5).

The rules for evaluating folded diagrams are as follows:

(1") The overall sign factor is $(-)^{f+\ell+h}$ where f is the number of folds (*not* the number of valence lines contained in each fold), ℓ is the number of closed loops, and h the number of hole line segments in the 'closed' form of the 'unfolded' version of the diagram. Closure lines as well as holes in the core contribute to h.

For the folded diagram expansion of a transition operator $\mathscr{F}$(see eqn (5.53) and the subsequent discussion), this rule (1") is modified slightly as enumerated at the end of section 5.7.

(2") Include a factor of (1/2) for each 'equivalent pair' of lines in the unfolded version of the diagram. This rule applies to the folded valence lines between blocks as well as to lines within a single block.

(3") For each intermediate state include an energy denominator obtained as follows. Take the folded version of the diagram and 'close' it, labelling the closure lines with the label that came from the bottom of the original diagram. Then the energy denominators are given by the Goldstone prescription, namely the sum of all down-going line energies minus the sum of all up-going line energies. Note that the folded valence lines may be upgoing or downgoing depending on the particular diagram.

(4") The two-body matrix elements are read off the unfolded version of the diagram according to rule (4') in section B.2.

(5") Within each block, sum each up-going line over all valence orbitals and over all higher-lying orbitals. Sum each down-going line over all core orbitals. Between blocks, sum each folded valence line over all valence orbitals. Each summation is to be done independently, without regard to exclusion, except that no intermediate states within a block may be in the degenerate model space, D.

As an example, we evaluate folded diagrams (d) and (e) in Fig. 5.8, both of which originate from the same unfolded version, namely diagram (c). We use the following notation in labelling the lines: a,b,c, ... represent valence orbitals in the degenerate model space; i,j,k, ... represent hole orbitals in the closed-shell core; and m,n,p, ... represent unoccupied particle orbitals including the valence orbitals and higher-lying orbitals, subject to the condition that no intermediate state is degenerate with the states in the model space D. The values of diagrams (d) and (e) are given by the same expression:

$$-\sum_{\substack{mnp\\b'c'}} \frac{1}{e}\langle a''b''|V|mp\rangle_A \ \langle pc''|V|nc'\rangle_A \ \langle mn|V|ab'\rangle_A \ \langle b'c'|V|bc\rangle_A$$

where e is a product of three energy denominators. The overall minus sign comes from f=1, ℓ=3, h=3, where three 'closure' lines have to be drawn in closing diagram (c). Each closure line completes a closed loop, hence ℓ=3, and each closure line introduces a down-going line segment, hence h=3. There are no equivalent pairs; external lines do not count in rule (2"). Diagrams (d) and (e) only differ in their energy denominators. For diagram (d), we have

$$\begin{aligned} e &= (\varepsilon_a+\varepsilon_b+\varepsilon_c-\varepsilon_{c''}-\varepsilon_m-\varepsilon_p)(\varepsilon_a+\varepsilon_b+\varepsilon_c-\varepsilon_{c'}-\varepsilon_m-\varepsilon_n)(\varepsilon_a+\varepsilon_{b'}-\varepsilon_m-\varepsilon_n) \\ &= (2\varepsilon-\varepsilon_m-\varepsilon_p)(2\varepsilon-\varepsilon_m-\varepsilon_n)^2, \end{aligned}$$

and for diagram (e),

$$e = (\varepsilon_a+\varepsilon_b+\varepsilon_c-\varepsilon_{c''}-\varepsilon_m-\varepsilon_p)(\varepsilon_a+\varepsilon_{b'}+\varepsilon_{c'}-\varepsilon_{c''}-\varepsilon_m-\varepsilon_p)(\varepsilon_a+\varepsilon_{b'}-\varepsilon_m-\varepsilon_n)$$

$$= (2\varepsilon-\varepsilon_m-\varepsilon_p)^2\ (2\varepsilon-\varepsilon_m-\varepsilon_n)$$

where the single-particle energies for all particles in the valence orbitals have been set equal to ε. The sums over b' and c' are over valence orbitals only, and the sums over m, n, and p are over valence and higher-lying orbitals providing no term with e=zero is retained.

REFERENCES AND AUTHOR INDEX

AHRENS, J., BORCHERT, H., EPPLER, H.B., GIMM, H., GUNDRUM, H., RIEHN, P., SITA RAM, G., ZIEGLER, A., KRONING, M., and ZIEGLER, B. (1972). *Proc. Intern. Conf. on Nuclear Structure Studies, Sendai, Japan.* (115)

AKIYAMA, Y., ARIMA, A., and SEBE, T. (1969). *Nucl. Phys.* **A138**, 273. (273)

AJZENBERG-SELOVE, F. (1971). *Nucl. Phys.* **A166**, 1. (19)

——— (1972). *Nucl. Phys.* **A190**, 1. (231,236)

ARIMA, A., COHEN, S., LAWSON, R.D., and MacFARLANE, M.H. (1968). *Nucl. Phys.* **A108**, 94. (276)

——— and HORIE, H. (1954). *Prog. theor. Phys.* **11**, 509. (202)

AUSTERN, N., DRISKO, R.M., HALBERT, E.C., and SATCHLER, G.R. (1964). *Phys. Rev.* **133B**, 3. (283,286)

BALASHOV, V.V. (1959). *JETP* **36**, 988. (226)

BAND, I.M., and KHARITONOV, Yu.I. (1971). *Nucl. Data* **A10**, 107. (234,236)

BARANGER, M. (1963). In *Cargese lectures in theoretical physics* (ed. M. Levy). Benjamin Press, New York. (50)

——— (1967). In *Nuclear structure and nuclear reactions, Proceedings of the international school of physics 'Enrico Fermi' course XL, Varenna* (eds. M. Jean and R.A. Ricci). Academic Press Inc., New York. (76,78,81,93)

——— and DAVIES, K.T.R. (1966). *Nucl. Phys.* **79**, 403. (351)

BARRETT, B.R. (1974). *Nucl. Phys.* **A221**, 299. (185)

——— HEWITT, R.G.L., and McCARTHY, R.J. (1970). *Phys. Rev.* **C2**, 1199. (92,101)

——— ——— ——— (1971). *Phys. Rev.* **C3**, 1137. (101,202,203)

——— and KIRSON, M.W. (1970). *Nucl. Phys.* **A148**, 145. (187,212)

——— ——— (1973). In *Advances in nuclear physics,* Vol. 6. (eds. M. Baranger and E. Vogt), p. 219. Plenum Press, New York. (157,199,200,205)

BASSICHIS, W.H., KERMAN, A.K., and SVENNE, J.P. (1967). *Phys. Rev.* **160**, 746. (59)

BAYMAN, B.F. (1957). *Lectures on groups and their applications to spectroscopy.* Nordita, Copenhagen. (239)

——— and KALLIO, A. (1967). *Phys. Rev.* **156**, 1121. (351)

——— and LANDE, A. (1966). *Nucl. Phys.* **77**, 1. (219,226)

BECKER, R.L., MacKELLAR, A.D., and MORRIS, B.M. (1968). *Phys. Rev.* **174**, 1264. (80)

BERTSCH, G. (1965). *Nucl. Phys.* **74**, 234. (186)

BETHE, H.A. (1965). *Phys. Rev.* **138B**, 804. (76)

——— (1967). *Phys. Rev.* **158**, 941. (77)

——— (1971). *A. Rev. nucl. Sci.* **21**, 93. (81,99)

——— BRANDOW, B.H., and PETSCHEK, A.G. (1963). *Phys. Rev.* **129**, 225. (70,80,89,91)

BLOCH, C. and HOROWITZ, J. (1958). *Nucl. Phys.* **8**, 91. (157)

BOHR, A. and MOTTELSON, B.R. (1953). *K. Danske. Vidensk. Selsk. Mat.-Fys. Medd.* **27**, 1. (340)

BOUTEN, M. (1970). In *Theory of nuclear structure: Trieste lectures 1969.* p. 361. International Atomic Energy Agency, Vienna. (50,55,57)

BRANDOW, B.H. (1965). In *Many-body description of nuclear structure and reactions, Proceedings of the international school of physics 'Enrico Fermi', Course XXXVI, Varenna* (ed. C. Bloch). Academic Press Inc., New York. (178,181)

——— (1967). *Rev. mod. Phys.* **39**, 771. (63,157,166,178,181,195,198,199,360)

——— (1969). In *Lectures in theoretical physics*, Vol. XIB, p. 55 (eds. K.T. Mahanthappa and W.E. Brittin). Gordon and Breach, New York. (157,166)

BRINK, D.M. and BOEKER, E. (1967). *Nucl. Phys.* **91**, 1. (59)

——— and SATCHLER, G.R. (1968). *Angular momentum.* Clarendon Press, Oxford. (317,319,324,325)

BRODY, T.A. and MOSHINSKY, M. (1960). *Tables of transformation brackets.* Monografias del Instituto de Fisica, Mexico. (351)

BROWN, G.E. (1967). *Unified theory of nuclear models and forces.* North-Holland, Amsterdam. (119,128,148)

——— and BOLSTERLI, M. (1959). *Phys. Rev. Lett.* **3**, 472. (118)

——— CASTILLEJO, L., and EVANS, J.A. (1961). *Nucl. Phys.* **22**, 1. (118)

——— EVANS, J.A., and THOULESS, D.J. (1961). *Nucl. Phys.* **24**, 1. (152)

BRUECKNER, K.A. and GOLDMAN, D.T. (1960). *Phys. Rev.* **117**, 207. (69)

BUCK, B. and HILL, A.D. (1967). *Nucl. Phys.* **A95**, 271. (154)

BUTTLE, P.J.A. and GOLDFARB, L.J.B. (1966). *Nucl. Phys.* **78**, 409. (288)

CHI, B.E. (1970). *Nucl. Phys.* **A146**, 449. (141)

CLEMENT, C.F. (1973). *Nucl. Phys.* **A213**, 469 and 493. (313)

——— and PEREZ, S.M. (1973). *Nucl. Phys.* **A213**, 510. (313,314)

CLEMENT, D.M. and BARANGER, E.U. (1968). *Nucl. Phys.* **A108**, 27. (101)

COHEN, S. and KURATH, D. (1965). *Nucl. Phys.* **73**, 1. (274,275)

——— ——— (1967). *Nucl. Phys.* **A101**, 1. (275)

——— ——— (1970). *Nucl. Phys.* **A141**, 145. (275)

——— LAWSON, R.D., MacFARLANE, M.H., and SOGA, M. (1966). In *Methods of computational physics*, Vol. 6, p. 235. (eds. B. Alder, S. Fernbach and M. Rotenberg). Academic Press, New York. (277)

——— LAWSON, R.D., and SOPER, J.M. (1966). *Phys. Lett.* **21**, 306. (277)

COLE, B.J., WATT, A., and WHITEHEAD, R.R. (1973). *Phys. Lett.* **45B**, 429. (280)

CONDON, E.U. and SHORTLEY, G.H. (1951). *The theory of atomic spectra* Cambridge Univ. Press., New York and London. (278)

CUSSON, R.Y. and LEE, H.C. (1973). *Nucl. Phys.* **A211**, 429. (207)

DAVIES, K.T.R., BARANGER, M., TARBUTTON, R.M., and KUO, T.T.S. (1969). *Phys. Rev.* **177**, 1519. (80)

——— ——— (1970). *Phys. Rev.* **C1**, 1640. (80)

——— KRIEGER, S.J., and BARANGER, M. (1966). *Nucl. Phys.* **84**, 545. (50)

——— and McCARTHY, R.J. (1971). *Phys. Rev.* **C4**, 81. (80)

DAY, B.D. (1967). *Rev. mod. Phys.* **39**, 719. (63,67,88,91)

DES CLOIZEAUX, J. (1960). *Nucl. Phys.* **20**, 321. (191,199)

De SHALIT, A. and TALMI, I. (1963). *Nuclear shell theory*. Academic Press, New York and London. (220,226,241,245,248,319,321)

De VOIGT, M.J.A., GLAUDEMANS, P.W.M., De BOER, J., and WILDENTHAL, B.H. (1972). *Nucl. Phys.* **A186**, 365. (279)

——— and WILDENTHAL, B.H. (1973). *Nucl. Phys.* **A206**, 305. (279)

DeVRIES, R.M. (1973). *Phys. Rev.* **C8**, 951. (288)

EDMONDS, A.R. and FLOWERS, B.H. (1952). *Proc. R. Soc.* **A214**, 515. (218,226)

EISENBERG, J.M. and GREINER, W. (1970). *Excitation mechanisms of the nucleus*. North-Holland, Amsterdam. (337)

EISENSTEIN, I. and KIRSON, M.W. (1973). *Phys. Lett.* **B47**, 315. (247)

ELLIOTT, J.P. and FLOWERS, B.H. (1955). *Proc. R. Soc.* **A229**, 536. (273)

——— ——— (1957). *Proc. R. Soc.* **A242**, 57. (117)

——— HOPE, J. and JAHN, H.A. (1953). *Phil. Trans. R. Soc.* **A246**, 241. (226)

——— JACKSON, A.D., MAVROMATIS, H.A., SANDERSON, E.A., and SINGH, B. (1968). *Nucl. Phys.* **A121**, 241. (101,184,206)

ELLIS, P.J. and MAVROMATIS, H.A. (1971). *Nucl. Phys.* **A175**, 309. (188,205,206)

——— and SIEGEL, S. (1971). *Phys. Lett.* **34B**, 177. (202,203)

ENDT, P.M. and VAN DER LEUN, C. (1973). *Nucl. Phys.* **A214**, 1. (24,231,253)

ENGEL, M.Y. and UNNA, I. (1968). *Phys. Lett.* **28B**, 12. (277)

FEDERMAN, P. and PITTEL, S. (1970). *Nucl. Phys.* **A155**, 161. (237)

——— and ZAMICK, L. (1969). *Phys. Rev.* **177**, 1534. (200,202)

FEENBERG, E. and PHILLIPS, M. (1937). *Phys. Rev.* **51**, 597. (272)

FINK, M., GARI, M., and HEBACH, H. (1974). *Phys. Lett.* **49B**, 20. (117)

——— ——— ——— and ZABOLITZKY, J.G. (1974). *Phys. Lett.* **51B**, 320. (117)

FLOWERS, B.H. (1952). *Proc. R. Soc.* **A215**, 398. (218,225,248)

FORD, K.W. and KONOPINSKI, E.J. (1959). *Nucl. Phys.* **9**, 218. (347)

FRENCH, J.B. (1965a). In *Many-body description of nuclear structure and reactions, Proceedings of the international school of physics 'Enrico Fermi', Course XXXVI, Varenna.* (ed. C. Bloch). Academic Press Inc., New York. (266,317)

——— (1965b). *Phys. Lett.* **15**, 327. (313)

——— and MacFARLANE, M.H. (1961). *Nucl. Phys.* **26**, 168. (306,310)

——— HALBERT, E.C., McGRORY, J.B., and WONG, S.S.M. (1969). In *Advances in nuclear physics*, Vol. 3, p. 193. (eds. M. Baranger and E. Vogt). Plenum Press, New York. (258,277)

FULLER, G.H. and COHEN, V.W. (1969). *Nucl. Data* **A5**, 433. (256)

GILLET, V. (1964). *Nucl. Phys.* **51**, 410. (152,153)

——— GREEN, A.M. and SANDERSON, E.A. (1966). *Nucl. Phys.* **88**, 321. (152,153)

——— and SANDERSON, E.A. (1964). *Nucl. Phys.* **54**, 472. (152)

——— and MELKANOFF, M.A. (1964). *Phys. Rev.* **133B**, 1190. (152)

——— and VINH MAU, N. (1964). *Nucl. Phys.* **54**, 321. (152)

GLAUDEMANS, P.W.M., WIECHERS, G., and BRUSSAARD, P.J. (1964). *Nucl. Phys.* **56**, 529 and 548. (266,261,274,276,308)

GLENDENNING, N.K. (1965). *Phys. Rev.* **137B**, 102. (294,295)

GOLDSTONE, J. (1957). *Proc. R. Soc.* **A239**, 267. (63)

GOODE, P. and KOLTUN, D.S. (1972). *Phys. Lett.* **39B**, 159. (188)

——— ——— (1975). *Nucl. Phys.* **A243**, 44. (188)

GOSWAMI, A. and PAL, M.K. (1962). *Nucl. Phys.* **35**, 544. (125)

GOUDSMIT, S. and BACHER, R.F. (1934). *Phys. Rev.* **46**, 948. (235)

HAFTEL, M.I. and TABAKIN, F. (1970). *Nucl. Phys.* **A158**, 1. (99)

——— ——— (1971). *Phys. Rev.* **C3**, 921. (99)

HALBERT, E.C., McGRORY, J.B., WILDENTHAL, B.H., and PANDYA, S.P. (1971). In *Advances in nuclear physics*, Vol. 4, p. 315. (eds M. Baranger and E. Vogt). Plenum Press, New York. (186,278)

HARVEY, M. (1968). In *Advances in nuclear physics*, Vol. 1, p. 67. (eds M. Baranger and E. Vogt). Plenum Press, New York. (104)

——— (1974). In *Nuclear spectroscopy and nuclear reactions with heavy ions, Proceedings of the international school of physics 'Enrico Fermi', course LXII, Varenna* (eds H. Faraggi and R.A. Ricci). North-Holland, Amsterdam. (207)

——— (1975). *Ann. Phys.* **94**, 47. (207)

——— and KHANNA, F.C. (1970). *Nucl. Phys.* **A155**, 337. (207)

HAMADA, T. and JOHNSTON, I.D. (1962). *Nucl. Phys.* **34**, 383. (99)

HODGSON, P.E. and MILLENER, D.J. (1972). *Revta bras. Fis.* **2**, 87. (313)

HUBBARD, L.B. (1971). *Nucl. Data* **A9**, 85. (226)

HUGENHOLTZ, N.M. (1957). *Physica* **23**, 533. (65)

ICHIMURA, M., ARIMA, A., HALBERT, E.C., and TERASAWA, T. (1973). *Nucl. Phys.* **A204**, 225. (289,291,292)

INGLIS, D.R. (1953). *Rev. mod. Phys.* **25**, 390. (272)

JAHN, H.A. (1954). *Phys. Rev.* **96**, 989. (218)

——— and VAN WIERINGEN H. (1951). *Proc. R. Soc.* **A209**, 502. (218,226)

JOHNSON, M.B. and BARANGER, M. (1971). *Ann. Phys.* **62**, 172. (157)

KAHANA, S., LEE, H.C., and SCOTT, C.K. (1969). *Phys. Rev.* **185**, 1378. (101)

KALLIO, A. and KOLLTVEIT, K. (1964). *Nucl. Phys.* **53**, 87. (153,203)

KASSIS, N.I. (1972). *Nucl. Phys.* **A194**, 205. (184,188,212)

KERMAN, A.K. and PAL, M.K. (1967). *Phys. Rev.* **162**, 970. (59,101)

——— SVENNE, J.P., and VILLARS, F.M.H. (1966). *Phys. Rev.* **147**, 710. (59)

KIRSON, M.W. (1967). *Nucl. Phys.* **A99**, 353. (71)

——— (1968). *Nucl. Phys.* **A115**, 49. (70,71)

——— (1969). *Nucl. Phys.* **A139**, 57. (70,71)

——— (1971). *Ann. Phys.* **66**, 624. *Erratum Ann. Phys.* **68**, 556. (205)

——— (1974). *Ann. Phys.* **82**, 345. (205)

KÖHLER, H.S. (1961). *Ann. Phys.* **16**, 375. (94)

——— (1971). *Nucl. Phys.* **A162**, 385. (80)

——— and McCARTHY, R.J. (1968). *Nucl. Phys.* **A106**, 313. (92)

KOLTUN, D.S. (1973). *A. Rev. nucl. Sci.* **23**, 163. (247,254)

KUO, T.T.S. (1967). *Nucl. Phys.* **A103**, 71. (100,185,186,205,278,281)

——— (1968). *Nucl. Phys.* **A122**, 325. (153)

——— (1974). *A. Rev. nucl. Sci.* **24**, 101. (157)

——— BLOMQVIST, J., and BROWN, G.E. (1970). *Phys. Lett.* **31B**, 93. (153)

——— and BROWN, G.E. (1966). *Nucl. Phys.* **85**, 40. (100,153,186,278)

——— ——— (1968). *Nucl. Phys.* **A114**, 241. (237)

——— LEE, S.Y., and RATCLIFF, K.F. (1971). *Nucl. Phys.* **A176**, 65. (157)

——— and OSNES, E. (1973). *Nucl. Phys.* **A205**, 1. (204,205)

KURATH, D. (1952). *Phys. Rev.* **88**, 804. (272)

——— (1956). *Phys. Rev.* **101**, 216. (273)

KURATH, D. (1973). *Phys. Rev.* **C7**, 1390. (275)

KUTSCHERA, W., BROWN, B.A., IKEZOE, H., SPROUSE, G.D., YAKAZAKI, Y., YOSHIDA, Y., NOMURA, T. and OHNUMA, H. (1975). *Phys. Rev.* **C12**, 813. (253)

LANE, A.M. (1964). *Nuclear theory*. Benjamin, New York. (125)

LANFORD, W.A. and WILDENTHAL, B.H. (1973). *Phys. Rev.* **C7**, 668. (279)

LASSILA, K.E., HULL, M.H., RUPPEL, H.M., McDONALD, F.A., and BREIT, G. (1962). *Phys. Rev.* **126**, 881. (99)

LEE, H.C. and CUSSON, R.Y. (1972). *Ann. Phys.* **72**, 353. (53,54,60)

LEVINGER, J.S. (1960). *Nuclear photo-disintegration*. Oxford University Press. (115)

——— and BETHE, H.A. (1950). *Phys. Rev.* **78**, 115. (115)

MACFARLANE, M.H. (1967). In *Nuclear structure and nuclear reactions, Proceedings of the international school of physics 'Enrico Fermi', Course XL, Varenna*. (eds. M. Jean and R.A. Ricci). Academic Press Inc., New York. (157,162,276)

——— and FRENCH, J.B. (1960). *Rev. mod. Phys.* **32**, 567. (226,259,298,305)

MAVROMATIS, H.A. (1973). *Nucl. Phys.* **A206**, 477. (63)

——— MARKIEWICZ, W., and GREEN, A.M. (1967). *Nucl. Phys.* **A90**, 101. (153)

McCARTHY, R.J. (1969). *Nucl. Phys.* **A130**, 305. (80,92,100)

——— and DAVIES, K.T.R. (1970). *Phys. Rev.* **C1**, 1644. (80)

MERCIER, R., BARANGER, E.U., and McCARTHY, R.J. (1969). *Nucl. Phys.* **A130**, 322. (100)

MESSIAH, A. (1964). *Quantum mechanics*. North-Holland, Amsterdam. (38,61,328)

MOSZKOWSKI, S.A. and SCOTT, B.L. (1960). *Ann. Phys.* **11**, 65. (81,88,93)

MUIRHEAD, H. (1965). *The physics of elementary particles*. Pergamon Press, Oxford. (28)

NANN, H., MUELLER, D., and KASHY, E. (1976). To be published. (253)

NEGELE, J.W. (1970). *Phys. Rev.* **C1**, 1260. (59,80)

NILSSON, S.G. (1955). *K. Danske Visensk. Selsk. Mat.-Fys. Medd.* **29**, 16. (52)

OHNUMA, H. and SOURKES, A.M. (1971). *Phys. Rev.* **C3**, 158. (313)

PANDYA, S. (1956). *Phys. Rev.* **103**, 956. (110)

PEREZ, S.M. (1969). *Nucl. Phys.* **A136**, 599. (153)

——— (1970). *Phys. Lett.* **33B**, 317. (153)

POLETTI, A.R., WARBURTON, E.K., OLNESS, J.W., KOLATA, J.J. and GORODETZKY, Ph. (1976). *Phys. Rev.* **C13**, 1180. (236,253)

PREEDOM, B.M. and WILDENTHAL, B.H. (1972). *Phys. Rev.* **C6**, 1633. (278)

QUESNE, C. (1970). *Phys. Lett.* **31B**, 7. (247)

RACAH, G. (1943). *Phys. Rev.* **63**, 367. (218,238)

RAJARAMAN, R. (1963a). *Phys. Rev.* **129**, 265. (76,77)

——— (1963b). *Phys. Rev.* **131**, 1244. (76,77,362)

——— and BETHE, H.A. (1967). *Rev. mod. Phys.* **39**, 745. (77)

RAPAPORT, J., SPERDUTO, A., and BUECHNER, W.W. (1966). *Phys. Rev.* **151**, 939. (314)

RAYNAL, J., MELKANOFF, M.A. and SAWADA, T. (1967). *Nucl. Phys.* **A101**, 369. (154)

REID, R.V. (1968). *Ann. Phys.* **50**, 411. (99,116)

RING, P. and SPETH, J. (1973). *Phys. Lett.* **44B**, 477. (154)

RIPKA, G. (1968). In *Advances in nuclear physics*, Vol. 1, p. 183. (eds M. Baranger and E. Vogt). Plenum Press, New York. (50,55,58)

ROSE, H.J. and BRINK, D.M. (1967). *Rev. mod. Phys.* **39**, 306. (333,334,337)

ROWE, D.J. (1970). *Nuclear collective motion*. Methuen, London. (128,140)

SACHS, R.G. (1953). *Nuclear theory*. Addison-Wesley, Reading, Massachusetts. (337)

SATCHLER, G.R. (1964). *Nucl. Phys.* **55**, 1. (283,287)

——— (1974). *Physics Reports* **14C**, 97. (124)

SAUNIER, G. and PEARSON, J.M. (1970). *Phys. Rev.* **C1**, 1353. (60)

SCHUCAN, T.H. and WEIDENMULLER, H.A. (1972). *Ann. Phys.* **73**, 108. (162)

——— ——— (1973). *Ann. Phys.* **76**, 483. (162)

SCHWEBER, S.S. (1962). *An introduction to relativistic quantum field theory*. Harper and Row, New York. (2)

SCOTT, B.L. and MOSZKOWSKI, S.A. (1961). *Ann. Phys.* **14**, 107. (93)

SIEGEL, S. and ZAMICK, L. (1969). *Phys. Lett.* **28B**, 450 and 453. (202)

——— ——— (1970). *Nucl. Phys.* **A145**, 89. (202,203,204)

SIGNELL, P. (1969). In *Advances in nuclear physics*, Vol. 2, p. 223. (eds M. Baranger and E. Vogt). Plenum Press, New York. (99)

SKYRME, T.H.R. (1959). *Nucl. Phys.* **9**, 615. (60)

SMIRNOV, Y.F. (1961). *Nucl. Phys.* **27**, 177. (291)

SMITH, W.R. (1971). *Phys. Lett.* **34B**, 252. (285)

SOYEUR, M. and ZUKER, A.P. (1972). *Phys. Lett.* **41B**, 135. (279)

SPRUNG, D.W.L. (1972). In *Advances in nuclear physics*, Vol. 5, p. 225

(eds M. Baranger and E. Vogt). Plenum Press, New York. (81)

TABAKIN, F. (1964). *Ann. Phys.* **30**, 51. (59,99)

TALMAN, J.D. (1970). *Nucl. Phys.* **A141**, 273. (351)

TALMI, I. (1962). *Rev. mod. Phys.* **34**, 704. (104,235)

——— and UNNA, I. (1960). *A. Rev. nucl. Sci.* **10**, 353. (235)

THOULESS, D.J. (1960). *Nucl. Phys.* **21**, 225. (47,132)

——— (1961). *Nucl. Phys.* **22**, 78. (134,139,146)

TOBOCMAN, W., RYAN, R., BALTZ, A.J., and KAHANA, S. (1973). *Nucl. Phys.* **A205**, 193. (285)

TOWNER, I.S. (1970). *Nucl. Phys.* **A151**, 97. (124,127)

——— and HARDY, J.C. (1967). *Phys. Lett.* **25B**, 577. (276)

——— ——— (1969). *Nucl. Data* **A6**, 153. (226)

ULLAH, N. and ROWE, D.J. (1971). *Nucl. Phys.* **A163**, 257. (141)

VARY, J.P., SAUER, P.U., and WONG, C.W. (1973). *Phys. Rev.* **C7**, 1776. (189)

VAUTHERIN, D. and BRINK, D.M. (1972). *Phys. Rev.* **C5**, 626. (60)

VIGNON, B., LONGEQUEUE, J.P., and TOWNER, I.S. (1972). *Nucl. Phys.* **A189**, 513. (308)

VILLARS, F.M.H. (1963). In *Nuclear physics, Proceedings of the international school of physics 'Enrico Fermi', Course XXIII, Varenna.* (ed. V.F. Weisskopf). Academic Press Inc., New York. (50)

VOLKOV, A.B. (1965). *Nucl. Phys.* **74**, 33. (59)

WAPSTRA, A.H. and GOVE, N.B. (1971). *Nucl. Data* **A9**, 267. (246)

WEGNER, F. (1970). *Nucl. Phys.* **A141**, 609. (347)

WEISSKOPF, V.F. (1951). *Phys. Rev.* **83**, 1073. (340)

WENG, W.T., KUO, T.T.S., and BROWN, G.E. (1973). *Phys. Lett.* **46B**, 329. (115)

WHITEHEAD, R.R. (1972). *Nucl. Phys.* **A182**, 290. (280)

——— and WATT, A. (1971). *Phys. Lett.* **35B**, 189. (280)

——— ——— (1972). *Phys. Lett.* **41B**, 7. (280)

WILDENTHAL, B.H., HALBERT, E.C., McGRORY, J.B., and KUO, T.T.S. (1971). *Phys. Rev.* **C4**, 1266. (278)

——— McGRORY, J.B., HALBERT, E.C., and GRABER, H.D. (1971). *Phys. Rev.* **C4**, 1708. (278)

——— and McGRORY, J.B. (1973). *Phys. Rev.* **C7**, 714. (278)

WILKINSON, D.H. (1960). In *Nuclear spectroscopy - Part B*, p. 852. (ed. F. Ajzenberg-Selove). Academic Press, New York and London. (340)

Wilkinson, J.H. (1969). *The algebraic eigenvalue problem*. Oxford University Press. (112,214,280)

YARIV, Y. (1974). *Nucl. Phys.* A225, 382. (247)

YUTSIS, A.P., LEVINSON, I.B., and VANAGAS, V.V. (1962). *Mathematical apparatus of the theory of angular momentum*. Israel Program for Scientific Translations, Jerusalem. (264)

SUBJECT INDEX